An Introduction
to the Mathematics of Finance

An Introduction to the Mathematics of Finance

A Deterministic Approach

Second Edition

S. J. Garrett

Published for the Institute
and Faculty of Actuaries (RC000243)
http://www.actuaries.org.uk

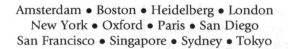

Amsterdam • Boston • Heidelberg • London
New York • Oxford • Paris • San Diego
San Francisco • Singapore • Sydney • Tokyo

Butterworth-Heinemann is an Imprint of Elsevier

Butterworth-Heinemann is an imprint of Elsevier
The Boulevard, Langford Lane, Kidlington, Oxford, OX5 1GB
225 Wyman Street, Waltham, MA 02451, USA

First edition 1989
Second edition 2013

Notices
Knowledge and best practice in this field are constantly changing. As new research and experience broaden our understanding, changes in research methods, professional practices, or medical treatment may become necessary.

Practitioners and researchers must always rely on their own experience and knowledge in evaluating and using any information, methods, compounds, or experiments described herein. In using such information or methods they should be mindful of their own safety and the safety of others, including parties for whom they have a professional responsibility.

To the fullest extent of the law, neither the Publisher nor the authors, contributors, or editors, assume any liability for any injury and/or damage to persons or property as a matter of products liability, negligence or otherwise, or from any use or operation of any methods, products, instructions, or ideas contained in the material herein.

British Library Cataloguing in Publication Data
A catalogue record for this book is available from the British Library

Library of Congress Cataloguing in Publication Data
A catalog record for this book is available from the Library of Congress

ISBN: 978-0-08-101302-1

For information on all Butterworth-Heinemann publications
visit our website at **store.elsevier.com**

Printed and bound in the United Kingdom

13 14 15 16   10 9 8 7 6 5 4 3 2 1

Working together
to grow libraries in
developing countries

www.elsevier.com • www.bookaid.org

Dedication

Dedicated to Adam and Matthew Garrett, my two greatest achievements.

Contents

Preface

This book is a revision of the original *An Introduction to the Mathematics of Finance* by J.J. McCutcheon and W.F. Scott. The subject of financial mathematics has expanded immensely since the publication of that first edition in the 1980s, and the aim of this second edition is to update the content for the modern audience. Despite the recent advances in stochastic models within financial mathematics, the book remains concerned almost entirely with deterministic approaches. The reason for this is twofold. Firstly, many readers will find a solid understanding of deterministic methods within the classical theory of compound interest entirely sufficient for their needs. This group of readers is likely to include economists, accountants, and general business practitioners. Secondly, readers intending to study towards an advanced understanding of financial mathematics need to start with the fundamental concept of compound interest. Such readers should treat this as an introductory text. Care has been taken to point towards areas where stochastic concepts will likely be developed in later studies; indeed, Chapters 10, 11, and 12 are intended as an introduction to the fundamentals and application of modern financial mathematics in the broader sense.

The book is primarily aimed at readers who are preparing for university or professional examinations. The material presented here now covers the entire *CT1* syllabus of the **Institute and Faculty of Actuaries** (as at 2013) and also some material relevant to the *CT8* and *ST5* syllabuses. This combination of material corresponds to the *FM-Financial Mathematics* syllabus of the **Society of Actuaries**. Furthermore, students of the **CFA Institute** will find this book useful in support of various aspects of their studies. With exam preparation in mind, this second edition includes many past examination questions from the *Institute and Faculty of Actuaries* and the *CFA Institute*, with worked solutions.

The book is necessarily mathematical, but I hope not too mathematical. It is expected that readers have a solid understanding of calculus, linear algebra, and probability, but to a level no higher than would be expected from a strong first year undergraduate in a numerate subject. That is not to say the material is easy, rather the difficulty arises from the sheer breath of application and the perhaps unfamiliar real-world contexts.

Where appropriate, additional material in this edition has been based on core reading material from the *Institute and Faculty of Actuaries*, and I am grateful to Dr. Trevor Watkins for permission to use this. I am also grateful to Laura Clarke and Sally Calder of the *Institute and Faculty of Actuaries* for their help, not least in sourcing relevant past examination questions from their archives. I am also grateful to Kathleen Paoni and Dr. J. Scott Bentley of *Elsevier* for supporting me in my first venture into the world of textbooks. I also wish to acknowledge the entertaining company of my good friend and colleague Dr. Andrew McMullan of the *University of Leicester* on the numerous coffee breaks between writing.

This edition has benefitted hugely from comments made by undergraduate and postgraduate students enrolled on my modules *An Introduction to Actuarial Mathematics* and *Theory of Interest* at the *University of Leicester* in 2012. Particular mention should be given to the eagle eyes of Fern Dyer, George Hodgson-Abbott, Hitesh Gohel, Prashray Khaire, Yueh-Chin Lin, Jian Li, and Jianjian Shao, who pointed out numerous typos in previous drafts. Any errors that remain are of course entirely my fault.

This list of acknowledgements would not be complete without special mention of my wife, Yvette, who puts up with my constant working and occasional grumpiness. Yvette is a constant supporter of everything I do, and I could not have done this, or indeed much else, without her.

Dr. Stephen J. Garrett
Department of Mathematics, University of Leicester
January 2013

CHAPTER 1

Introduction

1.1 THE CONCEPT OF *INTEREST*

Interest may be regarded as a reward paid by one person or organization (the *borrower*) for the use of an asset, referred to as *capital*, belonging to another person or organization (the *lender*). The precise conditions of any transaction will be mutually agreed. For example, after a stated period of time, the capital may be returned to the lender with the interest due. Alternatively, several interest payments may be made before the borrower finally returns the asset.

Capital and interest need not be measured in terms of the same commodity, but throughout this book, which relates primarily to problems of a financial nature, we shall assume that both are measured in the monetary units of a given currency. When expressed in monetary terms, capital is also referred to as *principal*.

If there is some risk of default (i.e., loss of capital or non-payment of interest), a lender would expect to be paid a higher rate of interest than would otherwise be the case; this additional interest is known as the *risk premium*. The additional interest in such a situation may be considered as a further reward for the lender's acceptance of the increased risk. For example, a person who uses his money to finance the drilling for oil in a previously unexplored region would expect a relatively high return on his investment if the drilling is successful, but might have to accept the loss of his capital if no oil were to be found. A further factor that may influence the rate of interest on any transaction is an allowance for the possible depreciation or appreciation in the value of the currency in which the transaction is carried out. This factor is obviously very important in times of high inflation.

It is convenient to describe the operation of interest within the familiar context of a savings account, held in a bank, building society, or other similar organization. An investor who had opened such an account some time ago with an initial deposit of £100, and who had made no other payments to or from the account, would expect to withdraw more than £100 if he were now to close the account. Suppose, for example, that he receives £106 on closing his account.

CONTENTS

An Introduction to the Mathematics of Finance. http://dx.doi.org/10.1016/B978-0-08-098240-3.00001-1

1

This sum may be regarded as consisting of £100 as the return of the initial deposit and £6 as interest. The interest is a payment by the bank to the investor for the use of his capital over the duration of the account.

The most elementary concept is that of *simple* interest. This naturally leads to the idea of *compound* interest, which is much more commonly found in practice in relation to all but short-term investments. Both concepts are easily described within the framework of a savings account, as described in the following sections.

1.2 SIMPLE INTEREST

Suppose that an investor opens a savings account, which pays simple interest at the rate of 9% per annum, with a single deposit of £100. The account will be credited with £9 of interest for each complete year the money remains on deposit. If the account is closed after 1 year, the investor will receive £109; if the account is closed after 2 years, he will receive £118, and so on. This may be summarized more generally as follows.

If an amount C is deposited in an account that pays simple interest at the rate of i per annum and the account is closed after n years (there being no intervening payments to or from the account), then the amount paid to the investor when the account is closed will be

$$C(1 + ni) \qquad (1.2.1)$$

This payment consists of a return of the initial deposit C, together with interest of amount

$$niC \qquad (1.2.2)$$

In our discussion so far, we have implicitly assumed that, in each of these last two expressions, n is an integer. However, the normal commercial practice in relation to fractional periods of a year is to pay interest on a pro rata basis, so that Eqs 1.2.1 and 1.2.2 may be considered as applying for *all* non-negative values of n.

Note that if the annual rate of interest is 12%, then $i = 0.12$ per annum; if the annual rate of interest is 9%, then $i = 0.09$ per annum; and so on.

Note that in the solution to Example 1.2.1, we have assumed that 6 months and 10 months are periods of 1/2 and 10/12 of 1 year, respectively. For accounts of duration less than 1 year, it is usual to allow for the actual number of *days* an account is held, so, for example, two 6-month periods are not necessarily regarded as being of equal length. In this case Eq. 1.2.1 becomes

---

EXAMPLE 1.2.1

Suppose that £860 is deposited in a savings account that pays simple interest at the rate of 5.375% per annum. Assuming that there are no subsequent payments to or from the account, find the amount finally withdrawn if the account is closed after

(a) 6 months,
(b) 10 months,
(c) 1 year.

Solution

The interest rate is given as a per annum value; therefore, n must be measured in years. By letting $n = 6/12$, $10/12$, and

1 in Eq. 1.2.1 with $C = 860$ and $i = 0.05375$, we obtain the answers

(a) £883.11,
(b) £898.52,
(c) £906.23.

In each case we have given the answer to two decimal places of one pound, rounded down. This is quite common in commercial practice.

EXAMPLE 1.2.2

Calculate the price of a 30-day £2,000 **treasury bill** issued by the government at a simple rate of discount of 5% per annum.

Solution

By issuing the treasury bill, the government is borrowing an amount equal to the price of the bill. In return, it pays £2,000 after 30 days. The price is given by

$$£2,000\left(1 - \frac{30}{365} \times 0.05\right) = £1,991.78$$

The investor has received interest of £8.22 under this transaction.

$$C\left(1 + \frac{mi}{365}\right) \tag{1.2.3}$$

where m is the duration of the account, measured in days, and i is the annual rate of interest.

The essential feature of simple interest, as expressed algebraically by Eq. 1.2.1, is that interest, once credited to an account, does not itself earn further interest. This leads to inconsistencies that are avoided by the application of *compound interest* theory, as discussed in Section 1.3.

As a result of these inconsistencies, simple interest has limited practical use, and this book will, necessarily, focus on compound interest. However, an important commercial application of simple interest is **simple discount**, which is commonly used for short-term loan transactions, i.e., up to 1 year. Under simple discount, the amount lent is determined by subtracting a *discount* from the amount due at the later date. If a lender bases his short-term transactions on a simple rate of discount d, then, in return for a repayment of X after a period t (typically $t < 1$), he will lend $X(1 - td)$ at the start of the period. In this situation, d is also known as a *rate of commercial discount*.

1.3 COMPOUND INTEREST

Suppose now that a certain type of savings account pays simple interest at the rate of i per annum. Suppose further that this rate is guaranteed to apply throughout the next 2 years and that accounts may be opened and closed at any time. Consider an investor who opens an account at the present time ($t = 0$) with an initial deposit of C. The investor may close this account after 1 year ($t = 1$), at which time he will withdraw $C(1 + i)$ (see Eq. 1.2.1). He may then place this sum on deposit in a new account and close this second account after one further year ($t = 2$). When this latter account is closed, the sum withdrawn (again see Eq. 1.2.1) will be

$$[C(1 + i)] \times (1 + i) \; = \; C(1 + i)^2 \; = \; C(1 + 2i + i^2)$$

If, however, the investor chooses not to switch accounts after 1 year and leaves his money in the original account, on closing this account after 2 years, he will receive $C(1 + 2i)$. Therefore, simply by switching accounts in the middle of the 2-year period, the investor will receive an *additional* amount i^2C at the end of the period. This extra payment is, of course, equal to $i(iC)$ and arises as interest paid (at $t = 2$) on the interest credited to the original account at the end of the first year.

From a practical viewpoint, it would be difficult to prevent an investor switching accounts in the manner described here (or with even greater frequency). Furthermore, the investor, having closed his second account after 1 year, could then deposit the entire amount withdrawn in yet another account. Any bank would find it administratively very inconvenient to have to keep opening and closing accounts in the manner just described. Moreover, on closing one account, the investor might choose to deposit his money elsewhere. Therefore, partly to encourage long-term investment and partly for other practical reasons, it is common commercial practice (at least in relation to investments of duration greater than 1 year) to pay *compound interest* on savings accounts. Moreover, the concepts of compound interest are used in the assessment and evaluation of investments as discussed throughout this book.

The essential feature of compound interest is that *interest itself earns interest*. The operation of compound interest may be described as follows: consider a savings account, which pays compound interest at rate i per annum, into which is placed an initial deposit C at time $t = 0$. (We assume that there are no further payments to or from the account.) If the account is closed after 1 year $(t = 1)$ the investor will receive $C(1 + i)$. More generally, let A_n be the amount that will be received by the investor if he closes the account after n years $(t = n)$. It is clear that $A_1 = C(1 + i)$. *By definition,* the amount received by the investor on closing the account at the end of any year is equal to the amount he would have received if he had closed the account 1 year previously plus further interest of i times this amount. The interest credited to the account up to the start of the final year itself earns interest (at rate i per annum) over the final year. Expressed algebraically, this definition becomes

$$A_{n+1} = A_n + iA_n$$

or

$$A_{n+1} = (1 + i)A_n \qquad n \geq 1 \qquad (1.3.1)$$

Since, by definition, $A_1 = C(1 + i)$, Eq.1.3.1 implies that, for $n = 1, 2, \ldots,$

$$A_n = C(1 + i)^n \qquad (1.3.2)$$

Therefore, if the investor closes the account after n years, he will receive

$$C(1 + i)^n \qquad (1.3.3)$$

This payment consists of a return of the initial deposit C, together with ***accumulated*** interest (i.e., interest which, if $n > 1$, has itself earned further interest) of amount

$$C[(1 + i)^n - 1] \qquad (1.3.4)$$

In our discussion so far, we have assumed that in both these last expressions n is an integer. However, in Chapter 2 we will widen the discussion and show that, under very general conditions, Eqs 1.3.3 and 1.3.4 remain valid for *all* non-negative values of n.

Since

$$[C(1 + i)^{t_1}](1 + i)^{t_2} = C(1 + i)^{t_1 + t_2}$$

an investor who is able to switch his money between two accounts, both of which pay compound interest at the same rate, is *not* able to profit by such action. This is in contrast with the somewhat anomalous situation, described at the beginning of this section, which may occur if simple interest is paid.

Equations 1.3.3 and 1.3.4 should be compared with the corresponding expressions under the operation of simple interest (i.e., Eqs 1.2.1 and 1.2.2). If interest *compounds* (i.e., earns further interest), the effect on the accumulation of an account can be very significant, especially if the duration of the account or

EXAMPLE 1.3.1

Suppose that £100 is deposited in a savings account. Construct a table to show the accumulated amount of the account after 5, 10, 20, and 40 years on the assumption that compound interest is paid at the rate of

(a) 4% per annum,
(b) 8% per annum.

Give also the corresponding figures on the assumption that only simple interest is paid at the same rate.

Solution

From Eqs 1.2.1 and 1.3.3 we obtain the following values. The reader should verify the figures by direct calculation and also by the use of standard compound interest tables.

Term (Years)	Annual Rate of Interest 4%		Annual Rate of Interest 8%	
	Simple	Compound	Simple	Compound
5	£120	£121.67	£140	£146.93
10	£140	£148.02	£180	£215.89
20	£180	£219.11	£260	£466.10
40	£260	£480.10	£420	£2,172.45

the rate of interest is great. This is revisited mathematically in Section 2.1, but is illustrated by Example 1.3.1.

Note that in Example 1.3.1, compound interest over 40 years at 8% per annum accumulates to more than five times the amount of the corresponding account with simple interest. The exponential growth of money under compound interest and its linear growth under simple interest are illustrated in Figure 1.3.1 for the case when $i = 0.08$.

As we have already indicated, compound interest is used in the assessment and evaluation of investments. In the final section of this chapter, we describe briefly several kinds of situations that can typically arise in practice. The analyses of these types of problems are among those discussed later in this book.

1.4 SOME PRACTICAL ILLUSTRATIONS

As a simple illustration, consider an investor who is offered a contract with a financial institution that provides £22,500 at the end of 10 years in return for a single payment of £10,000 now. If the investor is willing to tie up this amount of capital for 10 years, the decision as to whether or not he enters into the contract will depend upon the alternative investments available. For example, if the investor can obtain elsewhere a guaranteed compound rate of interest for the next 10 years of 10% per annum, then he should not enter into the contract

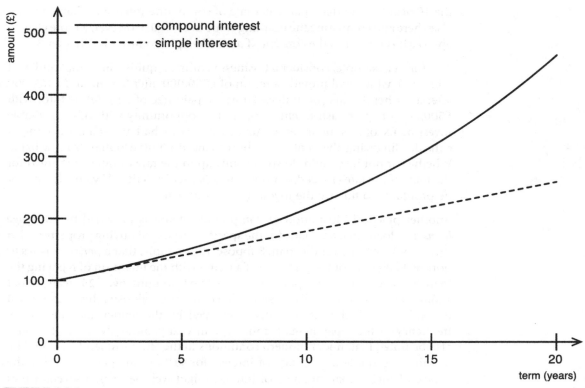

FIGURE 1.3.1
Accumulation of £100 with interest at 8% per annum

as, from Eq. 1.3.3, $£10,000 \times (1 + 10\%)^{10} = £25,937.42$, which is greater than £22,500.

However, if he can obtain this rate of interest with certainty only for the next 6 years, in deciding whether or not to enter into the contract, he will have to make a judgment about the rates of interest he is likely to be able to obtain over the 4-year period commencing 6 years from now. (Note that in these illustrations we ignore further possible complications, such as the effect of taxation or the reliability of the company offering the contract.)

Similar considerations would apply in relation to a contract which offered to provide a specified lump sum at the end of a given period in return for the payment of a series of premiums of stated (and often constant) amount at regular intervals throughout the period. Would an investor favorably consider a contract that provides £3,500 tax free at the end of 10 years in return for ten annual premiums, each of £200, payable at the start of each year? This question can be answered by considering the growth of each individual premium to the end of the 10-year term under a particular rate of compound interest available to him elsewhere and comparing the resulting value to £3,500. However, a more elegant approach is related to the concept of *annuities* as introduced in Chapter 3.

As a further example, consider a business venture, requiring an initial outlay of £500,000, which will provide a return of £550,000 after 5 years and £480,000 after a further 3 years (both these sums are paid free of tax). An investor with £500,000 of spare cash might compare this opportunity with other available investments of a similar term. An investor who had no spare cash might consider financing the venture by borrowing the initial outlay from a bank. Whether or not he should do so depends upon the rate of interest charged for the loan. If the rate charged is more than a particular "critical" value, it will not be profitable to finance the investment in this way.

Another practical illustration of compound interest is provided by *mortgage loans*, i.e., loans that are made for the specific purpose of buying property which then acts as security for the loan. Suppose, for example, that a person wishes to borrow £200,000 for the purchase of a house with the intention of repaying the loan by regular periodic payments of a fixed amount over 25 years. What should be the amount of each regular repayment? Obviously, this amount will depend on both the rate of interest charged by the lender and the precise frequency of the repayments (monthly, half-yearly, annually, etc.). It should also be noted that, under modern conditions in the UK, most lenders would be unwilling to quote a fixed rate of interest for such a long period. During the course of such a loan, the rate of interest might well be revised several times (according to market conditions), and on each revision there would be a corresponding change in either the amount of the borrower's regular repayment or in the outstanding term of the loan. Compound interest techniques enable the revised amount of the repayment or the new outstanding term to be found in such cases. Loan repayments are considered in detail in Chapter 5.

One of the most important applications of compound interest lies in the analysis and evaluation of investments, particularly *fixed-interest* securities. For example, assume that any *one* of the following series of payments may be purchased for £1,000 by an investor who is not liable to tax:

(i)   Income of £120 per annum payable in arrears at yearly intervals for 8 years, together with a payment of £1,000 at the end of 8 years;

(ii) Income of £90 per annum payable in arrears at yearly intervals for 8 years, together with a payment of £1,300 at the end of 8 years;

(iii) A series of eight payments, each of amount £180, payable annually in arrears.

The first two of the preceding may be considered as typical fixed-interest securities. The third is generally known as a level *annuity* (or, more precisely, a level *annuity certain*, as the payment timings and amounts are *known* in advance), payable for 8 years, in this case. In an obvious sense, the *yield* (or return) on the first investment is 12% per annum. Each year the investor receives an income of 12% of his outlay until such time as this outlay is repaid. However, it is less clear what is meant by the yield on the second or third investments. For the second investment, the annual income is 9% of the purchase price, but the final payment after 8 years exceeds the purchase price. Intuitively, therefore, one would consider the second investment as providing a yield greater than 9% per annum. How much greater? Does the yield on the second investment exceed that on the first? Furthermore, what is the yield on the third investment? Is the investment with the highest yield likely to be the most profitable? The appraisal of investment and project opportunities is considered in Chapter 6 and fixed-interest investments in particular are considered in detail in Chapters 7 and 8.

In addition to considering the theoretical analysis and numerous practical applications of compound interest, this book provides an introduction to *derivative pricing* in Chapters 10 and 11. In particular, we will demonstrate that compound interest plays a crucial role at the very heart of modern financial mathematics. Furthermore, despite this book having a clear focus on *deterministic* techniques, we end with a description of *stochastic* modeling techniques in Chapter 12.

SUMMARY

- *Interest* is the reward paid by the *borrower* for the use of money, referred to as *capital* or *principal*, belonging to the *lender*.
- Under the action of *simple interest*, interest is paid only on the principal amount and previously earned interest *does not* earn interest itself. A principal amount of C invested under simple interest at a rate of i per annum for n years will accumulate to

$$C(1 + in)$$

- Under the action of *compound interest*, interest is paid on previously earned interest. A principal amount of C invested under compound interest at a rate of i per annum for years will accumulate to

$$C(1 + i)^n$$

- Compound interest is used in practice for all but very short-term investments.

CHAPTER 2

Theory of Interest Rates

In this chapter we introduce the standard notation and concepts used in the study of compound interest problems throughout this book. We discuss the fundamental concepts of *accumulation, discount,* and *present values* in the context of discrete and continuous cash flows. Much of the material presented here will be considered in more detail in later chapters of this book; this chapter should therefore be considered as fundamental to all that follows.

2.1 THE RATE OF INTEREST

We begin by considering investments in which capital and interest are paid at the end of a fixed term, there being no intermediate interest or capital payments. This is the simplest form of a *cash flow*. An example of this kind of investment is a short-term deposit in which the lender invests £1,000 and receives a return of £1,035 6 months later; £1,000 may be considered to be a repayment of capital and £35 a payment of interest, i.e., the reward for the use of the capital for 6 months.

It is essential in any compound interest problem to define the unit of time. This may be, for example, a month or a year, the latter period being frequently used in practice. In certain situations, however, it is more appropriate to choose a different period (e.g., 6 months) as the basic time unit. As we shall see, the choice of time scale often arises naturally from the information one has.

Consider a unit investment (i.e., of 1) for a period of 1 time unit, commencing at time t, and suppose that $1 + i(t)$ is returned at time $t + 1$. We call $i(t)$ the *rate of interest* for the period t to $t + 1$. One sometimes refers to $i(t)$ as the *effective rate of interest* for the period, to distinguish it from *nominal* and *flat* rates of interest, which will be discussed later. If it is assumed that the rate of interest does not depend on the amount invested, the cash returned at time $t + 1$ from an investment of C at time t is $C[1 + i(t)]$. (Note that in practice a higher rate of interest may be obtained from a large investment than from a small one, but we ignore this point here and throughout this book.)

CONTENTS

11

An Introduction to the Mathematics of Finance. http://dx.doi.org/10.1016/B978-0-08-098240-3.00002-3

Recall from Chapter 1 that the defining feature of *compound interest* is that it is earned on previously earned interest; with this in mind, the **accumulation** of C from time $t = 0$ to time $t = n$ (where n is some positive integer) is

$$C[1 + i(0)][1 + i(1)] \cdots [1 + i(n - 1)] \qquad (2.1.1)$$

This is true since proceeds $C[1 + i(0)]$ at time 1 may be invested at this time to produce $C[1 + i(0)][1 + i(1)]$ at time 2, and so on.

Rates of interest are often quoted as percentages. For example, we may speak of an effective rate of interest (for a given period) of 12.75%. This means that the effective rate of interest for the period is 0.1275. As an example, £100 invested at 12.75% per annum will accumulate to £100 × (1 + 0.1275) = £112.75 after 1 year. Alternatively, £100 invested at 12.75% per 2-year period would have accumulated to £112.75 after 2 years. Computing the equivalent rate of return over different units of time is an essential skill that we will return to later in this chapter.

If the rate of interest per period does not depend on the time t at which the investment is made, we write $i(t) = i$ for all t. In this case the accumulation of an investment of C for *any* period of length n time units is, by Eq. 2.1.1,

$$C(1 + i)^n \qquad (2.1.2)$$

This formula, which will be shown later to hold (under particular assumptions) even when n is not an integer, is referred to as the **accumulation** of C for n time units under *compound interest* at rate i per time unit.

The corresponding accumulation under **simple interest** at rate i per time unit is defined, as in Chapter 1, as

$$C(1 + in) \qquad (2.1.3)$$

This last formula may also be considered to hold for any positive n, not necessarily an integer.

It is interesting to note the connection between the Taylor expansion of the formula for an n-year accumulation of a unit investment under compound interest, Eq. 2.1.2, and that for an accumulation under simple interest, Eq. 2.1.3

$$C(1 + i)^n = C(1 + in + O(i^2))$$

In particular, we see that, for *small* compound interest rates, the higher order terms are negligible and the two expressions are approximately equal. This reflects that for small interest rates the interest earned on interest would be negligible. A comparison of the accumulations under simple and compound interest was given in Example 1.3.1.

EXAMPLE 2.1.1

The rate of compound interest on a certain bank deposit account is 4.5% per annum effective. Find the accumulation of £5,000 after 7 years in this account.

Solution

The interest rate is fixed for the period. By Eq. 2.1.2, the accumulation is

$$5,000(1.045)^7 = 5,000 \times 1.36086 = £6,804.31$$

EXAMPLE 2.1.2

The effective compounding rate of interest per annum on a certain building society account is currently 7%, but in 2 years' time it will be reduced to 6%. Find the accumulation in 5 years' time of an investment of £4,000 in this account.

Solution

The interest rate is fixed at 7% per annum for the first 2 years and then fixed at 6% per annum for the following 3 years. It is

necessary to consider accumulations over these two periods separately, and, by Eq. 2.1.1, the total accumulation is

$$4,000(1.07)^2(1.06)^3 = £5,454.38$$

The approach taken in Example 2.1.2 is standard practice where interest rates are fixed within two or more subintervals within the period of the investment. It is an application of the *principle of consistency*, introduced in Section 2.3.

2.2 NOMINAL RATES OF INTEREST

Now consider transactions for a term of length h time units, where $h > 0$ and need not be an integer. We define $i_h(t)$, the *nominal rate of interest* per unit time on transactions of term h beginning at time t, to be such that the effective rate of interest for the period of length h beginning at time t is $h i_h(t)$. Therefore, if the sum of C is invested at time t for a term h, the sum to be received at time $t + h$ is, *by definition*,

$$C[1 + h i_h(t)] \tag{2.2.1}$$

If $h = 1$, the nominal rate of interest coincides with the effective rate of interest for the period to $t = 1$, so

$$i_1(t) = i(t) \tag{2.2.2}$$

In many practical applications, $i_h(t)$ does not depend on t, in which case we may write

$$i_h(t) = i_h \quad \text{for all } t \tag{2.2.3}$$

EXAMPLE 2.2.1

If the nominal rate of interest is 12% per annum on transactions of term a month, calculate the accumulation of £100 invested at this rate after 1 month.

Solution

We have $h = 1/12$ and $i_{1/12}(t) = 12\%$ per annum. The effective monthly rate is therefore 1%, and the accumulation of £100 invested over 1 month is then $£100(1 + 1\%) = £101$.

EXAMPLE 2.2.2

If the nominal rate of interest is 12% per annum on transactions of term 2 years, calculate the accumulation of £100 invested at this rate over 2 years.

Solution

We have $h = 2$ and $i_2(t) = 12\%$ per annum. The effective rate of interest over a 2-year period is therefore 24%, and the accumulation of £100 invested over 2 years is then $£100(1 + 24\%) = £124$.

If, in this case, we also have $h = 1/p$, where p is a positive integer (i.e., h is a simple fraction of a time unit), it is more usual to write $i^{(p)}$ rather than $i_{1/p}$. We therefore have

$$i^{(p)} = i_{1/p} \tag{2.2.4}$$

It follows that a unit investment for a period of length $1/p$ will produce a return of

$$1 + \frac{i^{(p)}}{p} \tag{2.2.5}$$

Note that $i^{(p)}$ is often referred to as a nominal rate of interest per unit time *payable pthly*, or *convertible pthly*, or *with pthly rests*. In Example 2.2.1, $i_{1/12} = i^{(12)} = 12\%$ is the yearly rate of *nominal interest converted monthly*, such that the effective rate of interest is $i = \dfrac{12\%}{12} = 1\%$ per month. See Chapter 4 for a fuller discussion of this topic.

Nominal rates of interest are often quoted in practice; however, it is important to realize that these need to be converted to effective rates to be used in calculations, as was done in Examples 2.2.1 and 2.2.2. This is further demonstrated in Example 2.2.3.

EXAMPLE 2.2.3

The nominal rates of interest per annum quoted in the financial press for local authority deposits on a particular day are as follows:

Term	Nominal Rate of Interest (%)
1 day	11.75
2 days	11.625
7 days	11.5
1 month	11.375
3 months	11.25

(Investments of term 1 day are often referred to as **overnight money**.) Find the accumulation of an investment at this time of £1,000 for

(a) 1 week,
(b) 1 month.

Solution

To express the preceding information in terms of our notation, we draw up the following table in which the unit of time is 1 year and the particular time is taken as t_0:

Term h	1/365	2/365	7/365	1/12	1/4
$i_h(t_0)$	0.1175	0.11625	0.115	0.11375	0.1125

By Eq. 2.2.1, the accumulations are $1,000[1 + h i_h(t_0)]$ where a) $h = 7/365$ and b) $h = 1/12$. This gives the answers

(a) $1,000\left(1 + \dfrac{7}{365} \times 0.115\right) = £1,002.21$,

(b) $1,000\left(1 + \dfrac{1}{12} \times 0.11375\right) = £1,009.48$.

Note that the nominal rates of interest for different terms (as illustrated by Example 2.2.3) are liable to vary from day to day: they should not be assumed to be fixed. If they were constant with time and equal to the above values, an investment of £1,000,000 for two successive 1-day periods would

accumulate to $£1,000,000 \times \left[1 + 0.1175 \times \dfrac{1}{365}\right]^2 = £1,000,644$, whereas an investment for a single 2-day term would give $£1,000,000 \times$

$\left[1 + 0.11625 \times \dfrac{2}{365}\right] = £1,000,637$. This apparent inconsistency may be

explained (partly) by the fact that the market expects interest rates to change in the future. These ideas are related to the *term structure of interest rates*, which will be discussed in detail in Chapter 9. We return to nominal rates of interest in Chapter 4.

2.3 ACCUMULATION FACTORS

As has been implied so far, investments are made in order to exploit the growth of money under the action of compound interest as time goes forward. In order to quantify this growth, we introduce the concept of *accumulation factors*.

Let time be measured in suitable units (e.g., years); for $t_1 \leq t_2$ we define $A(t_1, t_2)$ to be the accumulation at time t_2 of a unit investment made at time t_1 for a term of $(t_2 - t_1)$. It follows by the definition of $i_h(t)$ that, for all t and for all $h > 0$, the accumulation over a time unit of length h is

$$A(t, t + h) = 1 + h i_h(t) \qquad (2.3.1)$$

and hence that

$$i_h(t) = \frac{A(t, t + h) - 1}{h} \qquad h > 0 \qquad (2.3.2)$$

The quantity $A(t_1, t_2)$ is often called an *accumulation factor*, since the accumulation at time t_2 of an investment of the sum C at time t_1 is

$$CA(t_1, t_2) \qquad (2.3.3)$$

We define $A(t, t) = 1$ for all t, reflecting that the accumulation factor must be unity over zero time.

In relation to the past, i.e., when the present moment is taken as time 0 and t and $t + h$ are both less than or equal to 0, the factors $A(t, t + h)$ and the nominal rates of interest $i_h(t)$ are a matter of recorded fact in respect of any given transaction. As for their values in the future, estimates must be made (unless one invests in fixed-interest securities with guaranteed rates of interest applying both now and in the future).

Now let $t_0 \leq t_1 \leq t_2$ and consider an investment of 1 at time t_0. The proceeds at time t_2 will be $A(t_0, t_2)$ if one invests at time t_0 for term $t_2 - t_0$, or $A(t_0, t_1) \times A(t_1, t_2)$ if one invests at time t_0 for term $t_1 - t_0$ and then, at time t_1, reinvests the proceeds for term $t_2 - t_1$. In a consistent market, these proceeds should not depend on the course of action taken by the investor. Accordingly, we say that under the **principle of consistency**

$$A(t_0, t_2) = A(t_0, t_1) A(t_1, t_2) \qquad (2.3.4)$$

for all $t_0 \leq t_1 \leq t_2$. It follows easily by induction that, if the principle of consistency holds,

$$A(t_0, t_n) = A(t_0, t_1) A(t_1, t_2) \cdots A(t_{n-1}, t_n) \qquad (2.3.5)$$

for any n and any increasing set of numbers $t_0, t_1, ..., t_n$.

Unless it is stated otherwise, one should assume that the principle of consistency holds. In practice, however, it is unlikely to be realized exactly because of dealing expenses, taxation, and other factors. Moreover, it is sometimes true that the accumulation factors implied by certain mathematical models do not in general satisfy the principle of consistency. It will be shown in Section 2.4 that, under very general conditions, accumulation factors satisfying the principle of consistency must have a particular form (see Eq. 2.4.3).

> **EXAMPLE 2.3.1**
>
> Let time be measured in years, and suppose that, for all $t_1 \leq t_2$,
>
> $$A(t_1, t_2) = \exp[0.05(t_2 - t_1)]$$
>
> Verify that the principle of consistency holds and find the accumulation 15 years later of an investment of £600 made at any time.
>
> **Solution**
> Consider $t_1 \leq s \leq t_2$: from the principle of consistency, we expect
>
> $$A(t_1, t_2) = A(t_1, s) \times A(s, t_2)$$
>
> The right side of this expression can be written as
>
> $$\exp[0.05(s - t_1)] \times \exp[0.05(t_2 - s)] = \exp[0.05(t_2 - t_1)]$$
> $$= A(t_1, t_2)$$
>
> which equals the left side, as required.
>
> By Eq. 2.3.3, the accumulation is $600e^{0.05 \times 15} = £1,270.20$

2.4 THE FORCE OF INTEREST

Equation 2.3.2 indicates how $i_h(t)$ is defined in terms of the accumulation factor $A(t, t + h)$. In Example 2.2.3 we gave (in relation to a particular time t_0) the values of $i_h(t_0)$ for a series of values of h, varying from 1/4 (i.e., 3 months) to 1/365 (i.e., 1 day). The trend of these values should be noted. In practical situations, it is not unreasonable to assume that, as h becomes smaller and smaller, $i_h(t)$ tends to a limiting value. In general, of course, this limiting value will depend on t. We therefore assume that for each value of t there is a number $\delta(t)$ such that

$$\lim_{h \to 0^+} i_h(t) = \delta(t) \tag{2.4.1}$$

The notation $h \to 0^+$ indicates that the limit is considered as h tends to zero "from above", i.e., through positive values. This is, of course, always true in the limit of a time interval tending to zero.

It is usual to call $\delta(t)$ the *force of interest per unit time at time t*. In view of Eq. 2.4.1, $\delta(t)$ is sometimes called the *nominal rate of interest per unit time at time t convertible momently*. Although it is a mathematical idealization of reality, the force of interest plays a crucial role in compound interest theory. Note that by combining Eqs 2.3.2 and 2.4.1, we may define $\delta(t)$ directly in terms of the accumulation factor as

$$\delta(t) = \lim_{h \to 0^+} \left[\frac{A(t, t + h) - 1}{h} \right] \tag{2.4.2}$$

The force of interest function $\delta(t)$ is defined in terms of the accumulation function $A(t_1, t_2)$, but when the principle of consistency holds, it is possible, under very general conditions, to express the accumulation factor in terms of the force of interest. This result is contained in Theorem 2.4.1.

THEOREM 2.4.1

If $\delta(t)$ and $A(t_0, t)$ are continuous functions of t for $t \geq t_0$, and the principle of consistency holds, then, for $t_0 \leq t_1 \leq t_2$

$$A(t_1, t_2) = \exp\left[\int_{t_1}^{t_2} \delta(t)dt\right] \qquad (2.4.3)$$

The proof of this theorem is given in Appendix 1, but essentially relies on the fact that Eq. 2.4.2 is the derivative of A with respect to time.

Equation 2.4.3 indicates the vital importance of the force of interest. As soon as $\delta(t)$, the force of interest per unit time, is specified, the accumulation factors $A(t_1, t_2)$ can be determined by Eq. 2.4.3. We may also find $i_h(t)$ by Eqs 2.4.3 and 2.3.2, and so

$$i_h(t) = \frac{\exp\left[\int_t^{t+h} \delta(s)ds\right] - 1}{h} \qquad (2.4.4)$$

The following examples illustrate the preceding discussion.

EXAMPLE 2.4.1

Assume that $\delta(t)$, the force of interest per unit time at time t, is given by

(a) $\delta(t) = \delta$ (where δ is some constant),
(b) $\delta(t) = a + bt$ (where a and b are some constants).

Find formulae for the accumulation of a unit investment from time t_1 to time t_2 in each case.

Solution
In case (a), Eq. 2.4.3 gives

$$A(t_1, t_2) = \exp[\delta \times (t_2 - t_1)]$$

and in case (b), we have

$$A(t_1, t_2) = \exp\left[\int_{t_1}^{t_2} (a + bt)dt\right]$$
$$= \exp\left[\left(at_2 + \tfrac{1}{2}bt_2^2\right) - \left(at_1 + \tfrac{1}{2}bt_1^2\right)\right]$$
$$= \exp\left[a(t_2 - t_1) + \frac{b}{2}\left(t_2^2 - t_1^2\right)\right]$$

The particular case that $\delta(t) = \delta$ for all t is of significant practical importance. It is clear that in this case

$$A(t_0, t_0 + n) = e^{\delta n} \qquad (2.4.5)$$

for *all* t_0 and $n \geq 0$. By Eq. 2.4.4, the effective rate of interest per time unit is

$$i = e^{\delta} - 1 \qquad (2.4.6)$$

and hence

$$e^{\delta} = 1 + i \qquad (2.4.7)$$

The accumulation factor $A(t_0, t_0 + n)$ may therefore be expressed in the alternative form

$$A(t_0, t_0 + n) = (1 + i)^n \qquad (2.4.8)$$

We therefore have a generalization of Eq. 2.1.2 to *all* $n \geq 0$, not merely the positive integers. Notation and theory may be simplified when $\delta(t) = \delta$ for all t. This case will be considered in detail in Chapter 3.

EXAMPLE 2.4.2

The force of interest per unit time, $\delta(t) = 0.12$ per annum for all t. Find the nominal rate of interest per annum on deposits of term (a) 7 days, (b) 1 month, and (c) 6 months.

Solution

Note that the natural time scale here is a year. Using Eq. 2.4.4 with $\delta(t) = 0.12$ for all t, we obtain, for all t,

$$i_h = i_h(t) = \frac{\exp(0.12h) - 1}{h}$$

Substituting (a) $h = \frac{7}{365}$, (b) $h = \frac{1}{12}$, and (c) $h = \frac{6}{12}$, we obtain the nominal rates of interest (a) 12.01%, (b) 12.06%, and (c) 12.37% per annum.

Let us now define

$$F(t) = A(t_0, t) \qquad (2.4.9)$$

where t_0 is fixed and $t_0 \leq t$. Therefore, $F(t)$ is the accumulation at time t of a unit investment at time t_0. By Eq. 2.4.3,

$$\ln F(t) = \int_{t_0}^{t} \delta(s)\,ds \qquad (2.4.10)$$

and hence we can express the force of interest in terms of the derivative of the accumulation factor, for $t > t_0$,

$$\delta(t) = \frac{d}{dt} \ln F(t) = \frac{F'(t)}{F(t)} \qquad (2.4.11)$$

Although we have assumed so far that $\delta(t)$ is a continuous function of time t, in certain practical problems we may wish to consider rather more general functions. In particular, we sometimes consider $\delta(t)$ to be piecewise. In such cases, Theorem 2.4.1 and other results are still valid. They may be established by considering $\delta(t)$ to be the limit, in a certain sense, of a sequence of continuous functions.

EXAMPLE 2.4.3

Measuring time in years, the force of interest paid on deposits to a particular bank are assumed to be

$$\delta(t) = \begin{cases} 0.06 & \text{for} \quad t < 5 \\ 0.05 & \text{for } 5 \leq t < 10 \\ 0.03 & \text{for} \quad t \geq 10 \end{cases}$$

Calculate the accumulated value after 12 years of an initial deposit of £1,000.

Solution

Using Eq. 2.4.3 and the principle of consistency, we can divide the investment term into subintervals defined by periods of constant interest. The accumulation is then considered as resulting from 5 years at 6%, 5 years at 5%, and 2 years at 3%, and we compute

$$£1,000e^{0.06\times5}e^{0.05\times5}e^{0.03\times2} = £1,840.43$$

2.5 PRESENT VALUES

In Section 2.3, accumulation factors were introduced to quantify the growth of an initial investment as time moves forward. However, one can consider the situation in the opposite direction. For example, if one has a future liability of known amount at a known future time, how much should one invest now (at known interest rate) to cover this liability when it falls due? This leads us to the concept of *present values*.

Let $t_1 \leq t_2$. It follows by Eq. 2.3.3 that an investment of $\frac{C}{A(t_1,t_2)}$, i.e., $C\exp(-\int_{t_1}^{t_2}\delta(t)dt)$, at time t_1 will produce a return of C at time t_2. We therefore say that the *discounted value* at time t_1 of C due at time t_2 is

$$C\exp\left[-\int_{t_1}^{t_2}\delta(t)dt\right] \tag{2.5.1}$$

This is the sum of money which, if invested at time t_1, will give C at time t_2 under the action of the known force of interest, $\delta(t)$. In particular, the discounted value at time 0 of C due at time $t \geq 0$ is called its *discounted present value* (or, more briefly, its *present value*); it is equal to

$$C\exp\left[-\int_{0}^{t}\delta(s)ds\right] \tag{2.5.2}$$

We now define the function

$$v(t) = \exp\left[-\int_0^t \delta(s)ds\right] \qquad (2.5.3)$$

When $t \geq 0$, $v(t)$ is the **(discounted) present value** of 1 due at time t. When $t < 0$, the convention $\int_0^t \delta(s)ds = -\int_t^0 \delta(s)ds$ shows that $v(t)$ is the **accumulation** of 1 from time t to time 0. It follows by Eqs 2.5.2 and 2.5.3 that the discounted present value of C due at a non-negative time t is

$$Cv(t) \qquad (2.5.4)$$

In the important practical case in which $\delta(t) = \delta$ for all t, we may write

$$v(t) = v^t \qquad \text{for all } t \qquad (2.5.5)$$

where $v = v(1) = e^{-\delta}$. Expressions for v can be easily related back to the interest rate quantities i and $i^{(p)}$; this is further discussed in Chapter 3. The values of $v^t (t = 1, 2, 3, ...)$ at various interest rates are included in standard compound interest tables, including those at the end of this book.

EXAMPLE 2.5.1

What is the present value of £1,000 due in 10 years' time if the effective interest rate is 5% per annum?

Solution

Using Eq. 2.5.4 and that $e^\delta = 1 + i$, the present value is

$$£1,000 \times v^{10} = 1,000 \times (1.05)^{-10} = £613.91$$

EXAMPLE 2.5.2

Measuring time in years from the present, suppose that $\delta(t) = 0.06 \times (0.9)^t$ for all t. Find a simple expression for $v(t)$, and hence find the discounted present value of £100 due in 3.5 years' time.

Solution

By Eq. 2.5.3,

$$v\left(t\right) = \exp\left[-\int_0^t 0.06(0.9)^s ds\right]$$

$$= \exp[-0.06(0.9^t - 1)/\ln(0.9)]$$

Hence, the present value of £100 due in 3.5 years' time is, by Eq. 2.5.4,

$$100\exp\left[-0.06(0.9^{3.5} - 1)/\ln\left(0.9\right)\right] = £83.89$$

EXAMPLE 2.5.3

Suppose that

$$\delta(t) = \begin{cases} 0.09 & \text{for } 0 \le t < 5 \\ 0.08 & \text{for } 5 \le t < 10 \\ 0.07 & \text{for } \quad t \ge 10 \end{cases}$$

Find simple expressions for $v(t)$ when $t \ge 0$.

Solution

The piecewise function for $\delta(t)$ means that $v(t)$ is also piece-wise. Note that for $5 \le t < 10$, we may evaluate $\int_0^t \delta(s)\mathrm{d}s$ as

$\left[\int_0^5 \delta(s)\mathrm{d}s + \int_5^t \delta(s)\mathrm{d}s \right]$ and that for $t \ge 10$ we may use the form $\left[\int_0^{10} \delta(s)\mathrm{d}s + \int_{10}^t \delta(s)\mathrm{d}s \right]$, with the appropriate formula for $\delta(s)$ in each case. This immediately gives

$$v(t) = \begin{cases} \exp(-0.09t) & \text{for } 0 \le t < 5 \\ \exp(-0.05 - 0.08t) & \text{for } 5 \le t < 10 \\ \exp(-0.15 - 0.07t) & \text{for } \quad t \ge 10 \end{cases}$$

2.6 PRESENT VALUES OF CASH FLOWS

In many compound interest problems, one is required to find the discounted present value of cash payments (or, as they are often called, *cash flows*) due in the future. It is important to distinguish between *discrete* and *continuous* payments.

Discrete Cash Flows

The present value of the sums $C_{t_1}, C_{t_2}, \ldots, C_{t_n}$ due at times $t_1, t_2, \ldots, t_n$ (where $0 \le t_1 < t_2 < \ldots < t_n$) is, by Eq. 2.5.4,

$$c_{t_1} v(t_1) + c_{t_2} v(t_2) + \cdots + c_{t_n} v(t_n) = \sum_{j=1}^{n} c_{t_j} v(t_j) \qquad (2.6.1)$$

If the number of payments is infinite, the present value is defined to be

$$\sum_{j=1}^{\infty} c_{t_j} v(t_j) \qquad (2.6.2)$$

provided that this series converges, which it usually will, in practical problems.

The process of finding discounted present values may be illustrated as in Figure 2.6.1. The discounting factors $v(t_1), v(t_2)$, and $v(t_3)$ are applied to bring each cash payment "back to the present time".

Continuously Payable Cash Flows (Payment Streams)

The concept of a continuously payable cash flow, although essentially theoretical, is important. For example, for many practical purposes, a pension that is payable weekly may be considered as payable continuously over an extended time period. Suppose that $T > 0$ and that between times 0 and T an investor

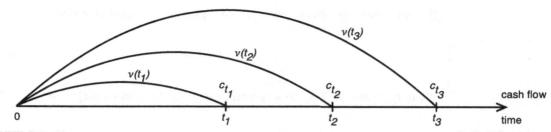

FIGURE 2.6.1
Discounted cash flow

will be paid money continuously, the rate of payment at time t being £$\rho(t)$ per unit time. What is the present value of this cash flow?

In order to answer this question, one needs to understand what is meant by the *rate of payment* of the cash flow at time t. If $M(t)$ denotes the *total* payment made between time 0 and time t, then, *by definition*,

$$\rho(t) = M'(t) \quad \text{for all } t \tag{2.6.3}$$

where the prime denotes differentiation with respect to time. Then, if $0 \le \alpha < \beta \le T$, the total payment received between time α and time β is

$$M\left(\beta\right) - M\left(\alpha\right) = \int_{\alpha}^{\beta} M'\left(t\right)dt$$

$$= \int_{\alpha}^{\beta} \rho(t)dt \tag{2.6.4}$$

The *rate of payment* at any time is therefore simply the derivative of the *total* amount paid up to that time, and the total amount paid between any two times is the integral of the *rate of payment* over the appropriate time interval.

Between times t and $t + dt$, the total payment received is $M(t + dt) - M(t)$. If dt is very small, this is approximately $M'(t)dt$ or $\rho(t)dt$. Theoretically, therefore, we may consider the present value of the money received between times t and $t + dt$ as $v(t)\rho(t)dt$. The present value of the entire cash flow is then obtained by integration as

$$\int_{0}^{T} v(t)\rho(t)dt \tag{2.6.5}$$

A rigorous proof of this result is given in textbooks on elementary analysis but is not necessary here; $\rho(t)$ will be assumed to satisfy an appropriate condition (e.g., that it is piecewise continuous).

If T is infinite, we obtain, by a similar argument, the present value

$$\int_0^\infty v(t)\rho(t)\mathrm{d}t \qquad (2.6.6)$$

We may regard Eq. 2.6.5 as a special case of Eq. 2.6.6 when $\rho(t) = 0$ for $t > T$.

EXAMPLE 2.6.1

Assume that time is measured in years, and that

$$\delta(t) = \begin{cases} 0.04 & \text{for } t < 10 \\ 0.03 & \text{for } t \geq 10 \end{cases}$$

Find $v(t)$ for all t, and hence find the present value of a continuous payment stream at the rate of 1 per annum for 15 years, beginning at time 0.

Solution

By Eq. 2.5.3,

$$v(t) = \begin{cases} \exp\left(-\int_0^t 0.04\mathrm{d}s\right) & \text{for } t < 10 \\ \exp\left(-\int_0^{10} 0.04\mathrm{d}s - \int_{10}^t 0.03\mathrm{d}s\right) & \text{for } t \geq 10 \end{cases}$$

i.e.,

$$v(t) = \begin{cases} \exp(-0.04t) & \text{for } t < 10 \\ \exp(-0.1 - 0.03t) & \text{for } t \geq 10 \end{cases}$$

The present value of the payment stream is, by Eq. 2.6.5 with $\rho(t) = 1$,

$$\int_0^{15} 1 \times v(t)\mathrm{d}t = \int_0^{10} \exp(-0.04t)\mathrm{d}t + \int_{10}^{15} \exp(-0.1 - 0.03t)\mathrm{d}t$$

$$= \frac{1 - \exp(-0.4)}{0.04} + \exp(-0.1)$$

$$\times \frac{\exp(-0.3) - \exp(-0.45)}{0.03}$$

$$= \pounds 11.35$$

By combining the results for discrete and continuous cash flows, we obtain the formula

$$\sum c_t v(t) + \int_0^\infty v(t)\rho(t)\mathrm{d}t \qquad (2.6.7)$$

for the present value of a general cash flow (the summation being over those values of t for which C_t, the discrete cash flow at time t, is non-zero).

So far we have assumed that all payments, whether discrete or continuous, are positive. If one has a series of incoming payments (which may be regarded as positive) and a series of outgoings (which may be regarded as negative), their *net present value* is defined as the difference between the value of the positive cash flow and the value of the negative cash flow. These ideas are further developed in the particular case that $\delta(t)$ is constant in later chapters.

2.7 VALUING CASH FLOWS

Consider times t_1 and t_2, where t_2 is not necessarily greater than t_1. The *value at time t_1 of the sum C due at time t_2* is defined as follows.

(a) if $t_1 \geq t_2$, the accumulation of C from time t_2 until time t_1, or
(b) if $t_1 < t_2$, the discounted value at time t_1 of C due at time t_2.

It follows by Eqs 2.4.3 and 2.5.1 that in both cases the value at time t_1 of C due at time t_2 is

$$C \exp\left[-\int_{t_1}^{t_2} \delta(t)dt \right] \qquad (2.7.1)$$

Note the convention that, if $t_1 > t_2$, $\int_{t_1}^{t_2} \delta(t)dt = -\int_{t_2}^{t_1} \delta(t)dt$.

Since

$$\int_{t_1}^{t_2} \delta(t)dt = \int_{0}^{t_2} \delta(t)dt - \int_{0}^{t_1} \delta(t)dt$$

it follows immediately from Eqs 2.5.3 and 2.7.1 that the value at time t_1 of C due at time t_2 is

$$C\frac{v(t_2)}{v(t_1)} \qquad (2.7.2)$$

The value at a general time t_1, of a discrete cash flow of c_t at time t (for various values of t) and a continuous payment stream at rate $\rho(t)$ per time unit, may now be found, by the methods given in Section 2.6, as

$$\sum c_t \frac{v(t)}{v(t_1)} + \int_{-\infty}^{\infty} \rho(t) \frac{v(t)}{v(t_1)} dt \qquad (2.7.3)$$

where the summation is over those values of t for which $c_t \neq 0$. We note that in the special case when $t_1 = 0$ (the present time), the value of the cash flow is

$$\sum c_t v(t) + \int_{-\infty}^{\infty} \rho(t)v(t)dt \qquad (2.7.4)$$

where the summation is over those values of t for which $c_t \neq 0$. This is a generalization of Eq. 2.6.7 to cover past, as well as present or future, payments.

If there are incoming and outgoing payments, the corresponding *net value* may be defined, as in Section 2.6, as the difference between the value of the *positive* and the *negative* cash flows. If all the payments are due at or after time t_1, their value at time t_1 may also be called their *discounted value*, and if they are due at or before time t_1, their value may be referred to as their *accumulation*. It follows that any value may be expressed as the sum of a discounted value and an accumulation; this fact is helpful in certain problems. Also, if $t_1 = 0$, and all the payments are due at or after the present time, their value may also be described as their (*discounted*) *present value*, as defined by Eq. 2.6.7.

It follows from Eq. 2.7.3 that the value at any time t_1 of a cash flow may be obtained from its value at another time t_2 by applying the factor $\dfrac{v(t_2)}{v(t_1)}$, i.e.,

$$\begin{bmatrix} \text{value at time } t_1 \\ \text{of cash flow} \end{bmatrix} = \begin{bmatrix} \text{value at time } t_2 \\ \text{of cash flow} \end{bmatrix}\begin{bmatrix} \dfrac{v(t_2)}{v(t_1)} \end{bmatrix} \quad (2.7.5)$$

or

$$\begin{bmatrix} \text{value at time } t_1 \\ \text{of cash flow} \end{bmatrix}\begin{bmatrix} v(t_1) \end{bmatrix} = \begin{bmatrix} \text{value at time } t_2 \\ \text{of cash flow} \end{bmatrix}\begin{bmatrix} v(t_2) \end{bmatrix} \quad (2.7.6)$$

Each side of Eq. 2.7.6 is the value of the cash flow at the present time (time $t = 0$).

In particular, by choosing time t_2 as the present time and letting $t_1 = t$, we obtain the result

$$\begin{bmatrix} \text{value at time } t \\ \text{of cash flow} \end{bmatrix} = \begin{bmatrix} \text{value at the present} \\ \text{time of cash flow} \end{bmatrix}\begin{bmatrix} \dfrac{1}{v(t)} \end{bmatrix} \quad (2.7.7)$$

As we shall see later in this book, these results are extremely useful in practical examples.

EXAMPLE 2.7.1

A businessman is owed the following amounts: £1,000 on 1 January 2013, £2,500 on 1 January 2014, and £3,000 on 1 July 2014. Assuming a constant force of interest of 0.06 per annum, find the value of these payments on

(a) 1 January 2011,
(b) 1 March 2012.

(b) The value of the same debts at 1 March 2012 is found by advancing the present value forwards by 14 months, by Eq. 2.7.7,

$$5,406.85 \, \exp\left(0.06 \times \frac{14}{12}\right) = £5,798.89$$

Solution
(a) Let time be measured in years from 1 January 2011. The value of the debts at that date is, by Eq. 2.7.1,

$$1,000v(2) + 2,500v(3) + 3,000v(3.5)$$
$$= 1,000\exp(-0.12) + 2,500\exp(-0.18)$$
$$+ 3,000\exp(-0.21) = £5,406.85$$

Note that the approach taken in Example 2.7.1 (b) is quicker than performing the calculation at 1 March 2012 from first principles.

EXAMPLE 2.7.2

The force of interest at any time t, measured in years, is given by

$$
\delta(t) = \begin{cases}
0.04 + 0.005t & \text{for } 0 \le t < 6 \\
0.16 - 0.015t & \text{for } 6 \le t < 8 \\
0.04 & \text{for } t \ge 8
\end{cases}
$$

(a) Calculate the value at time 0 of £100 due at time $t = 8$.
(b) Calculate the accumulated value at time $t = 10$ of a payment stream of rate $\rho(t) = 16 - 1.5t$ paid continuously between times $t = 6$ and $t = 8$.

Solution

(a) We need the present value of £100 at time 8, i.e., $100/A(0,8)$ with the accumulation factor

$$
\begin{aligned}
A(0,8) &= A(0,6) \times A(6,8) \\
&= \exp\left(\int_0^6 0.04 + 0.005t \, dt \right) \\
&\quad \times \exp\left(\int_6^8 0.16 - 0.015t \, dt \right) \\
&= \exp(0.44)
\end{aligned}
$$

leading to the present value of $£100/e^{0.44} = £64.40$

(b) The accumulated value is given by the accumulation of each payment element $\rho(t)\mathrm{d}t$ from time t to 10

$$
\int_6^8 A(t,10).\rho(t)\mathrm{d}t
$$

Using the principle of consistency, we can express the accumulation factor as easily found quantities, $A(0,10)$ and $A(0,t)$ as

$$
A(t,10) = \frac{A(0,10)}{A(0,t)} = e^{0.88 - 0.16t + 0.0075t^2}
$$

for $6 \le t \le 8$. The required present value is then obtained via integration by parts as £12.60.

2.8 INTEREST INCOME

Consider now an investor who wishes not to accumulate money but to receive an income while keeping his capital fixed at C. If the rate of interest is fixed at i per time unit, and if the investor wishes to receive his income at the end of each time unit, it is clear that his income will be iC per time unit, payable in arrears, until such time as he withdraws his capital.

More generally, suppose that $t > t_0$ and that an investor wishes to deposit C at time t_0 for withdrawal at time t. Suppose further that $n > 1$ and that the investor wishes to receive interest on his deposit at the n equally spaced times

$t_0 + h, t_0 + 2h, \ldots, t_0 + nh$, where $h = (t - t_0)/n$. The interest payable at time $t_0 + (j + 1)h$, for the period $t_0 + jh$ to $t_0 + (j + 1)h$, will be

$$Chi_h(t_0 + jh)$$

where $i_h(t)$ is the nominal rate over the period h starting at time t. The total interest income payable between times t_0 and t will then be

$$C\sum_{j=0}^{n-1} hi_h(t_0 + jh) \tag{2.8.1}$$

Since, by assumption, $i_h(t)$ tends to $\delta(t)$ as h tends to 0, it is fairly easily shown (provided that $\delta(t)$ is continuous) that as n increases (so that h tends to 0) the total interest received between times t_0 and t converges to

$$I(t) = C\int_{t_0}^{t} \delta(s)\mathrm{d}s \tag{2.8.2}$$

Hence, in the limit, the *rate of payment of interest income* per unit time at time t, $I'(t)$, equals

$$C\delta(t) \tag{2.8.3}$$

The position is illustrated in Figure 2.8.1. The cash C in the "tank" remains constant at C, while interest income is decanted continuously at the instantaneous rate $C\delta(t)$ per unit time at time t. If interest is paid very frequently from a variable-interest deposit account, the position may be idealized to that

FIGURE 2.8.1 Interest income flow

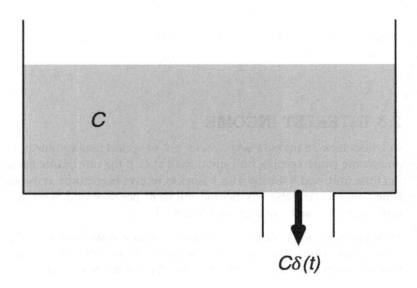

shown in the figure, which depicts a continuous flow of interest income. Of course, if $\delta(t) = \delta$ for all t, interest is received at the constant rate $C\delta$ per time unit.

If the investor withdraws his capital at time T, the present values of his income and capital are, by Eqs 2.5.4 and 2.6.5,

$$C \int_0^T \delta(t)v(t)\mathrm{d}t \tag{2.8.4}$$

and

$$Cv(T) \tag{2.8.5}$$

Since

$$\int_0^T \delta(t)v(t)\mathrm{d}t = \int_0^T \delta(t)\exp\left[-\int_0^t \delta(s)\mathrm{d}s\right]\mathrm{d}t$$

$$= \left[-\exp\left(-\int_0^t \delta(s)\mathrm{d}s\right)\right]_0^T$$

$$= 1 - v(T)$$

we obtain

$$C = C \int_0^T \delta(t)v(t)\mathrm{d}t + Cv(T) \tag{2.8.6}$$

as one would expect by general reasoning. In the case when $T = \infty$ (in which the investor never withdraws his capital), a similar argument gives the result that

$$C = C \int_0^\infty \delta(t)v(t)\mathrm{d}t \tag{2.8.7}$$

where the expression on the right side is the present value of the interest income. The case when $\delta(t) = \delta$ for all t is discussed further in Chapter 3.

2.9 CAPITAL GAINS AND LOSSES, AND TAXATION

So far we have described the difference between money returned at the end of the term and the cash originally invested as "interest". In practice, however, this quantity may be divided into *interest income* and *capital gains* (the term *capital loss* being used for a negative capital gain). Some investments, known as zero-coupon bonds, bear no interest income. Many other securities provide both interest income and capital gains; these will be considered later in this book. Since the basis of taxation of capital gains is usually different from that of interest income, the distinction between interest income and capital gains is of importance for tax-paying investors.

The theory developed in the preceding sections is unaltered if we replace the term *interest* by *interest and capital gains less any income and capital gains taxes*. The force of interest (which, to avoid any confusion of terminology, should perhaps be called the *force of growth*) will include an allowance for capital appreciation or depreciation, as well as interest income, and will also allow for the incidence of income and capital gains taxes on the investor.

Both income and capital gains tax are considered more fully in Chapters 7 and 8.

SUMMARY

- The *accumulation factor*, $A(t, T)$, gives the value, at time T, of a unit investment made at time $t < T$. If the investment is subject to an *effective rate* of compound interest i, then

$$A(t, T) = (1 + i)^{T-t}$$

- The *discount factor*, v^t, gives the *present value* at time zero of an investment that has unit value at time $t > 0$.

$$v^t = (1 + i)^{-t} = A(0, t)^{-1} = A(t, 0)$$

- The *principle of consistency* states that
  $A(t_0, t_n) = A(t_0, t_1)A(t_1, t_2)...A(t_{n-1}, t_n)$ for all $t_0 < t_1 < ... < t_{n-1} < t_n$. It is a common assumption on consistent markets.
- The *nominal rate of interest converted p*thly, $i^{(p)}$, is defined such that the effective rate of interest is $i = i^{(p)}/p$ per period of length $1/p$.
- The *force of interest* at time t can be defined by the expression
  $\delta(t) = \lim_{p \to \infty} i^{(p)}(t)$, i.e., is the nominal rate converted momentarily.
- The accumulation factor under the action of a force of interest between times t_1 and t_2 is

$$A(t_1, t_2) = e^{\int_{t_1}^{t_2} \delta(t)dt}$$

- The present value at $t = 0$ of a cash flow consisting of discrete payments C_{t_i} made at times t_i and a continuous payment stream of rate $\rho(t)$ is given by

$$\sum C_{t_i} v(t_i) + \int_0^\infty v(t)\rho(t)\mathrm{d}t$$

- The value of a cash flow at times t_1 and t_2 are connected by

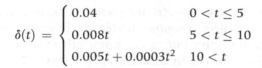

$$\begin{bmatrix} \text{value at time } t_1 \\ \text{of cash flow} \end{bmatrix}[v(t_1)] = \begin{bmatrix} \text{value at time } t_2 \\ \text{of cash flow} \end{bmatrix}[v(t_2)]$$

EXERCISES

2.1 Calculate the time in days for £1,500 to accumulate to £1,550 at
  (a) Simple rate of interest of 5% per annum,
  (b) A force of interest of 5% per annum.

Institute
and Faculty
of Actuaries

2.2 The force of interest $\delta(t)$ is a function of time and at any time t, measured in years, is given by the formula

Institute
and Faculty
of Actuaries

$$\delta(t) = \begin{cases} 0.04 & 0 < t \le 5 \\ 0.008t & 5 < t \le 10 \\ 0.005t + 0.0003t^2 & 10 < t \end{cases}$$

  (a) Calculate the present value of a unit sum of money due at time $t = 12$.
  (b) Calculate the effective annual rate of interest over the 12 years.
  (c) Calculate the present value at time $t = 0$ of a continuous payment stream that is paid at the rate of $e^{-0.05t}$ per unit time between time $t = 2$ and time $t = 5$.

2.3 Over a given year the force of interest per annum is a linear function of time, falling from 0.15 at the start of the year to 0.12 at the end of the year. Find the value at the start of the year of the nominal rate of interest per annum on transactions of term
  (a) 3 months,
  (b) 1 month,
  (c) 1 day.
  Find also the corresponding values midway through the year. (Note how these values tend to the force of interest at the appropriate time.)

2.4 A bank credits interest on deposits using a variable force of interest. At the start of a given year, an investor deposited £20,000 with the bank. The accumulated amount of the investor's account was £20,596.21 midway through the year and £21,183.70 at the end of the year. Measuring time in years from the start of the given year and assuming that over the year the force of interest per annum was a linear function of time, derive an expression for the force of interest per annum at time $t (0 \leq t \leq 1)$ and find the accumulated amount of the account three-quarters of the way through the year.

2.5 A borrower is under an obligation to repay a bank £6,280 in 4 years' time, £8,460 in 7 years' time, and £7,350 in 13 years' time. As part of a review of his future commitments the borrower now offers either

  (a) To discharge his liability for these three debts by making an appropriate single payment 5 years from now, or

  (b) To repay the total amount owed (i.e., £22,090) in a single payment at an appropriate future time.

On the basis of a constant force of interest per annum of $\delta = \ln 1.08$, find the appropriate single payment if offer (a) is accepted by the bank, and the appropriate time to repay the entire indebtedness if offer (b) is accepted.

2.6 Assume that $\delta(t)$, the force of interest per annum at time t (years), is given by the formula

$$\delta(t) \begin{cases} 0.08 & \text{for } 0 \leq t < 5 \\ 0.06 & \text{for } 5 \leq t < 10 \\ 0.04 & \text{for } \quad t \geq 10 \end{cases}$$

  (a) Derive expressions for $v(t)$, the present value of 1 due at time t.

  (b) An investor effects a contract under which he will pay 15 premiums annually in advance into an account which will accumulate according to the above force of interest. Each premium will be of amount £600, and the first premium will be paid at time 0. In return, the investor will receive either

   (i) The accumulated amount of the account 1 year after the final premium is paid; or

   (ii) A level annuity payable annually for 8 years, the first payment being made 1 year after the final premium is paid.

Find the lump sum payment under option (i) and the amount of the annual annuity under option (ii).

2.7 Suppose that the force of interest per annum at time t years is

$$\delta(t) = ae^{-bt}$$

  (a) Show that the present value of 1 due at time t is

$$v(t) = \exp\left[\frac{a}{b}\left(e^{-bt} - 1\right)\right]$$

  (b)   **(i)** Assuming that the force of interest per annum is as given above and that it will fall by 50% over 10 years from the value 0.10 at time 0, find the present value of a series of four annual payments, each of amount £1,000, the first payment being made at time 1.

(ii) At what *constant* force of interest per annum does this series of payments have the same present value as that found in (i)?

2.8 Suppose that the force of interest per annum at time t years is

$$\delta(t) = r + se^{-rt}$$

(a) Show that the present value of 1 due at time t is

$$v(t) = \exp\left(\frac{-s}{r}\right)\exp(-rt)\exp\left(\frac{s}{r}e^{-rt}\right)$$

(b) **(i)** Hence, show that the present value of an n-year continuous payment stream at a constant rate of £1,000 per annum is

$$\frac{1,000}{s}\left\{1 - \exp\left[\frac{s}{r}\left(e^{-rn} - 1\right)\right]\right\}$$

(ii) Evaluate the last expression when $n = 50$, $r = \ln 1.01$, and $s = 0.03$.

The Basic Compound Interest Functions

In this chapter we consider the particular case that the force of interest and therefore other interest rate quantities are independent of time. We define standard actuarial notation for the present values of simple payment streams called *annuities*, which can be used to construct more complicated payment streams in practical applications. Closed-form expressions to evaluate the present values and accumulations of various types of annuity are derived. The concept of an *equation of value* is discussed, which is of fundamental importance to the analysis of cash flows in various applications throughout the remainder of this book. Finally, we briefly discuss approaches to incorporating *uncertainty* into the analysis of cash flow streams.

CONTENTS

3.1 INTEREST RATE QUANTITIES

The particular case in which $\delta(t)$, the force of interest per unit time at time t, does *not* depend on t is of special importance. In this situation we assume that, for all values of t,

$$\delta(t) = \delta \tag{3.1.1}$$

where δ is some constant. Throughout this chapter, we shall assume that Eq. 3.1.1 is valid, unless otherwise stated.

The value at time s of 1 due at time $s + t$ is (see Eq. 2.7.1)

$$\exp\left[- \int_s^{s+t} \delta(r)dr \right] = \exp\left(- \int_s^{s+t} \delta dr \right)$$
$$= \exp(-\delta t)$$

which does *not* depend on s, only the time interval t. Therefore, the value at *any* given time of a unit amount due after a further period t is

$$v(t) = e^{-\delta t} \tag{3.1.2}$$

$$= v^t \tag{3.1.3}$$

$$= (1 - d)^t \tag{3.1.4}$$

35

where v and d are defined in terms of δ by the equations

$$v = e^{-\delta} \tag{3.1.5}$$

and

$$1 - d = e^{-\delta} \tag{3.1.6}$$

Then, in return for a repayment of a unit amount at time 1, an investor will lend an amount $(1 - d)$ at time 0. The sum of $(1 - d)$ may be considered as a loan of 1 (to be repaid after 1 unit of time) on which interest of amount d is payable *in advance*. For this reason, d is called the **rate of discount** per unit time. Sometimes, in order to avoid confusion with nominal rates of discount (see Chapter 4), d is called the **effective rate of discount** per unit time.

Similarly, it follows immediately from Eq. 2.4.9 that the accumulated amount at time $s + t$ of 1 invested at time s does *not* depend on s and is given by

$$F(t) = e^{\delta t} \tag{3.1.7}$$

$$= (1 + i)^t \tag{3.1.8}$$

where i is defined by the equation

$$1 + i = e^{\delta} \tag{3.1.9}$$

Therefore, an investor will lend a unit amount at time $t = 0$ in return for a repayment of $(1 + i)$ at time $t = 1$. Accordingly, i is called the **rate of interest** (or the **effective rate of interest**) per unit time.

Although we have chosen to define i, v, and d in terms of the force of interest δ, any three of i, v, d, and δ are uniquely determined by the fourth. For example, if we choose to regard i as the basic parameter, then it follows from Eq. 3.1.9 that

$$\delta = \ln(1 + i)$$

In addition, Eqs 3.1.5 and 3.1.9 imply that

$$v = (1 + i)^{-1}$$

while Eqs 3.1.6 and 3.1.9 imply that

$$d = 1 - (1 + i)^{-1}$$

$$= \frac{i}{1 + i}$$

These last three equations define δ, v, and d in terms of i.

The last equation may be written as

$$d = iv$$

which confirms that an interest payment of i at time $t = 1$ has the same value as a payment of d at time $t = 0$. But what sum paid *continuously* (at a constant rate) over the time interval $[0, 1]$ has the same value as either of these payments? Let the required amount be σ such that the amount paid in time increment dt is σdt. Then, taking values at time 0, we have

$$d = \int_0^1 \sigma e^{-\delta t} dt$$

$$= \sigma\left(\frac{1 - e^{-\delta}}{\delta}\right) \quad \text{(if } \delta \neq 0) \qquad \text{(by Eq. 3.1.6)}$$

$$= \sigma\left(\frac{d}{\delta}\right)$$

Hence $\sigma = \delta$. This result is also true, of course, when $\delta = 0$. This establishes the important fact that a payment of δ made continuously over the period $[0,1]$ has the same value as a payment of d at time 0 or a payment of i at time 1. Each of the three payments may be regarded as alternative methods of paying interest on a unit loan over the period.

In certain situations, it may be natural to regard the force of interest as the basic parameter, with implied values for i, v, and d. In other cases, it may be preferable to assume a certain value for i (or d or v) and to calculate, if necessary, the values implied for the other three parameters. Note that standard compound interest tables (e.g., those given in this book) give the values of δ, v, and d for given values of i. It is left as a simple, but important, exercise for the reader to verify the relationships summarized here.

Value Of	δ	i	v	d
In Terms Of				
δ		$e^\delta - 1$	$e^{-\delta}$	$1 - e^{-\delta}$
i	$\ln(1+i)$		$(1+i)^{-1}$	$i(1+i)^{-1}$
v	$-\ln v$	$v^{-1} - 1$		$1 - v$
d	$-\ln(1-d)$	$(1-d)^{-1} - 1$	$1 - d$	

EXAMPLE 3.1.1

Calculate the values of i, v, and d that are implied by $\delta = 6\%$ per annum.

$i = e^\delta - 1 = e^{0.06} - 1 = 6.184\%$
$v = e^{-\delta} = e^{-0.06} = 0.9418$
$d = 1 - e^{-\delta} = 5.824\%$

Solution
Using the expressions summarized previously, we have

When i is small, approximate formulae for d and δ in terms of i may be obtained from well-known series by neglecting the remainder after a small number of terms. For example, since

$$\delta = \ln\ (1+i)$$
$$= i - \tfrac{1}{2}i^2 + \tfrac{1}{3}i^3 - \tfrac{1}{4}i^4 + \cdots \qquad (\text{if } |i| < 1)$$

it follows that, for small values of i,

$$\delta \approx i - \frac{1}{2}i^2$$

Similarly

$$d = i(1+i)^{-1}$$
$$= i(1 - i + i^2 - i^3 + \cdots) \qquad (\text{if}|i| < 1)$$
$$= i - i^2 + i^3 - i^4 + \cdots$$

so, if i is small,

$$d \approx i - i^2$$

We note that if i is small, then i, δ, and d are all of the same order of magnitude. Similar expressions can be derived, which give approximate relations between any of combination of d, δ, and i.

3.2 THE EQUATION OF VALUE

Consider a transaction under which, in return for outlays of amount $a_{t_1}, a_{t_2}, \ldots, a_{t_n}$ at times $t_1, t_2, \ldots, t_n$, an investor will receive payments of $b_{t_1}, b_{t_2}, \ldots, b_{t_n}$ at these times, respectively. (In most situations, only *one* of a_{t_r} and b_{t_r} will be non-zero.) At what force or rate of interest does the series of outlays have the same present value as the series of receipts?

At force of interest δ, the two series are of equal present value if and only if

$$\sum_{r=1}^{n} a_{t_r} e^{-\delta t_r} = \sum_{r=1}^{n} b_{t_r} e^{-\delta t_r} \qquad (3.2.1)$$

This equation may be written as

$$\sum_{r=1}^{n} c_{t_r} e^{-\delta t_r} = 0 \qquad (3.2.2)$$

where

$$c_{t_r} = b_{t_r} - a_{t_r}$$

is the amount of the *net cash flow* at time t_r. Note that we adopted the standard convention that a negative cash flow corresponds to a payment *by* the investor and a positive cash flow represents a payment *to* him.

Equation 3.2.2 expresses algebraically the condition that, at force of interest δ, the total value of the net cash flows is 0; it is called the ***equation of value*** for the force of interest implied by the transaction. If we let $e^\delta = 1 + i$, the equation may be written as

$$\sum_{r=1}^{n} c_{t_r}(1+i)^{-t_r} = 0 \qquad (3.2.3)$$

The latter form is known as the *equation of value for the rate of interest* or the ***yield equation***. Alternatively, the equation may be written as

$$\sum_{r=1}^{n} c_{t_r} v^{t_r} = 0$$

Note that in the preceding equations n may be infinite.

In relation to continuous payment streams, if we let $\rho_1(t)$ and $\rho_2(t)$ be the rates of paying and receiving money at time t, respectively, we call $\rho(t) = \rho_2(t) - \rho_1(t)$ the ***net rate of cash flow*** at time t. The equation of value, corresponding to Eq. 3.2.2, for the force of interest is

$$\int_0^\infty \rho(t) e^{-\delta t}\, dt = 0 \qquad (3.2.4)$$

When both discrete and continuous cash flows are present, the equation of value is

$$\sum_{r=1}^{n} c_{t_r} e^{-\delta t_r} + \int_0^\infty \rho(t) e^{-\delta t}\, dt = 0 \qquad (3.2.5)$$

and the equivalent yield equation is

$$\sum_{r=1}^{n} c_{t_r}(1+i)^{-t_r} + \int_0^\infty \rho(t)(1+i)^{-t}\, dt = 0 \qquad (3.2.6)$$

For any given transaction, Eq. 3.2.5 may have no roots, a unique root, or several roots in δ. We consider only *real* roots as δ has a physical meaning. If there is a unique root, δ_0 say, it is known as the *force of interest implied by the transaction*, and the corresponding rate of interest $i_0 = e^{\delta_0} - 1$ is called the ***yield*** per unit time. Alternative terms for the yield are the *internal rate of return* and the *money-weighted rate of return* for the transaction, as discussed in Chapter 6.

Although for certain investments the yield does not exist (since the equation of value 3.2.2 has no roots or more than one root), there is one important class of transaction for which the yield always exists. This is described in Theorem 3.2.1.

THEOREM 3.2.1

For any transaction in which *all* the negative net cash flows
precede *all* the positive net cash flows (or vice versa), the yield
is well defined.

The proof of Theorem 3.2.1 is given in Appendix 1.

Although the yield is defined only when Eq. 3.2.3 has a unique root greater
than -1 (so that $e^{\delta} > 0$, by Eq. 3.1.9), it is sometimes of interest to consider
transactions in which the yield equation has a unique *positive* root (even though
there may also be negative roots). There is one easily described class of transaction for which the yield equation always has precisely one positive root, and
this is described in Theorem 3.2.2.

THEOREM 3.2.2

Suppose that $t_0 < t_1 < \ldots < t_n$ and consider a transaction for
which the investor's net cash flow at time t_i is of amount c_{t_i}.
(Some of the $\{c_{t_i}\}$ will be positive and some negative, according
to the convention described previously.) For $i = 0, 1, \ldots n$, let
$A_i = \sum_{r=0}^{i} c_{t_r}$, so that A_i denotes the *cumulative* total amount
received by the investor after the cash flow at time t_i has
occurred. Suppose that A_0 and A_n are both non-zero and
that, when any zero values are excluded, the sequence $\{A_0,
A_1, \ldots , A_n\}$ contains precisely one change of sign. Then the
yield equation has exactly one positive root.

It should be noted that Theorem 3.2.2 gives no information on the existence of
negative roots. We omit the proof of the theorem as it is beyond the scope of
the book.

One particular example of this situation is provided by a transaction in which
all the investor's outlays precede all his receipts and the total amount received
exceeds the total outlays. The existence of the yield for this type of transaction
has been established by Theorem 3.2.1, the proof of which shows that the yield
is positive.

The analysis of the equation of value for a given transaction may be somewhat
complex. (See Appendix 2 for possible methods of solution.) However, when
the equation $f(i) = 0$ is such that f is a *monotonic* function, its analysis is
particularly simple. The equation has a root if and only if we can find i_1 and i_2
with $f(i_1)$ and $f(i_2)$ of opposite sign. In this case, the root is unique and lies
between i_1 and i_2. By choosing i_1 and i_2 sufficiently close to each other, we may
determine the yield to any desired degree of accuracy(see Figure 3.2.1).

EXAMPLE 3.2.1

An investor is able to receive returns of 1, 8, and 4 at times 1, 3, and 4, respectively, in return for payment amounts 5 and 3 at times 0 and 2, respectively. Demonstrate that a unique positive yield exists for this transaction and verify that it is 22.107%.

Solution

Net cash flows are given chronologically by the sequence

$$\{c_{t_i}\} = \{-5, \ 1, \ -3, \ 8, \ 4\}$$

and the *cumulative* total cash flows (in chronological order) by the sequence

$$\{A_i\} = \{-5, \ -4, \ -7, \ 1, \ 5\}$$

Since the latter sequence contains only one sign change (in this case from negative to positive), the yield equation has only one positive root by Theorem 3.2.2.

The equation of value for the transaction is

$$f(i) = -5 + 1(1+i)^{-1} - 3(1+i)^{-2} + 8(1+i)^{-3} + 4(1+i)^{-4}$$

and we note that $f(0.22107) = 0$.

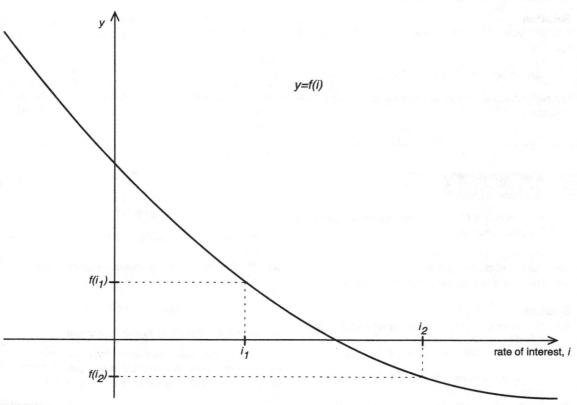

FIGURE 3.2.1
Equation of value

It should be noted that, after multiplication by $(1+i)^{t_0}$, Eq. 3.2.3 takes the equivalent form

$$\sum_{r=1}^{n} c_{t_r}(1+i)^{t_0-t_r} = 0 \qquad (3.2.7)$$

This slightly more general form may be called the *equation of value at time t_0*. It is, of course, directly equivalent to the original equation (which is now seen to be the equation of value at time 0), as expected from Eq. 2.7.7.

EXAMPLE 3.2.2

In return for an immediate payment of £500 and a further payment of £200 2 years from now, an investor will receive £1,000 after 5 years. Find the yield for the transaction.

Solution

Choose 1 year as the unit of time. The equation of value at time 0 is

$$f(i) = -500 - 200(1+i)^{-2} + 1,000(1+i)^{-5} = 0$$

Our earlier discussion indicates that there is a unique root by Theorem 3.2.2.

Since $f(0.08) = 9.115$ and $f(0.09) = -18.405$, the yield is between 8% and 9% per annum. A first approximation for the yield, obtained by linear interpolation, is

$$0.08 + (0.09 - 0.08)\frac{9.115 - 0}{9.115 - (-18.405)} = 0.0833$$

i.e., $i \approx 8.33\%$ per annum.

If the yield were required to a greater degree of accuracy, one might evaluate $f(0.085)$ and interpolate between this and $f(0.08)$. The yield to four decimal places is, in fact, 8.3248% per annum.

EXAMPLE 3.2.3

In return for a loan of £100, a borrower agrees to repay £110 after 7 months. Calculate

(a) The rate of interest per annum,
(b) The rate of discount per annum,
(c) The force of interest per annum for the transaction.

Solution

(a) The rate of interest per annum is given by the equation

$$100(1+i)^{7/12} = 110$$

from which it follows that $i = 0.17749$ or 17.749% per annum.

(b) The rate of discount per annum d is given by the equation

$$100 = 110(1 - d)^{7/12}$$

from which we obtain $d = 0.15074$ or 15.074% per annum.

(c) The force of interest per annum δ is given by the equation

$$100e^{(7/12)\delta} = 110$$

so that $\delta = 0.16339$ or 16.339% per annum.

Note that, for illustrative purposes, we have found each of i, d, and δ from first principles. It would also be possible find just one value, say i, and compute the other values from Eq. 3.1.9, etc.

EXAMPLE 3.2.4

Shortly after receiving the loan in Example 3.2.3, the borrower requests that he be allowed to repay the loan by a payment of £50 on the original settlement date and a second payment 6 months after this date. Assuming that the lender agrees to the request, and that the calculation is made on the original interest basis, find the amount of the second payment under the revised transaction.

Solution

Let the amount of the second payment required under the revised transaction be £X. Then, applying the equation of value at the final repayment date (which is 13 months after the loan was made), we obtain

$$100e^{\frac{13}{12}\delta} - 50e^{\frac{1}{2}\delta} - X = 0$$

so that

$$X = 100e^{\frac{13}{12}\delta} - 50e^{\frac{1}{2}\delta}$$

Since it is assumed that the original basis applies, $\delta = 0.16339$ (as found above) and $X = $ £65.11.

The interpretation of the yield in Example 3.2.2 is simple: if the outlays of £500 and £200 were to be deposited in an account on which interest is payable at the rate of 8.3248% per annum, at the end of 5 years the accumulated amount of the account would be £1,000. If the investor expects to be able to make deposits over the next 5 years at a greater rate of interest than 8.3248% per annum, he will not choose the investment.

Note that an equivalent approach to Example 3.2.2 would be to work with the equation of value at time 5. This is

$$-500(1 + i)^5 - 200(1 + i)^3 + 1,000 = 0$$

which is, of course, also solved by $i = 8.3248\%$ per annum.

An alternative approach to finding the value of X in Example 3.2.4 is provided by noting that the borrower wishes to replace the originally agreed payment of £110 by a payment of £50; i.e., he wishes to defer a payment of £60 due on the original repayment date by a period of 6 months. The deferred payment (of

EXAMPLE 3.2.5

A lender bases his short-term transactions on a *rate of commercial discount D*, where $0 < D < 1$. This means that, if $0 < t \leq 1$, in return for a repayment of X after a period t, he will lend $X(1 - Dt)$ at the start of the period. For such a transaction over an interval of length t $(0 < t \leq 1)$, derive an expression in terms of D and t for d, the *effective rate of discount per unit time.*

Solution

Using the definition of the rate of commercial discount from Section 1.2, we may write the equation of value for the transaction at the time the loan is made in the form $(1 - d)^t = 1 - Dt$, and so

$$d = 1 - (1 - Dt)^{\frac{1}{t}}$$

amount £X) must have the same value as the £60 it replaces. Equating these values at the final repayment date, we obtain

$$X = 60e^{\frac{1}{2}\delta}$$

from which $X = £65.11$, as before. Loan repayments are considered in detail in Chapter 5.

EXAMPLE 3.2.6

An investor takes a loan at a commercial rate of discount of 18% per annum that is settled with a payment of £1,000 after

(a) 3 months,
(b) 9 months.

Calculate the annual rate of discount and the effective rate of interest implied in each case.

Solution
(a) $-1-1 = 20.223\%$ per annum.
(b) 9/12. i.e., $d = 17.582\%$ per annum and $i = 21.333\%$ per annum.

(c) The initial loan was $1{,}000\left(1 - \frac{1}{4} \times 0.18\right) = £955$, and the annual rate of discount is therefore given from $955 = 1{,}000(1 - d)^{\frac{1}{4}}$ i.e. $d = 16.821\%$ per annum. We than have $i = (1 - d)^{-1} - 1 = 20.223\%$ per annum.

(d) The initial loan was $1{,}000\left(1 - \frac{3}{4} \times 0.18\right) = £865$, and the annual rate of discount is therefore given from $865 = 1{,}000(1 - d)^{9/12}$ i.e. $d = 17.582\%$ per annum and $i = 21.333\%$ per annum.

It should be noted that the longer loan has the greater effective annual rate of discount and interest.

3.3 ANNUITIES-CERTAIN: PRESENT VALUES AND ACCUMULATIONS

Consider a series of n payments, each of amount 1, to be made at time intervals of one unit, the first payment being made at time $t + 1$. Such a sequence of payments is illustrated in Figure 3.3.1, in which the rth payment is made at time $t + r$.

The present value of this series of payments *one unit of time before the first payment is made* is denoted by $a_{\overline{n}|}$. For the series of payments illustrated in Figure 3.3.1, the value relates to time t. Clearly, if $i = 0$, then $a_{\overline{n}|} = n$; otherwise,

$$
\begin{aligned}
a_{\overline{n}|} &= v + v^2 + v^3 + \ldots + v^n \\
&= \frac{v(1 - v^n)}{1 - v} \\
&= \frac{1 - v^n}{v^{-1} - 1} \\
&= \frac{1 - v^n}{i}
\end{aligned}
\tag{3.3.1}
$$

If $n = 0$, $a_{\overline{n}|}$ is defined to be zero, as no payments will be made.

In general, the quantity $a_{\overline{n}|}$ is the present value at the start of any period of length n of a series of n payments, each of unit amount, to be made *in arrears* at unit time intervals over the period. It is common to refer to such a series of

5-year annuity in arrears:

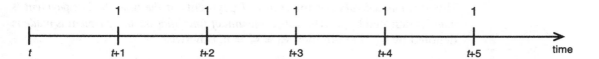

5-year annuity in advance:

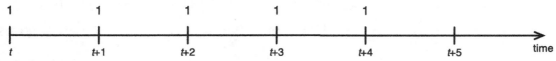

FIGURE 3.3.1
Cash flow diagram for unit annuities paid in arrears and advance

payments, made in arrears, as an *immediate annuity-certain* and to call $a_{\overline{n}|}$ the
present value of the immediate annuity-certain. When there is no possibility of
confusion with a life annuity, i.e., a series of payments dependent on the
survival of one or more human lives, the term *annuity* may be used as an
alternative to annuity-certain, and $a_{\overline{n}|}$ simply may be termed the *present value of
an n-year annuity paid in arrears*.

The value of this series of payments *at the time the first payment is made* is
denoted by $\ddot{a}_{\overline{n}|}$. If $i = 0$, then $\ddot{a}_{\overline{n}|} = n$; otherwise,

$$\ddot{a}_{\overline{n}|} = 1 + v + v^2 + \ldots + v^{n-1}$$
$$= \frac{1 - v^n}{1 - v} \qquad (3.3.2)$$
$$= \frac{1 - v^n}{d}$$

In general, the quantity $\ddot{a}_{\overline{n}|}$ is the value at the start of any given period of length
n of a series of n payments, each of unit amount, to be made *in advance* at unit
time intervals over the period. It is common to refer to such a series of
payments, made in advance, as an *annuity-due* and to call $\ddot{a}_{\overline{n}|}$ the *present value of
the annuity-due* or simply the *present value of an n-year paid in advance*. Again, if
$n = 0$, $\ddot{a}_{\overline{n}|}$ is defined to be zero.

It follows directly from the preceding definitions that, for $n \geq 2$,

$$\left. \begin{array}{l} \ddot{a}_{\overline{n}|} = (1 + i)a_{\overline{n}|} \\ \ddot{a}_{\overline{n}|} = 1 + a_{\overline{n-1}|} \end{array} \right\} \qquad (3.3.3)$$

The reader should verify these relationships algebraically and by general reasoning.

The *accumulated value* of the series of payments *at the time the last payment is made* is denoted by $s_{\overline{n}|}$. The value *one unit of time after the last payment is made* is denoted by $\ddot{s}_{\overline{n}|}$. If $i = 0$, then $s_{\overline{n}|} = \ddot{s}_{\overline{n}|} = n$, otherwise,

$$
\begin{aligned}
s_{\overline{n}|} &= (1+i)^{n-1} + (1+i)^{n-2} + (1+i)^{n-3} + \cdots + 1 \\
&= (1+i)^n a_{\overline{n}|} \\
&= \frac{(1+i)^n - 1}{i}
\end{aligned}
\tag{3.3.4}
$$

and

$$
\begin{aligned}
\ddot{s}_{\overline{n}|} &= (1+i)^n + (1+i)^{n-1} + (1+i)^{n-2} + \cdots + (1+i) \\
&= (1+i)^n \ddot{a}_{\overline{n}|} \\
&= \frac{(1+i)^n - 1}{d}
\end{aligned}
\tag{3.3.5}
$$

It is clear that $s_{\overline{n}|}$ and $\ddot{s}_{\overline{n}|}$ are the values at the end of any period of length n (i.e., at time $t = n$) of a series of n payments, each of amount 1, made at unit time intervals over the period, where the payments are made in arrears and in advance, respectively. Sometimes $s_{\overline{n}|}$ and $\ddot{s}_{\overline{n}|}$ are called the *accumulation* (or the *accumulated amount*) of an immediate annuity and an annuity-due, respectively. When $n = 0$, $s_{\overline{n}|}$ and $\ddot{s}_{\overline{n}|}$ are defined to be zero.

It is an immediate consequence of the preceding definition that

$$
\left.
\begin{aligned}
\ddot{s}_{\overline{n}|} &= (1+i)s_{\overline{n}|} \\
s_{\overline{n+1}|} &= 1 + \ddot{s}_{\overline{n}|} \\
\ddot{s}_{\overline{n}|} &= s_{\overline{n+1}|} - 1
\end{aligned}
\right\}
\tag{3.3.6}
$$

The reader should verify these relationships algebraically and by general reasoning.

Equations 3.3.1, 3.3.2, 3.3.4, and 3.3.5 may be expressed in the form

$$
\left.
\begin{aligned}
1 &= i a_{\overline{n}|} + v^n \\
1 &= d \ddot{a}_{\overline{n}|} + v^n \\
(1+i)^n &= i s_{\overline{n}|} + 1 \\
(1+i)^n &= d \ddot{s}_{\overline{n}|} + 1
\end{aligned}
\right\}
\tag{3.3.7}
$$

respectively. The reader should be able to write down these four expressions of Eq. 3.3.7 immediately. The first equation is simply the equation of value at time 0 for a unit loan over the period from time 0 to time n, when interest is payable in arrears. The other three equations may be similarly interpreted, the last two being equations of value at time n.

As the rate of interest i increases, v decreases, so $\sum_{r=1}^{n} v^r$ decreases. Therefore, for a fixed value of n, $a_{\overline{n}|}$ is a decreasing function of i. Similarly, $\ddot{a}_{\overline{n}|}$ is a decreasing function of i, while $s_{\overline{n}|}$ and $\ddot{s}_{\overline{n}|}$ are increasing functions of i.

For a fixed rate of interest, $a_{\overline{n}|}$, $\ddot{a}_{\overline{n}|}$, $s_{\overline{n}|}$, and $\ddot{s}_{\overline{n}|}$ are all increasing functions of n. When n becomes infinite, the corresponding annuity (or annuity-due) is known as a *perpetuity* (or *perpetuity-due*). The notations $a_{\overline{\infty}|}$ and $\ddot{a}_{\overline{\infty}|}$ are used to denote the corresponding present values; if $i > 0$,

$$a_{\overline{\infty}|} = \lim_{n \to \infty} a_{\overline{n}|} = \frac{1}{i} \qquad (3.3.8)$$

and

$$\ddot{a}_{\overline{\infty}|} = \lim_{n \to \infty} \ddot{a}_{\overline{n}|} = \frac{1}{d} \qquad (3.3.9)$$

These expressions follow directly from Eqs 3.3.1 and 3.3.2. Note that if $i \leq 0$, both $a_{\overline{\infty}|}$ and $\ddot{a}_{\overline{\infty}|}$ are infinite.

It is convenient to have standard tables of annuity and accumulation values at various rates of interest. In view of the relationship 3.3.3, it is not necessary to give the values of both $a_{\overline{n}|}$ and $\ddot{a}_{\overline{n}|}$; similarly, the relationship 3.3.6 removes the need to tabulate both $s_{\overline{n}|}$ and $\ddot{s}_{\overline{n}|}$. Such tables can be found later in this book.

Considering the quantity $a_{\overline{n}|}$ as the value of an n-year payment stream (made in arrears) at time t, and $s_{\overline{n}|}$ as the value of the same stream at time $t + n$, it is clear that

$$s_{\overline{n}|} = (1 + i)^n a_{\overline{n}|} \qquad (3.3.10)$$

Similarly,

$$\ddot{s}_{\overline{n}|} = (1 + i)^n \ddot{a}_{\overline{n}|} \qquad (3.3.11)$$

The reader should verify these relationships algebraically and by general reasoning.

EXAMPLE 3.3.1

Working in years, evaluate $a_{\overline{10}|}$, $\ddot{a}_{\overline{10}|}$, $s_{\overline{10}|}$, and $\ddot{s}_{\overline{10}|}$ for $i = 2\%$ per annum and $i = 5\%$ per annum. Confirm Eqs 3.3.10 and 3.3.11 and also confirm you answers using standard tables.

Solution

Using Eqs. 3.3.1, 3.3.2, 3.3.4, and 3.3.5, we see that

i	$a_{\overline{10}|}$	$\ddot{a}_{\overline{10}|}$	$s_{\overline{10}|}$	$\ddot{s}_{\overline{10}|}$
2%	8.9826	9.1622	10.9497	11.1687
5%	7.7217	8.1078	12.5779	13.2068

We note that $8.9826 \times (1.02)^{10} = 10.9497$ and $9.1622 \times (1.02)^{10} = 11.1687$, which confirm Eqs 3.3.10 and 3.3.11 at 2%. Similarly for 5%.

Standard tables give values of $a_{\overline{10}|}$ and $s_{\overline{10}|}$ equal to those calculated at each i.

EXAMPLE 3.3.2

A loan of £2,400 is to be repaid by 20 equal annual install-ments. The rate of interest for the transaction is 10% per annum. Find the amount of each annual repayment, assuming that payments are made in arrears and in advance.

Solution

Let the annual repayment in arrears be £X; therefore,

$$2,400 = X(v + v^2 + \cdots + v^{20})$$
$$= Xa_{\overline{20}|} \quad \text{at } 10\%$$

so $X = 2,400/a_{\overline{20}|} = 2,400/8.5136 = £281.90$

Let the annual repayment in advance be £Y; therefore,

$$2,400 = Y(1 + v + \cdots + v^{19})$$
$$= Y\ddot{a}_{\overline{20}|} \quad \text{at } 10\%$$

so $Y = 2,400/(1 + a_{\overline{19}|}) = 2,400/9.3649 = £256.28$

EXAMPLE 3.3.3

On 15 November in each of the years 1994 to 2009 inclusive, an investor deposited £500 in a special savings account. Find the sum that can be withdrawn by the investor on 15 November 2013 if the bank interest rate was 7% per annum for the entire period.

Solution

Two alternative solutions are considered:

(a) The investor made 16 deposits in his account. On 15 November 2009 (i.e., the date of the final deposit), the amount of the account was therefore $500\,s_{\overline{16}|}$ at 7%, i.e., £500 × 27.88805 = £13,944.03. Four years later, i.e., on 15 November 2013, the amount of the account was £13,944.03 × $(1.07)^4$ = £18,277.78.

(b) Alternatively, although the investor made no deposits in the years 2010 to 2013, we shall value his account on the basis that the payments of £500 continued in these years. We then deduct from the accumulated account, calculated on this basis, the accumulated value of the payments which he did *not* make in the years 2010 to 2013. This leads to

$$500(s_{\overline{20}|} - s_{\overline{4}|}) \quad \text{at } 7\%$$
$$= 500(40.995\ 49 - 4.439\ 94) = £18,277.78$$

EXAMPLE 3.3.4

A borrower agrees to repay a loan of £3,000 by 15 annual repayments of £500, the first repayment being due after 5 years. Find the annual yield for this transaction.

Solution

The equation of value may be written as

$$3,000 = 500(v^5 + v^6 + \cdots + v^{19})$$
$$= 500[(v + v^2 + \cdots + v^{19}) - (v + v^2 + v^3 + v^4)]$$
$$= 500(a_{\overline{19}|} - a_{\overline{4}|})$$

We are required to solve this equation for the rate of interest *i*. Our remarks in Section 3.2 indicate that there is a unique root. Since the right side of the equation is a monotonic

function of *i*, the solution may be found quite simply to any desired degree of accuracy.

At 8%, the right side has value 500 × (9.6036 − 3.3121) = 3,145.75. At 9%, its value is 500 × (8.9501 − 3.2397) = 2,855.20. Since 3,145.75 > 3,000 > 2,855.20, the value of *i* is between 8% and 9%. We estimate *i* by linear interpolation as

$$i \approx 0.08 + (0.09 - 0.08)\frac{3,145.75 - 3,000}{3,145.75 - 2,855.20} = 0.08502$$

Note that the solution, to five decimal places, is, in fact, $i = 0.08486$, say, 8.49%.

Note that loan repayment schedules are considered in more detail in Chapter 5.

Example 3.3.4 motivates the use of *deferred annuities*, which we consider now.

3.4 DEFERRED ANNUITIES

Suppose that m and n are non-negative integers. The value at time $t = 0$ of a series of n payments, each of amount 1, due at times $(m + 1)$, $(m + 2)$,..., $(m + n)$ is denoted by $_m|a_{\overline{n}|}$. This is illustrated in Figure 3.4.1.

3-year annuity in arrears deferred for 2 years:

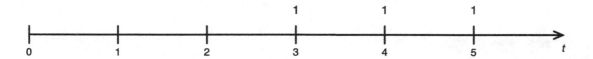

3-year annuity-due deferred for 2 years:

FIGURE 3.4.1
Cash flow diagrams for deferred annuities

Such a series of payments may be considered as an immediate annuity, *deferred* for m time units. When $n > 0$, this is denoted by

$$
\begin{aligned}
_m|a_{\overline{n}|} &= v^{m+1} + v^{m+2} + v^{m+3} + \cdots + v^{m+n} \\
&= \left(v + v^2 + v^3 + \cdots + v^{m+n}\right) - \left(v + v^2 + v^3 + \cdots + v^m\right) \quad (3.4.1) \\
&= v^m\left(v + v^2 + v^3 + \cdots + v^n\right)
\end{aligned}
$$

The last two equations show that

$$
_m|a_{\overline{n}|} = a_{\overline{m+n}|} - a_{\overline{m}|} \quad (3.4.2)
$$

$$
= v^m a_{\overline{n}|} \quad (3.4.3)
$$

Either of these two equations may be used to determine the value of a deferred immediate annuity. Together, they imply that

$$a_{\overline{m+n}|} = a_{\overline{m}|} + v^m a_{\overline{n}|} \qquad (3.4.4)$$

which is often a useful representation.

At this stage it is perhaps worth pointing out that the Eq. 3.4.1 may be used for $_m|a_{\overline{m}|}$ when m is any non-negative number, not only an integer. In this case, Eq. 3.4.3 is valid, but Eqs. 3.4.2 and 3.4.4 have, as yet, no meaning, since $a_{\overline{k}|}$ has been defined only when k is an integer. Later we shall extend the definition of $a_{\overline{k}|}$ to non-integral values of k, and it will be seen that Eqs 3.4.2 and 3.4.4 are always valid.

We may define the corresponding deferred annuity-due as

$$_m|\ddot{a}_{\overline{m}|} = v^m \ddot{a}_{\overline{m}|} \qquad (3.4.5)$$

EXAMPLE 3.4.1

Evaluate the present value of a unit 10-year annuity paid in arrears but deferred for 5 years at an interest rate of 5% per annum. Use two different approaches.

$$a_{\overline{15}|} - a_{\overline{5}|} = 10.3767 - 4.3295 = 6.0502$$
$$v^5 a_{\overline{10}|} = 0.7835 3.7.7217 = 6.0502$$

Solution

We require $_5|a_{\overline{10}|}$ $i = 15\%$. This can be evaluated as

---

3.5 CONTINUOUSLY PAYABLE ANNUITIES

Let n be a non-negative number. The value at time 0 of an annuity payable continuously between time 0 and time n, where the rate of payment per unit time is constant and equal to 1, is denoted by $\bar{a}_{\overline{n}|}$.

It is straightforward to demonstrate that

$$\bar{a}_{\overline{n}|} = \int_0^n e^{-\delta t} dt$$

$$= \frac{1 - e^{-\delta n}}{\delta} \qquad (3.5.1)$$

$$= \frac{1 - v^n}{\delta} \qquad (\text{if} \quad \delta \neq 0)$$

The justification of the first line is similar to that expressed in Section 2.6 with respect to continuously paid cash flows: During an increment of length dt at time t, the payment element made is dt since $\rho(t) = 1$. The present value of this element at time 0 is then $e^{-\delta t}dt$, and the entire present value of the stream is obtained from the integral of this between $t = 0$ and $t = n$. Note that $\bar{a}_{\overline{n}|}$ is defined even for non-integral values of n. If $\delta = 0$ (or, equivalently, $i = 0$), $\bar{a}_{\overline{n}|}$ is, of course, equal to n.

If m is a non-negative number, we use the symbol $_m|\bar{a}_{\overline{n}|}$ to denote the present value of a continuously payable annuity of 1 per unit time for n time units, deferred for m time units

$$_m|\bar{a}_{\overline{n}|} = \int_{m}^{m+n} e^{-\delta t}dt$$
$$= e^{-\delta m}\int_{0}^{n} e^{-\delta s}ds$$
$$= \int_{0}^{m+n} e^{-\delta t}dt - \int_{0}^{m} e^{-\delta t}dt$$

Hence,

$$_m|\bar{a}_{\overline{n}|} = \bar{a}_{\overline{m+n}|} - \bar{a}_{\overline{m}|} \qquad (3.5.2)$$
$$= v^{m}\,\bar{a}_{\overline{n}|} \qquad (3.5.3)$$

The reader will note that Eqs 3.5.2 and 3.5.3 are analogous to Eqs 3.4.2 and 3.4.3. In Chapter 4 we shall show that, with the appropriate definition of $a_{\overline{n}|}$ for non-integral n, Eq. 3.5.4 is valid for all non-negative n.

Since Eq. 3.5.1 may be written as

$$\bar{a}_{\overline{n}|} = \frac{i}{\delta}\left(\frac{1 - v^{n}}{i}\right)$$

it follows immediately that, if n is an integer,

$$\bar{a}_{\overline{n}|} = \frac{i}{\delta}a_{\overline{n}|} \qquad (\text{if } \delta \neq 0) \qquad (3.5.4)$$

The factor i/δ can be thought of as substituting the denominator of $a_{\overline{n}|}$, i.e., replacing i with δ, thereby transforming Eq. 3.3.1 into Eq. 3.5.1. Formulae like Eq. 3.5.4 and the equivalent expressions that link $a_{\overline{n}|}$ and $\ddot{a}_{\overline{n}|}$ through an effective substitution of the denominators are very useful when transferring between evaluations of the present values defined under payments made in arrears, advance or continuously.

Note similar methods to those used in Example 3.5.1 can be employed when transferring between evaluations of accumulations.

Given that at $a_{\overline{10}|} = 8.5302$ at $i = 3\%$ per annum, evaluate $\ddot{a}_{\overline{10}|}$ and $\overline{a}_{\overline{10}|}$.

Solution

Since $i = 3\%$, we know that $d = 0.029126$ and $\delta = 0.029559$ from standard compound interest tables or the conversion formulae, and so

$$\ddot{a}_{\overline{10}|} = \frac{0.03}{0.029126} a_{\overline{10}|} = 8.7862$$

$$\overline{a}_{\overline{10}|} = \frac{0.03}{0.029559} a_{\overline{10}|} = 8.6575$$

3.6 VARYING ANNUITIES

Until now we have considered annuities for which the amount of each payment is constant. For an annuity in which the payments are not all of an equal amount, it is a simple matter to find the present (or accumulated) value from first principles. For example, the present value of such an annuity may always be evaluated as

$$\sum_{i=1}^{n} X_i v^{t_i}$$

where the ith payment, of amount X_i, is made at time t_i.

In the particular case when $X_i = t_i = i$, the annuity is known as an *increasing annuity*, and its present value is denoted by $(Ia)_{\overline{n}|}$ with

$$(Ia)_{\overline{n}|} = v + 2v^2 + 3v^3 + \cdots + nv^n \qquad (3.6.1)$$

Hence,

$$(1 + i)(Ia)_{\overline{n}|} = 1 + 2v + 3v^2 + \cdots + nv^{n-1}$$

By subtraction, we obtain

$$i(Ia)_{\overline{n}|} = 1 + v + v^2 + \cdots + v^{n-1} - nv^n$$
$$= \ddot{a}_{\overline{n}|} - nv^n$$

and so

$$(Ia)_{\overline{n}|} = \frac{\ddot{a}_{\overline{n}|} - nv^n}{i} \qquad (3.6.2)$$

The last equation need *not* be memorized, as it may be rapidly derived from first principles. A simple way of recalling Eq. 3.6.2 is to express it in the form

$$\ddot{a}_{\overline{n}|} = i(Ia)_{\overline{n}|} + nv^n \qquad (3.6.3)$$

This equation is simply the equation of value for a transaction in which an investor lends 1 at the start of each year for n years in return for interest at the end of each year of amount i times the outstanding loan and a repayment of the total amount lent (i.e., n) after n years. The two sides of the equation represent the value (at the start of the transaction) of the payments made by the lender and the borrower, respectively. Numerical values of the function $(Ia)_{\overline{n}|}$ are included in standard compound interest tables for various i and n.

The present value of any annuity payable in arrears for n time units for which the amounts of successive payments form an arithmetic progression can be expressed in terms of $Ia_{\overline{n}|}$ and $(Ia)_{\overline{n}|}$. If the first payment of such an annuity is P and the second payment is $(P + Q)$, the tth payment is $(P - Q) + Qt$, and the present value of the annuity is therefore

$$(P - Q)a_{\overline{n}|} + Q(Ia)_{\overline{n}|}$$

Alternatively, the present value of the annuity can be derived from first principles.

EXAMPLE 3.6.1

Measuring time in years, calculate the present value of a cash flow stream £4, £6, £8, £10, and £12 paid at times 1, 2, 3, 4, and 5 if $i = 7\%$ per annum.

Solution

From first principles, let the present value be £X, then

$$X = 4v^1 + 6v^2 + 8v^3 + 10v^4 + 12v^5$$

which can be evaluated as

$$X = 4 \times 0.9346 + 6 \times 0.8734 + 8 \times 0.8163 + 10$$
$$\times 0.7629 + 12 \times 0.7130 = £31.69$$

Alternatively, using increasing annuity functions, we can write the present value as

$$X = 2a_{\overline{5}|} + 2(Ia)_{\overline{5}|}$$
$$= 2 \times 4.1002 + 2 \times 11.7469$$
$$= £31.69$$

The notation $(I\ddot{a})_{\overline{n}|}$ is used to denote the present value of an increasing annuity-due payable for n time units, the tth payment (of amount t) being made at time $t - 1$.

$$(I\ddot{a})_{\overline{n}|} = 1 + 2v + 3v^2 + \cdots + nv^{n-1}$$
$$= (1 + i)(Ia)_{\overline{n}|} \tag{3.6.4}$$

$$= 1 + a_{\overline{n-1}|} + (Ia)_{\overline{n-1}|} \tag{3.6.5}$$

For increasing annuities that are payable continuously, it is important to distinguish between annuities which have a constant rate of payment r (per unit time) throughout the rth period and annuities which have a rate of payment t at time t. For the former, the rate of payment is a step function, taking the discrete values 1, 2,.... For the latter, the rate of payment itself increases continuously. If the annuities are payable for n time units, their present values are denoted by $(\overline{Ia})_{\overline{n}|}$ and $(\overline{I}\overline{a})_{\overline{n}|}$, respectively.

Clearly

$$(I\bar{a})_{\overline{n}|} = \sum_{r=1}^{n}\left(\int_{r-1}^{r} rv^t dt\right)$$

and

$$(\bar{I}\bar{a})_{\overline{n}|} = \int_{0}^{n} tv^t dt$$

Using integration by parts in the second case, the reader should verify that

$$(I\bar{a})_{\overline{n}|} = \frac{\ddot{a}_{\overline{n}|} - nv^n}{\delta} \tag{3.6.6}$$

and

$$(\bar{I}\bar{a})_{\overline{n}|} = \frac{\bar{a}_{\overline{n}|} - nv^n}{\delta} \tag{3.6.7}$$

Each of the last two equations, expressed in a form analogous to Eq. 3.6.3, may be easily written down as the equation of value for an appropriate transaction.

Corresponding to the present values at time $t=0$, $(Ia)_{\overline{n}|}$, $(I\ddot{a})_{\overline{n}|}$, $(I\bar{a})_{\overline{n}|}$, and $(\bar{I}\bar{a})_{\overline{n}|}$ are the accumulations at time $t=n$ of the relevant series of payments. These accumulations are denoted by $(Is)_{\overline{n}|}$, $(I\ddot{s})_{\overline{n}|}$, $(I\bar{s})_{\overline{n}|}$, and $(\bar{I}\bar{s})_{\overline{n}|}$, respectively. It follows that

$$\left.\begin{array}{l}(Is)_{\overline{n}|} = (1+i)^n(Ia)_{\overline{n}|} \\ (I\ddot{s})_{\overline{n}|} = (1+i)^n(I\ddot{a})_{\overline{n}|} \\ (I\bar{s})_{\overline{n}|} = (1+i)^n(I\bar{a})_{\overline{n}|} \\ (\bar{I}\bar{s})_{\overline{n}|} = (1+i)^n(\bar{I}\bar{a})_{\overline{n}|}\end{array}\right\} \tag{3.6.8}$$

The present values of deferred increasing annuities are defined in the obvious manner. For example,

$$_m|(Ia)_{\overline{n}|} = v^m(Ia)_{\overline{n}|}$$

It is important to realize that in general there is no one correct or "best" method of solution for many compound interest problems. Provided that the reader has a good grasp of the underlying principles, he will be able to use that method which is most suited to his own approach.

EXAMPLE 3.6.2

An annuity is payable annually in arrears for 20 years. The first payment is of amount £8,000, and the amount of each subsequent payment decreases by £300 each year. Find the present value of the annuity on the basis of an interest rate of 5% per annum using

(a) First principles,
(b) *Increasing* annuity functions.

Solution

Let the present value be £X

(a) From first principles, we can write both

$$X = 8,000v + 7,700v^2 + 7,400v^3 + \cdots + 2,300v^{20}$$
$$(1+i)X = 8,000 + 7,700v + 7,400v^2 + \cdots + 2,300v^{19}$$

By subtraction, we obtain

$$iX = 8,000 - 300\left(v + v^2 + \cdots + v^{19}\right) - 2,300v^{20}$$

and so $X = \dfrac{8,000 - 300a_{\overline{19}|} - 2,300v^{20}}{i}$ at 5%, which is evaluated to be £70,151.

(b) Using increasing annuity functions, we consider the annuity to be a level annuity of £8,300 per annum less an increasing annuity for which the rth payment is of amount £$300r$. Hence,

$$X = 8,300\left(v + v^2 + \cdots + v^{20}\right) - 300\left(v + 2v^2 + \cdots + 20v^{20}\right)$$
$$= 8,300\,a_{\overline{20}|} - 300(Ia)_{\overline{20}|}$$
$$= 8,300\,a_{\overline{20}|} - 300\frac{\ddot{a}_{\overline{20}|} - 20v^{20}}{i} = £70,151$$

It is very important in practical work to incorporate checks to ensure, for example, that one does not misplace a decimal point or make some other simple error. For example, in Example 3.6.2, the first annuity payment is £8,000 and the last payment is £2,300. The average payment is then £5,150. Accordingly, a rough approximation for the value of the annuity is $5,150a_{\overline{20}|} = 64,180$. Because of the increasing effect of discount with time, this

EXAMPLE 3.6.3

An annuity is payable half-yearly for 6 years, the first half-yearly payment of amount £1,800 being due after 2 years. The amount of subsequent payments decreases by £30 each half-year. On the basis of an interest rate of 5% per half-year, find the present value of the annuity.

Solution

As before, we give two solutions that illustrate different approaches. It is quite possible that the reader may prefer yet another method.

(a) Choose the *half-year* as our basic time unit and $i = 5\%$. There are 12 annuity payments, the last of amount £$[1,800 - (11 \times 30)]$, i.e., £1,470. From first principles, we can write

$$X = 1,800v^4 + 1,770v^5 + \cdots + 1,470v^{15}$$
$$(i+i)X = 1,800v^3 + 1,770v^5 + \cdots + 1,470v^{14}$$

Subtracting and rearranging, we find that

$$X = \frac{1,800v^3 - 30(a_{\overline{14}|} - a_{\overline{3}|}) - 1,470v^{15}}{i} = 12,651$$

(b) Alternately, we may write

$$X = v^3\left[1,830(v + v^2 + \cdots + v^{12})\right.$$
$$\left. - 30(v + 2v^2 + \cdots + 12v^{12})\right]$$
$$= v^3[1,830a_{\overline{12}|} - 30(Ia)_{\overline{12}|}] = 12,651$$

approximation *understates* the true value, but confirms the order of magnitude of our answer.

When there is no ambiguity as to the value of i, the rate of interest, functions such as $a_{\overline{n}|}$, $s_{\overline{n}|}$, and d are clearly defined. In this case an equation such as

$$X\, a_{\overline{n}|} = 1{,}000$$

may be solved immediately for X. If it is desired to emphasize the rate of interest, the equation may be written in the form

$$X\, a_{\overline{n}|} = 1{,}000 \text{ at rate } i$$

The solution of particular problems may involve more than one rate of interest, and in such cases there may be doubt as to the interest rate implicit in a particular function. In order to avoid ambiguity, we may attach the rate of interest as a suffix to the standard functions. For example, we may write $a_{\overline{n}|i}, v_i^n, s_{\overline{n}|i}$, and d_i for $a_{\overline{n}|}, v^n, s_{\overline{n}|}$ and d at rate i. With this notation, an equation such as

$$X s_{\overline{10}|0.04} = 100 a_{\overline{15}|0.03}$$

is quite precise. This notation is readily extended to the functions defined in later chapters.

3.7 UNCERTAIN PAYMENTS

It is clear that the concept of the equation of value is very important in practice; indeed, it forms the foundation of much of this book. At this stage it is worth emphasizing that the payments (either discrete or continuous) need not be certain for the equation of value to be formed and used. A detailed study of the allowance of uncertain cash flows is beyond the scope of this book, and much of this book involves certain cash flows. However, the allowance for uncertainty in the payments/receipts can be summarized as taking one of two forms:

1. By applying probabilities to each uncertain cash flow;
2. By using an increased rate of interest.

We now demonstrate how these two approaches can be considered equivalent.

The equation of value or yield equation as stated in Eq. 3.2.6 can be modified by the introduction of probabilities of the cash flows p_{t_r} and $p(t)$; these represent the probabilities relevant to the discrete payments at time t_r and continuous stream at time t, respectively. The resulting equation of value (or yield equation) is then formed from *expected present values* ("expected" in the probabilistic sense) and is written as

$$\sum_{r=1}^{n} p_{t_r} c_{t_r}(1+i)^{-t_r} + \int_0^\infty p(t)\rho(t)(1+i)^{-t}dt = 0 \qquad (3.7.1)$$

Equation 3.7.1 can be thought of as a more general form of Eq. 3.2.6 in that it also allows for certain cash flows by setting p_{t_r} and $p(t)$ equal to 1 for all time. The estimation of the probabilities in practice requires a considerable amount of skill and experience with the cash flows in question.

We can express the discounting factors in terms of the force of interest δ and introduce a new quantity, $\mu > 0$, representing the "force" of probability such that $p_{t_r} = e^{-\mu t_r}$ and $p(t) = e^{-\mu t}$. This form is consistent with the probability of payment reducing with time. Doing this allows Eq. 3.7.1, for example, to be rewritten as

$$\sum_{r=1}^{n} c_{t_r} e^{-\delta t_r} e^{-\mu t_r} + \int_0^\infty \rho(t) e^{-\delta t} e^{-\mu t} dt = 0 \qquad (3.7.2)$$

It is then clear that defining $\delta' = \delta + \mu$ enables Eq. 3.7.2 to be expressed in the form identical to Eq. 3.2.6, but at the higher force of interest, δ'.

The corresponding addition to the rate of interest, i, is often known as the **risk premium**, as introduced in Chapter 1. It reflects the greater return an investor would demand to compensate for the addition risk caused by the uncertainty in proceeds. Introducing a risk premium is particularly useful when it is not possible to calculate explicit probabilities for each cash flow.

EXAMPLE 3.7.1

An investor invests in a company for 3 years. The terms of the investment are such that the company pays the investor £100 at the end of each of the next 3 years and lump sum payment of £1,000 after 3 years.

(a) If the force of interest required by the investor is 4% per annum, calculate a fair price for the investment without allowance for default.

(b) The investor estimates that the probability the first payment will be made is 90%, the second payment is 60%, and the third payment (including capital redemption) is 50% (i.e., there is a 10%, 40%, and 50% chance of receiving *zero* payment at each time). Calculate the fair price for the investment on this basis.

(c) Rather than explicitly allowing for the probabilities of default, the investor decides to demand an increased return equivalent to $\delta = 26.7\%$ per annum. Calculate the fair price for the investment on this basis.

Solution

Let the fair price of the investment be P in each case.

(a) The equation of value is

$$P = 100e^{-0.04} + 100e^{-2\times0.04} + 1,100e^{-3\times0.04} = 1,164$$

(b) The equation of value is now written in terms of *expected* present values and

$$P = 0.90 \times 100e^{-0.04} + 0.60 \times 100e^{-2\times0.04} + 0.50$$
$$\times 1,100e^{-3\times0.04}$$
$$= £629.66$$

(c) The equation of value is

$$P = 100e^{-0.267} + 100e^{-2\times0.267} + 1,100e^{-3\times0.267} = £628.96$$

SUMMARY

- The *equation of value* is a mathematical expression that equates the present value (at a particular time) of the constituent cash flows of a transaction to zero. Discounting is expressed in terms of δ. The equivalent expression expressed in terms of i is called the *yield equation*.
- Standard notation exists for the present value of regular streams of cash flows called *annuities*. These can be used to construct the equations of value for more complicated cash flows.
- Annuities can be *immediate* or *deferred*, paid in *arrears*, in *advance* or *continuously*, and be *level* or *increasing*. Closed-form expressions can be derived to evaluate the present values of all such annuities.
- The accumulated value at time n of an n-year annuity can be obtained by accumulating the present value at $t = 0$ through n years. Standard notation exists for all such *accumulations*.
- Uncertainties in the timings and/or value of cash flows can be accounted for by the introduction of probabilities to form *expected present values*. Alternatively, a *risk premium* can be added to the discount rate. Both approaches act to reduce the present value of an uncertain cash flow stream.

EXERCISES

3.1 At retirement, a client has two payment options: a 20-year annuity at €50,000 per year starting after 1 year or a lump sum of €500,000 today. If the client's required rate of return on retirement fund investments is 6% per year, which plan has the higher present value and by how much?

3.2 Suppose you plan to send your daughter to college in 3 years. You expect her to earn two-thirds of her tuition payment in scholarship money, so you estimate that your payments will be $10,000 a year for 4 years. To estimate whether you have set aside enough money, you ignore possible inflation in tuition payments and assume that you can earn 8% annually on your investments. How much should you set aside now to cover these payments?

3.3 A computer manufacturer is to develop a new chip to be produced from 1 January 2008 until 31 December 2020. Development begins on 1 January 2006. The cost of development comprises £9 million payable on 1 January 2006 and £12 million payable continuously during 2007. Calculate the present value as at 1 January 2006 of the liabilities at an effective rate of interest of 9% per annum.

3.4 An investor is considering investing in a capital project. The project requires an outlay of £500,000 at outset and further payments at the end of each of the first 5 years, the first payment being £100,000 and each successive payment increasing by £10,000.

The project is expected to provide a continuous income at a rate of £80,000 in the first year, £83,200 in the second year, and so on, with income increasing each year by 4% per annum compound. The income is received for 25 years.

It is assumed that, at the end of 15 years, a further investment of £300,000 will be required and that the project can be sold to another investor for £700,000 at the end of 25 years.

(a) Calculate the net present value of the project at a rate of interest of 11% per annum effective.

(b) Without doing any further calculations, explain how the net present value would alter if the interest rate had been greater than 11% per annum effective.

3.5 In return for a single payment of £1,000 a building society offers the following alternative benefits:

(i) A lump sum of £1,330 after 3 years,

(ii) A lump sum of £1,550 after 5 years, or

(iii) Four annual payments, each of amount £425, the first payment being made after 5 years.

Any investor must specify which benefit he is choosing when he makes the single payment.

(a) Write down an equation of value for each savings plan and, hence, find the yield for each.

(b) Assume that an investor opts for plan (i), and that after 3 years, he invests the proceeds of the plan for a further 2 years at a fixed rate of interest. How large must this rate of interest be in order for him to receive £1,550 from this further investment?

(c) Assume that an investor opts for plan (ii), and that after 5 years, he uses the proceeds of the plan to buy a level annuity-due payable for 4 years, the amount of the annuity payment being calculated on the basis of a fixed interest rate. How large must this rate of interest be in order for the annuity payment to be £425?

3.6 An investor has the choice of either of the following savings plans:

(i) Ten annual premiums, each of £100 and payable in advance, will give £1,700 after 10 years, or

(ii) Fifteen annual premiums, each of £100 and payable in advance, will give £3,200 after 15 years.

The investor must declare which plan he is choosing when he pays the first premium.

(a) Find the yield on each plan.

(b) Assume that an investor has chosen plan (i). Assume further that after 10 years he deposits the proceeds of the plan in an account which will earn interest at a fixed rate and that he also makes five annual payments of £100 to this account, the first payment being made at the time the original savings plan matures. How large must the fixed rate of interest be in order that finally, after 15 years, the investor may receive £3,200?

3.7 (a) The manufacturer of a particular toy sells to retailers on either of the following terms:

   (i) Cash payment: 30% below recommended retail price,

   (ii) Six months' credit: 25% below recommended retail price.

   Find the effective annual rate of discount offered by the manufacturer to retailers who pay cash. Express this as an effective annual interest rate charged to those retailers who accept the credit terms.

(b) The manufacturer is considering changing his credit terms. Credit for 6 months will no longer be available, but for 3 months' credit, a discount of 27.5% below the recommended retail price will be allowed. The terms for cash payment will be unaltered.

   Does this new arrangement offer a greater or lower effective annual rate of discount to cash purchasers?

3.8 (a) Find the value at an interest rate of 4% of the following:

$$_{5}|a_{\overline{32}|}, \quad \ddot{a}_{\overline{62}|}, \quad \overline{a}_{\overline{62}|}, \quad _{12}|\ddot{a}_{\overline{50}|}, \quad s_{\overline{62}|}, \quad \ddot{s}_{\overline{61}|},$$
$$(I\ddot{a})_{\overline{62}|}, \quad _{5}|(Ia)_{\overline{20}|}, \quad (I\overline{a})_{\overline{25}|}, \quad (\overline{I}\overline{a})_{\overline{25}|}$$

(b) Given that $\ddot{a}_{\overline{n}|} = 7.029584$ and $\ddot{a}_{\overline{2n}|} = 10.934\ 563$, find the rate of interest and n.

3.9 An annuity-certain is payable annually for 20 years. The annual payment is £5 for 6 years, then £7 for 9 years, and, finally, £10 for 5 years.

(a) Show that the present value of the annuity at the time of the first payment may be expressed as

   (i) $5\ddot{a}_{\overline{6}|} + 7_{6}|\ddot{a}_{\overline{9}|} + 10_{15}|\ddot{a}_{\overline{5}|}$;

   (ii) $10\ddot{a}_{\overline{20}|} - 3\ddot{a}_{\overline{15}|} - 2\ddot{a}_{\overline{6}|}$; or

   (iii) $5 + 10a_{\overline{19}|} - 3a_{\overline{14}|} - 2a_{\overline{5}|}$.

(b) Show that the value of the annuity at the time of the last payment may be expressed as

   (i) $5(1+i)^{14}s_{\overline{6}|} + 7(1+i)^{5}s_{\overline{9}|} + 10s_{\overline{5}|}$; or

   (ii) $5s_{\overline{20}|} + 2s_{\overline{14}|} + 3s_{\overline{5}|}$.

3.10 An investor agrees to pay 20 premiums, annually in advance. At the end of 20 years the investor will receive the accumulated amount of his payments. This amount is calculated on the basis of an effective annual interest rate of 8% for the first 5 years, 6% for the next 7 years, and 5% for the final 8 years. Find the amount which the investor will receive in return for an annual premium of £100. Find also his yield per annum on the complete transaction.

CHAPTER 4

Further Compound Interest Functions

In this chapter we introduce variations on the compound interest and annuity notation we have seen in previous chapters. In particular, we consider the case where interest is received and payments are made at $p$thy intervals. The material is related to the nominal rate of interest, $i^{(p)}$, first introduced in Chapter 2.

4.1 INTEREST PAYABLE pTHLY

Suppose that, as in the preceding chapter, the force of interest per unit time is constant and equal to δ. Let i and d be the corresponding rates of interest and discount, respectively. In Section 3.1 we showed that d payable at time 0, i payable at time 1, and δ payable continuously at a constant rate over the time interval [0, 1] all have the same value (on the basis of the force of interest δ). Each of these payments may be regarded as the interest for the period [0, 1] payable on a loan of 1 made at time $t = 0$.

Suppose, however, that a borrower, who is lent 1 at time $t = 0$ for repayment at time $t = 1$, wishes to pay the interest on his loan in p equal installments over the interval. How much interest should he pay? This question motivates what follows.

We define $i^{(p)}$ to be that *total* amount of interest, payable in equal installments at the *end* of each pth subinterval (i.e., at times $1/p$, $2/p$, $3/p$, ..., 1), which has the same value as each of the interest payments just described. Likewise, we define $d^{(p)}$ to be that *total* amount of interest, payable in equal installments at the *start* of each pth subinterval (i.e., at times 0, $1/p$, $2/p$, ... $(p-1)/p$), which has the same value as each of these other payments.

We may easily express $i^{(p)}$ in terms of i. Since $i^{(p)}$ is the total interest paid, each interest payment is of amount $i^{(p)}/p$ and, when we consider the present value of the payments at the end of the interval, our definition implies the following

$$\sum_{t=1}^{p} \frac{i^{(p)}}{p}(1+i)^{(p-t)/p} = i \qquad (4.1.1)$$

61

An Introduction to the Mathematics of Finance. http://dx.doi.org/10.1016/B978-0-08-098240-3.00004-7

or, if $i \neq 0$,

$$\frac{i^{(p)}}{p} \left[\frac{(1+i) - 1}{(1+i)^{1/p} - 1} \right] = i$$

Hence,

$$i^{(p)} = p\left[(1+i)^{1/p} - 1\right] \tag{4.1.2}$$

and

$$\left[1 + \frac{i^{(p)}}{p}\right]^p = 1 + i \tag{4.1.3}$$

Note that the last two equations are valid even when $i = 0$.

Equations 4.1.2 and 4.1.3 are most important. Indeed, either equation may be regarded as providing a definition of $i^{(p)}$. If such a definition is used, it is a trivial matter to establish Eq. 4.1.1, which shows that $i^{(p)}$ may be interpreted as the total interest payable $p$thly in arrears in equal installments for a loan of 1 over one time unit.

Likewise, it is a consequence of our definition of $d^{(p)}$ that, when we consider the present value payments at the start of the interval, the following is true

$$\sum_{t=1}^{p} \frac{d^{(p)}}{p}(1-d)^{(t-1)/p} = d \tag{4.1.4}$$

or, if $d \neq 0$,

$$\frac{d^{(p)}}{p} \left[\frac{1 - (1-d)}{1 - (1-d)^{1/p}} \right] = d$$

Hence,

$$d^{(p)} = p\left[1 - (1-d)^{1/p}\right] \tag{4.1.5}$$

and

$$\left[1 - \frac{d^{(p)}}{p}\right]^p = 1 - d \tag{4.1.6}$$

Again, the last two equations are important and are valid even when $d = 0$. Either may be used to define $d^{(p)}$, in which case Eq. 4.1.4 is readily verified and

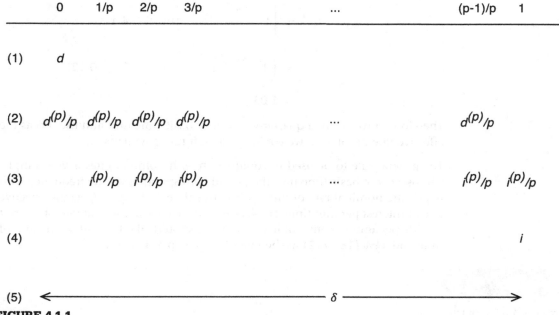

	0	1/p	2/p	3/p	...	(p-1)/p	1
(1)	d						
(2)	$d^{(p)}/p$	$d^{(p)}/p$	$d^{(p)}/p$	$d^{(p)}/p$	...	$d^{(p)}/p$	
(3)		$i^{(p)}/p$	$i^{(p)}/p$	$i^{(p)}/p$	...	$i^{(p)}/p$	$i^{(p)}/p$
(4)							i
(5)	⟵				δ		⟶

FIGURE 4.1.1
Equivalent payments

our original definition is confirmed. Note that $i^{(1)} = i$ and $d^{(1)} = d$. It is usual to include values of $i^{(p)}$ and $d^{(p)}$, at least for $p = 2$, 4, and 12, in standard compound interest tables. It is essential to appreciate that, at force of interest δ per unit time, the five series of payments illustrated in Figure 4.1.1 all have the same value.

If we choose to regard $i^{(p)}$ or $d^{(p)}$ as the basic quantity, Eqs 4.1.3 or 4.1.6 may be used to define i in terms of $i^{(p)}$ or d in terms of $d^{(p)}$. It is customary to refer to $i^{(p)}$ and $d^{(p)}$ as **nominal rates of interest** and **discount convertible pthly**. For example, if we speak of a rate of interest of 12% per annum convertible quarterly, we have $i^{(4)} = 0.12$ (with 1 year as the unit of time). Since $(1 + i) = \{1 + [i^{(4)}/4]\}^4$, this means that $i = 0.125509$. Therefore, the equivalent annual rate of interest is 12.5509%. As has been mentioned in Chapter 2, when interest rates are expressed in nominal terms, it is customary to refer to the equivalent rate per unit time as an *effective* rate. Therefore, if the nominal rate of interest convertible quarterly is $i^{(4)} = 12\%$ per annum, the effective rate per annum is $i = 12.5509\%$.

The treatment of problems involving nominal rates of interest (or discount) is almost always considerably simplified by an appropriate choice of the time unit. For example, on the basis of a nominal rate of interest of 12% per annum convertible quarterly, the present value of 1 due after t years is

$$(1+i)^{-t} = \left[1 + \frac{i^{(4)}}{4}\right]^{-4t} \qquad \text{(by Eq. 4.1.3)}$$

$$= \left(1 + \frac{0.12}{4}\right)^{-4t} \qquad \text{(since } i^{(4)} = 0.12\text{)}$$

$$= 1.03^{-4t}$$

Therefore, if we adopt a quarter-year as our basic time unit and use 3% as the effective rate of interest, we correctly value future payments.

The general rule to be used in conjunction with nominal rates is very simple. Choose as the basic time unit the period corresponding to the frequency with which the nominal rate of interest is convertible and use $i^{(p)}/p$ as the effective rate of interest per unit time. For example, if we have a nominal rate of interest of 18% per annum convertible monthly, we should take 1 month as the unit of time and $1\frac{1}{2}\%$ (18%/12) as the rate of interest per unit time.

EXAMPLE 4.1.1

Given that $\delta = 0.1$, find the values of

(i) i, $i^{(4)}$, $i^{(12)}$, $i^{(52)}$, $i^{(365)}$,
(ii) d, $d^{(4)}$, $d^{(12)}$, $d^{(52)}$, $d^{(365)}$.

Solution

$$i^{(p)} = p\left[(1+i)^{1/p} - 1\right]$$
$$= p(e^{\delta/p} - 1)$$
$$= p(e^{0.1/p} - 1) \qquad \text{(since } \delta = 0.1\text{)}$$

Also,

$$d^{(p)} = p\left[1 - (1-d)^{1/p}\right]$$
$$= p(1 - e^{-\delta/p})$$
$$= p(1 - e^{-0.1/p}) \qquad \text{(since } \delta = 0.1\text{)}$$

Hence, we have the following table for the required nominal rates of interest and discount when $\delta = 0.1$:

p	1	4	12	52	365
$i^{(p)}$	0.105 171	0.101 260	0.100 418	0.100 096	0.100 014
$d^{(p)}$	0.095 163	0.098 760	0.099 584	0.099 904	0.099 986

EXAMPLE 4.1.2

Given that $i = 0.08$, find the values of $i^{(12)}$, $d^{(4)}$, and δ.

Solution

$$i^{(12)} = 12\left[(1+i)^{1/12} - 1\right] = 0.077208$$

$$d^{(4)} = 4\left[1 - (1-d)^{1/4}\right] = 4\left[1 - (1+i)^{-1/4}\right]$$
$$= 0.076225$$

$$\delta = \ln(1+i) = 0.076961$$

EXAMPLE 4.1.3

Suppose that l and m are positive integers. Express $i^{(m)}$ in terms of l, m, and $d^{(l)}$. Hence, find $i^{(12)}$ when $d^{(4)} = 0.057\,847$.

Solution

$$\left[1 - \frac{d^{(l)}}{l}\right]^{l} = e^{-\delta} = \left[1 + \frac{i^{(m)}}{m}\right]^{-m}$$

Hence,

$$d^{(l)} = l\left\{1 - \left[1 + \frac{i^{(m)}}{m}\right]^{-m/l}\right\}$$

and

$$i^{(m)} = m\left\{\left[1 - \frac{d^{(l)}}{l}\right]^{-l/m} - 1\right\}$$

In particular,

$$i^{(12)} = 12\left\{\left[1 - \frac{d^{(4)}}{4}\right]^{-1/3} - 1\right\}$$

$$= 0.058\,411 \qquad (\text{when } d^{(4)} = 0.057847)$$

Alternatively, we may note that $(1 + i) = \{1 - [d^{(4)}/4]\}^{-4} = 1.06$, from which the value of $i^{(12)}$ follows immediately.

Note that $i^{(p)}$ and $d^{(p)}$ are given directly in terms of the force of interest δ by the equations

$$\left.\begin{array}{l} i^{(p)} = p\left(e^{\delta/p} - 1\right) \\ d^{(p)} = p\left(1 - e^{-\delta/p}\right) \end{array}\right\} \qquad (4.1.7)$$

Since

$$\lim_{x \to \infty} x\left(e^{\delta/x} - 1\right) = \lim_{x \to \infty} x\left(1 - e^{-\delta/x}\right) = \delta$$

it follows immediately from the Eq. 4.1.7 that

$$\lim_{p \to \infty} i^{(p)} = \lim_{p \to \infty} d^{(p)} = \delta \qquad (4.1.8)$$

This is intuitively obvious from our original definitions, since a continuous payment stream may be regarded as the limit, as p tends to infinity of a corresponding series of payments at intervals of time $1/p$ (see also Section 2.4).

Using the preceding definitions, we can easily establish that

$$i > i^{(2)} > i^{(3)} > \cdots > \delta$$

and

$$d < d^{(2)} < d^{(3)} < \cdots < \delta$$

so that the sequences $\{i^{(p)}\}$ and $\{d^{(p)}\}$ tend monotonically to the common limit δ from above and below, respectively.

Various approximations for $i^{(p)}$ and $d^{(p)}$ in terms of p and i, d, or δ may be obtained, as in Example 4.1.4.

EXAMPLE 4.1.4

Show that, if δ is small,

$$d^{(p)} \approx \delta - \frac{\delta^2}{2p}$$

and derive a corresponding approximation for $i^{(p)}$.

Solution

$$d^{(p)} = p(1 - e^{-\delta/p}) \quad \text{(by Eq. 4.1.7)}$$

$$= p\left[1 - \left(1 - \frac{\delta}{p} + \frac{\delta^2}{2p^2} - \frac{\delta^3}{6p^3} + \cdots\right)\right]$$

$$= \delta - \frac{\delta^2}{2p} + \frac{\delta^3}{6p^2} - \cdots$$

$$\approx \delta - \frac{\delta^2}{2p} \quad \left(\text{if } \delta \text{ is small}\right)$$

Similarly,

$$i^{(p)} = p(e^{\delta/p} - 1)$$

$$= p\left[\left(1 + \frac{\delta}{p} + \frac{\delta^2}{2p^2} + \frac{\delta^3}{6p^3} + \cdots\right) - 1\right]$$

$$= \delta + \frac{\delta^2}{2p} + \frac{\delta^3}{6p^2} + \cdots$$

$$\approx \delta + \frac{\delta^2}{2p} \quad \text{(if } \delta \text{ is small)}$$

4.2 ANNUITIES PAYABLE pTHLY: PRESENT VALUES AND ACCUMULATIONS

The nominal rates of interest and discount introduced in the preceding section are of particular importance in relation to annuities which are payable more frequently than once per unit time. We shall refer to an annuity which is payable p times per unit time as *payable pthly*.

If p and n are positive integers, the notation $a_{\overline{n}|}^{(p)}$ is used to denote the present value at time 0 of a level annuity payable $p$thly in arrears at the rate of 1 per unit time over the time interval $[0, n]$. For this annuity the payments are made at times $1/p$, $2/p$, $3/p$, ..., n, and the amount of each payment is $1/p$.

It is a simple matter to derive an expression for $a_{\overline{n}|}^{(p)}$ from first principles. However, the following argument, possibly less immediately obvious, is an important illustration of a kind of reasoning which has widespread application.

By definition, a series of p payments, each of amount $i^{(p)}/p$ in arrears at $p$thly subintervals over any unit time interval, has the same present value as a single payment of amount i at the end of the interval. By proportion, p payments, each of amount $1/p$ in arrears at $p$thly subintervals over any unit time interval, have the same present value as a single payment of amount $i/i^{(p)}$ at the end of the interval. Consider now that annuity for which the present value is $a_{\overline{n}|}^{(p)}$. The p payments after time $r - 1$ and not later than time r therefore have the same value as a single payment of amount $i/i^{(p)}$ at time r. This is true for $r = 1, 2, ..., n$, so the annuity has the same value as a series of n payments, each of amount $i/i^{(p)}$, at times $1, 2, ..., n$. This means that

$$a_{\overline{n}|}^{(p)} = \frac{i}{i^{(p)}} a_{\overline{n}|} \qquad (4.2.1)$$

The alternative approach, from first principles, is to write

$$
\begin{aligned}
a_{\overline{n}|}^{(p)} &= \sum_{t=1}^{np} \frac{1}{p} v^{t/p} \\
&= \frac{1}{p} \frac{v^{1/p}(1 - v^n)}{1 - v^{1/p}} \\
&= \frac{1 - v^n}{p\left[(1 + i)^{1/p} - 1\right]} \\
&= \frac{1 - v^n}{i^{(p)}}
\end{aligned}
\qquad (4.2.2)
$$

which confirms Eq. 4.2.1.

Likewise, we define $\ddot{a}_{\overline{n}|}^{(p)}$ to be the present value of a level annuity-due payable $p$thly at the rate of 1 per unit time over the time interval $[0, n]$. (The annuity payments, each of amount $1/p$, are made at times $0, 1/p, 2/p, ..., n - (1/p)$.) By definition, a series of p payments, each of amount $d^{(p)}/p$, in advance at $p$thly subintervals over any unit time interval has the same value as a single payment of amount i at the *end* of the interval. Hence, by proportion, p payments, each of amount $1/p$ in advance at $p$thly subintervals, have the same value as a single payment of amount $i/d^{(p)}$ at the *end* of the interval. This means (by an identical argument to that above) that

$$\ddot{a}_{\overline{n}|}^{(p)} \frac{i}{d^{(p)}} a_{\overline{n}|} \qquad (4.2.3)$$

It is usual to include the values of $i/i^{(p)}$ and $i/d^{(p)}$ in published tables. This enables the values of $a_{\overline{n}|}^{(p)}$ and $\ddot{a}_{\overline{n}|}^{(p)}$ to be calculated easily.

Alternatively, from first principles, we may write

$$
\begin{aligned}
\ddot{a}_{\overline{n}|}^{(p)} &= \sum_{t=1}^{np} \frac{1}{p} v^{(t-1)/p} \\
&= \frac{1 - v^n}{d^{(p)}}
\end{aligned}
\qquad (4.2.4)
$$

(on simplification), which confirms Eq. 4.2.3. Note that

$$a_{\overline{n}|}^{(p)} = v^{1/p} \ddot{a}_{\overline{n}|}^{(p)}$$

By combining Eqs 4.2.1 and 4.2.3, we obtain

$$i a_{\overline{n}|} = i^{(p)} a_{\overline{n}|}^{(p)} = d^{(p)} \ddot{a}_{\overline{n}|}^{(p)} = d\ddot{a}_{\overline{n}|} = \delta \bar{a}_{\overline{n}|} \qquad (4.2.5)$$

each expression being equal to $(1 - v^n)$.

In light of Eqs 4.2.2 and 4.2.4, Eqs 4.2.1 and 4.2.3 can be thought of as substituting the denominator in the expression for $a_{\overline{n}|}$ to form either $a_{\overline{n}|}^{(p)}$ or $\ddot{a}_{\overline{n}|}^{(p)}$. This approach was discussed in Section 3.5.

Note that since

$$\lim_{p \to \infty} i^{(p)} = \lim_{p \to \infty} d^{(p)} = \delta \qquad \text{(by Eq. 4.1.8)}$$

it follows immediately from Eqs 4.2.2 and 4.2.4 that

$$\lim_{p \to \infty} a_{\overline{n}|}^{(p)} = \lim_{p \to \infty} \ddot{a}_{\overline{n}|}^{(p)} = \bar{a}_{\overline{n}|}$$

These equations should be intuitively clear.

Similarly, we define $s_{\overline{n}|}^{(p)}$ and $\ddot{s}_{\overline{n}|}^{(p)}$ to be the accumulated amounts of the corresponding $p$thly immediate annuity and annuity-due, respectively. Therefore,

$$
\begin{aligned}
s_{\overline{n}|}^{(p)} &= (1+i)^n a_{\overline{n}|}^{(p)} \\
&= (1+i)^n \frac{i}{i^{(p)}} a_{\overline{n}|} \qquad \text{(by Eq. 4.2.1)} \qquad (4.2.6) \\
&= \frac{i}{i^{(p)}} s_{\overline{n}|}
\end{aligned}
$$

Also

$$
\begin{aligned}
\ddot{s}_{\overline{n}|}^{(p)} &= (1+i)^n \ddot{a}_{\overline{n}|}^{(p)} \\
&= (1+i)^n \frac{i}{d^{(p)}} a_{\overline{n}|} \qquad \text{(by Eq. 4.2.3)} \qquad (4.2.7) \\
&= \frac{i}{d^{(p)}} s_{\overline{n}|}
\end{aligned}
$$

The preceding proportional arguments may be applied to other varying series of payments. Consider, for example, an annuity payable annually in arrears for n years, the payment in the tth year being x_t. The present value of this annuity is

$$a = \sum_{t=1}^{n} x_t v^t \qquad (4.2.8)$$

EXAMPLE 4.2.1

Describe what the following quantities represent and compute their value for $i = 5\%$ per annum using values from standard tables.

(a) $50a_{\overline{10}|}^{(2)}$

(b) $60\ddot{s}_{\overline{2}|}^{(12)}$

Solution

(a) The quantity $50a_{\overline{10}|}^{(2)}$ represents the present value (at time $t = 0$) of an annuity paid for 10 years; in each year a total of 50 is paid in two equal installments in arrears.

$$50a_{\overline{10}|}^{(2)} = 50a_{\overline{10}|} \times \frac{i}{i^{(2)}}$$
$$= 50 \times 7.7217 \times 1.012348$$
$$= 390.85$$

(b) The quantity $60\ddot{s}_{\overline{2}|}^{(12)}$ represents the accumulated value (at time $t = 2$) of an annuity paid for 2 years; in each year a total of 60 is paid in 12 equal installments in advance.

$$60\ddot{s}_{\overline{2}|}^{(12)} = 60s_{\overline{2}|} \times \frac{i}{d^{(12)}}$$
$$= 60 \times 2.050 \times 1.026881$$
$$= 126.31$$

Consider now a second annuity, also payable for n years with the payment in the tth year, again of amount x_t, being made in p equal instalments in arrears over that year. If $a^{(p)}$ denotes the present value of this second annuity, by replacing the p payments for year t (each of amount x_t/p) by a single equivalent payment at the end of the year of amount $x_t[i/i^{(p)}]$, we immediately obtain

$$a^{(p)} = \frac{i}{i^{(p)}} a$$

where a is given by Eq. 4.2.8.

An annuity payable *pthly* in arrears, under which the payments continue indefinitely, is called a ***perpetuity payable pthly***. When the rate of payment is constant and equal to 1 per unit time, the present value of such a perpetuity is denoted by $a_{\overline{\infty}|}^{(p)}$. If the payments are in advance, we have a ***perpetuity-due***, with the corresponding present value denoted by $\ddot{a}_{\overline{\infty}|}^{(p)}$.

Since the payments differ only in the first payment at time 0, it is clear that

$$\ddot{a}_{\overline{\infty}|}^{(p)} = \frac{1}{p} + a_{\overline{\infty}|}^{(p)} \tag{4.2.9}$$

By letting n tend to infinity in Eqs 4.2.2 and 4.2.4, we obtain (if $i > 0$)

$$a_{\overline{\infty}|}^{(p)} = \frac{1}{i^{(p)}} \tag{4.2.10}$$

and

$$\ddot{a}_{\overline{\infty}|}^{(p)} = \frac{1}{d^{(p)}} \tag{4.2.11}$$

respectively.

The present values of an immediate annuity and an annuity-due, payable $p$thly at the rate of 1 per unit time for n time units and deferred for m time units, are denoted by

$$\left.\begin{array}{c} {}_{m|}a_{\overline{n}|}^{(p)} = v^m a_{\overline{n}|}^{(p)} \\[2mm] \text{and} \quad {}_{m|}\ddot{a}_{\overline{n}|}^{(p)} = v^m \ddot{a}_{\overline{n}|}^{(p)} \end{array}\right\}$$ (4.2.12)

respectively.

Finally, we remark that if $p = 1$, $a_{\overline{n}|}^{(p)}$, $\ddot{a}_{\overline{n}|}^{(p)}$, $s_{\overline{n}|}^{(p)}$, and $\ddot{s}_{\overline{n}|}^{(p)}$ are equal to $a_{\overline{n}|}$, $\ddot{a}_{\overline{n}|}$, $s_{\overline{n}|}$, and $\ddot{s}_{\overline{n}|}$, respectively.

4.3 ANNUITIES PAYABLE AT INTERVALS OF TIME r, WHERE $r > 1$

In Section 4.2 we showed how, by replacing a series of payments to be received by an equivalent series of payments of equal value, we could immediately write down an expression for the value of a $p$thly annuity. This technique of equivalent payments may be used to value a series of payments of constant amount payable at intervals of time length r, where r is some integer greater than 1.

For example, suppose that k and r are integers greater than 1 and consider a series of payments, each of amount X, due at times r, $2r$, $3r$, ..., kr. What is the value of this series at time 0 on the basis of an interest rate i per unit time?

The situation is illustrated in Figure 4.3.1, which shows the payments of amount X due at the appropriate times. Let us "replace" the payment of X due at time r by a series of r payments, each of amount Y, due at times 1, 2, ..., r, where Y is chosen to make these r equivalent payments of the same total value as the single payment they replace. This means that

$$Y s_{\overline{r}|} = X$$

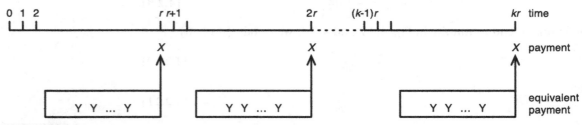

FIGURE 4.3.1
Annuity valuation through equivalent payments

at rate i, or

$$Y = \frac{X}{s_{\overline{r}|}} \qquad (4.3.1)$$

Similarly, *each* payment of amount X can be replaced by r equivalent payments of amount Y of the same value (see Figure 4.3.1). Then the original series of payments of X, due every rth time interval, has the same value as a series of kr payments of $Y = X/s_{\overline{r}|}$ due at unit time intervals. Hence, the value of the annuity is

$$\frac{X}{s_{\overline{r}|}} a_{\overline{kr}|} \qquad (4.3.2)$$

at rate i. (This result may also be obtained from first principles simply by summing the appropriate geometric progression.)

This technique is illustrated in Example 4.3.1. The reader should make no attempt to memorize this last result, which may be obtained immediately from first principles, provided that the underlying idea has been clearly understood.

EXAMPLE 4.3.1

An investor wishes to purchase a level annuity of £120 per annum payable quarterly in arrears for 5 years. Find the purchase price, given that it is calculated on the basis of an interest rate of 12% per annum

(a) Effective,
(b) Convertible half-yearly,
(c) Convertible quarterly,
(d) Convertible monthly.

Solution

(a) The value is

$$120a_{\overline{5}|}^{(4)} \text{ at } 12\% = 120\frac{i}{i^{(4)}} a_{\overline{5}|} = 451.583, \quad \text{say} \quad £451.58$$

(b) Since the rate of interest is nominal convertible half-yearly, we take the half-year as our unit of time and 6% as our rate of interest. The annuity is payable twice per half-year for 10 half-years at the rate of £60 per half-year. Hence, its value is

$$60a_{\overline{10}|}^{(2)} \text{ at } 6\% = 60\frac{i}{j^{(2)}} a_{\overline{10}|} = 448.134, \quad \text{say} \quad £448.13$$

(c) We take the quarter-year as the unit of time and 3% as the rate of interest. The value is then

$$30a_{\overline{20}|} \text{ at } 3\% = 446.324, \quad \text{say} \quad £446.32$$

(d) We take the month as the unit of time and 1% as the rate of interest. The annuity payments, of amount £30, at the end of every third month can be replaced by a series of equivalent monthly payments, each of $30/s_{\overline{3}|}$ (at 1%). The value is then

$$\frac{30}{s_{\overline{3}|}} a_{\overline{60}|} \text{ at } 1\% = 445.084, \quad \text{say} \quad £445.08$$

Alternatively, one may evaluate each of (b), (c), and (d) by converting from the given $i^{(p)}$ to the equivalent effective rate and its associated $i^{(4)}$, and proceeding to work in years, as in (a). For example, in (d), $i^{(12)} = 12\%$ per annum is equivalent to $i = 12.683\%$, and $i^{(4)} = 12.120\%$ and $120a_{\overline{5}|}^{(4)}$ can be evaluated directly as £445.08.

4.4 DEFINITION OF $a_{\overline{n}|}^{(p)}$ FOR NON-INTEGER VALUES OF n

Let p be a positive integer. Until now, the symbol $a_{\overline{n}|}^{(p)}$ has been defined only when n is a positive integer. For certain non-integral values of n, the symbol $a_{\overline{n}|}^{(p)}$ has an intuitively obvious interpretation. For example, it is not clear what meaning, if any, may be given to $a_{\overline{23.5}|}$, but the symbol $a_{\overline{23.5}|}^{(4)}$ ought to represent the present value of an immediate annuity of 1 per annum payable quarterly in arrears for 23.5 years (i.e., a total of 94 quarterly payments, each of amount 0.25). However, $a_{\overline{23.25}|}^{(2)}$ has no obvious meaning.

Suppose that n is an integer multiple of $1/p$, say $n = r/p$, where r is an integer. In this case we define $a_{\overline{n}|}^{(p)}$ to be the value at time $t = 0$ of a series of r payments, each of amount $1/p$, at times $1/p$, $2/p$, $3/p$, ..., $r/p = n$. If $i = 0$, then clearly $a_{\overline{n}|}^{(p)} = n$. If $i \neq 0$, then

$$a_{\overline{n}|}^{(p)} = \frac{1}{p}(v^{1/p} + v^{2/p} + v^{3/p} + \cdots + v^{r/p})$$

$$= \frac{1}{p}v^{1/p}\left(\frac{1 - v^{r/p}}{1 - v^{1/p}}\right) \tag{4.4.1}$$

$$= \frac{1}{p}\left[\frac{1 - v^{r/p}}{(1+i)^{1/p} - 1}\right]$$

and so

$$a_{\overline{n}|}^{(p)} = \begin{cases} \dfrac{1 - v^n}{i^{(p)}} & \text{if } i \neq 0 \\ n & \text{if } i = 0 \end{cases} \tag{4.4.2}$$

Note that, by working in terms of a new time unit equal to $1/p$ times the original time unit and with the equivalent effective interest rate of $i^{(p)}/p$ per new time unit, we see that

$$a_{\overline{n}|}^{(p)} \text{ at rate } i = \frac{1}{p}a_{\overline{np}|} \text{ at rate } i^{(p)}/p \tag{4.4.3}$$

The definition of $a_{\overline{n}|}^{(p)}$ given by Eq. 4.4.2 is mathematically meaningful for *all* non-negative values of n. For our present purpose, therefore, it is convenient to adopt Eq. 4.4.2 as a definition of $a_{\overline{n}|}^{(p)}$ for *all* n. If n is not an integer multiple of $1/p$, there is no universally recognized definition of $a_{\overline{n}|}^{(p)}$. For example, if $n = n_1 + f$, where n_1 is an integer multiple of $1/p$ and $0 < f < 1/p$, some writers define $a_{\overline{n}|}^{(p)}$ as

$$a_{\overline{n_1}|}^{(p)} + fv^n.$$

With this alternative definition

$$a_{\overline{23.75}|}^{(2)} = a_{\overline{23.5}|}^{(2)} + \frac{1}{4}v^{23.75}$$

which is the present value of an annuity of 1 per annum, payable half-yearly for 23.5 years, together with a final payment of 0.25 after 23.75 years. Note that this is *not* equal to the value obtained from Eq. 4.4.2 (see Example 4.4.1).

If $i \neq 0$, we define for all non-negative n

$$
\left.
\begin{aligned}
\ddot{a}_{\overline{n}|}^{(p)} &= (1+i)^{1/p}\, a_{\overline{n}|}^{(p)} = \frac{1 - v^n}{d^{(p)}} \\[2mm]
s_{\overline{n}|}^{(p)} &= (1+i)^n\, a_{\overline{n}|}^{(p)} = \frac{(1+i)^n - 1}{i^{(p)}} \\[2mm]
\ddot{s}_{\overline{n}|}^{(p)} &= (1+i)^n\, \ddot{a}_{\overline{n}|}^{(p)} = \frac{(1+i)^n - 1}{d^{(p)}}
\end{aligned}
\right\}
\qquad (4.4.4)
$$

where $i^{(p)}$ and $d^{(p)}$ are defined by Eqs 4.1.2 and 4.1.5, respectively. If $i = 0$, each of these last three functions is defined to equal n.

The reader should verify that, whenever n is an integer multiple of $1/p$, say $n = r/p$, then $\ddot{a}_{\overline{n}|}^{(p)}$, $s_{\overline{n}|}^{(p)}$, and $\ddot{s}_{\overline{n}|}^{(p)}$ are the values at different times of an annuity-certain of $r = np$ payments, each of amount $1/p$, at intervals of $1/p$ time unit.

As before, we use the simpler notations $a_{\overline{n}|}$, $\ddot{a}_{\overline{n}|}$, $s_{\overline{n}|}$, and $\ddot{s}_{\overline{n}|}$, to denote $a_{\overline{n}|}^{(1)}$, $\ddot{a}_{\overline{n}|}^{(1)}$, $s_{\overline{n}|}^{(1)}$, and $\ddot{s}_{\overline{n}|}^{(1)}$, respectively, therefore extending the definition of $a_{\overline{n}|}$, etc., given in Chapter 3 to all non-negative values of n. It is a trivial consequence of our definitions that the formulae

$$
\left.
\begin{aligned}
a_{\overline{n}|}^{(p)} &= \frac{i}{i^{(p)}}\, a_{\overline{n}|} \\[2mm]
\ddot{a}_{\overline{n}|}^{(p)} &= \frac{i}{d^{(p)}}\, a_{\overline{n}|} \\[2mm]
s_{\overline{n}|}^{(p)} &= \frac{i}{i^{(p)}}\, s_{\overline{n}|} \\[2mm]
\ddot{s}_{\overline{n}|}^{(p)} &= \frac{i}{d^{(p)}}\, s_{\overline{n}|}
\end{aligned}
\right\}
\qquad (4.4.5)
$$

(valid when $i \neq 0$) now hold for *all* values of n.

We may also extend the definitions of $_m|a_{\overline{n}|}^{(p)}$ and $_m|\ddot{a}_{\overline{n}|}^{(p)}$ to all values of n by the formulae

$$\left.\begin{array}{l}_m|a_{\overline{n}|}^{(p)} = v^m a_{\overline{n}|}^{(p)} \\[2mm] _m|\ddot{a}_{\overline{n}|}^{(p)} = v^m \ddot{a}_{\overline{n}|}^{(p)}\end{array}\right\}$$ (4.4.6)

The reader should verify that these definitions imply that

$$\left.\begin{array}{l}_m|a_{\overline{n}|}^{(p)} = a_{\overline{n+m}|}^{(p)} - a_{\overline{m}|}^{(p)} \\[2mm] _m|\ddot{a}_{\overline{n}|}^{(p)} = \ddot{a}_{\overline{n+m}|}^{(p)} - \ddot{a}_{\overline{m}|}^{(p)}\end{array}\right\}$$ (4.4.7)

EXAMPLE 4.4.1

Given that $i = 0.03$, evaluate (a) $a_{\overline{23.5}|}^{(4)}$, (b) $\ddot{a}_{\overline{23.75}|}^{(4)}$, (c) $_{1.5}|a_{\overline{5.25}|}^{(4)}$, (d) $\ddot{s}_{\overline{6.5}|}^{(2)}$, and (e) $a_{\overline{23.75}|}^{(2)}$.

Note that the last value should be calculated from Eq. 4.4.2 and your answer compared with that obtained from the alternative definition just described.

Solution

(a) $a_{\overline{23.5}|}^{(4)} = \dfrac{1 - v^{23.5}}{i^{(4)}} = 16.8780$

(b) $\ddot{a}_{\overline{23.75}|}^{(4)} = \dfrac{1}{4} + a_{\overline{23.5}|}^{(4)} = 17.1280$

This value may also be obtained from Eq. 4.4.4.

(c) $_{1.5}|a_{\overline{5.25}|}^{(4)} = v^{1.5} a_{\overline{5.25}|}^{(4)}$

$= v^{1.5}\left(\dfrac{1 - v^{5.25}}{i^{(4)}}\right) = 4.6349$

(d) $\ddot{s}_{\overline{6.5}|}^{(2)} = \dfrac{(1+i)^{6.5} - 1}{d^{(2)}} = 7.2195$

(e) $a_{\overline{23.75}|}^{(2)} = \dfrac{1 - v^{23.75}}{i^{(2)}} = 16.9391$

The alternative definition gives

$a_{\overline{23.75}|}^{(2)} = a_{\overline{23.5}|}^{(2)} + \dfrac{1}{4}v^{23.75} = 16.9395$

SUMMARY

- $i^{(p)}$ is the **nominal rate of interest converted pthly.** It is defined such that the effective rate of interest over a period of length $1/p$ is $i^{(p)}/p$. Therefore,

$$(1 + i) = \left(1 + \frac{i^{(p)}}{p}\right)^p$$

- $d^{(p)}$ is the **nominal rate of discount converted pthly.** It is defined such that the effective rate of discount over a period of length $1/p$ is $d^{(p)}/p$. Therefore,

$$(1 - d) = \left(1 - \frac{d^{(p)}}{p}\right)^p$$

- When n is an integer multiple of p, well-defined and standard notation exists for the present and accumulated values of n-year annuities paid $p$thly, with the usual variations of in arrears, in advance, and deferred payment.
- When n is not an integer multiple of p, an element of discretion exists in the interpretation of the standard notation and how to evaluate the present and accumulated values.

EXERCISES

4.1 **(a)** Calculate the present value of £100 over 10 years at the following rates of interest/discount:

Institute
and Faculty
of Actuaries

    (i) A rate of interest of 5% per annum convertible monthly;

    (ii) A rate of discount of 5% per annum convertible monthly;

    (iii) A force of interest of 5% per annum.

  (b) A 91-day treasury bill is bought for $98.91 and is redeemed at $100. Calculate the annual effective rate of interest obtained from the bill.

4.2 **(a)** Given that $i = 0.0625$, find the values of $i^{(4)}$, δ, and $d^{(2)}$.

  (b) Given that $i^{(2)} = 0.0625$, find the values of $i^{(12)}$, δ, and $d^{(4)}$.

  (c) Given that $d^{(12)} = 0.0625$, find the values of $i^{(2)}$, δ, and d.

  (d) Given that $\delta = 0.0625$, find the values of $i^{(4)}$ and $d^{(2)}$.

4.3 Find, on the basis of an effective interest rate of 4% per unit time, the values of

$$a^{(4)}_{\overline{67}|}, \quad \ddot{s}^{(12)}_{\overline{18}|}, \quad _{14|}\ddot{a}^{(2)}_{\overline{10}|}, \quad \bar{s}_{\overline{56}|}, \quad a^{(4)}_{\overline{16.5}|}, \quad \ddot{s}^{(12)}_{\overline{15.25}|}, \quad _{4.25|}a^{(4)}_{\overline{3.75}|}, \quad \bar{a}_{\overline{26/3}|}.$$

4.4 **(a)** Every 3 years £100 is paid into an account which earns interest at a constant rate. Find the accumulated amount of the account immediately before the sixth payment is made, given that the interest rate is

    (i) 10% per annum effective;

    (ii) 10% per annum convertible half-yearly.

  (b) Sixteen payments, each of amount £240, will be made at 3-yearly intervals, the first payment being made 1 year from now. Find the present value of this series of payments on the basis of an interest rate of 8% per annum effective.

4.5 **(a)** On the basis of an interest rate of 12% per annum effective, find (to the nearest pound) the present value of an annuity of £600 per annum for 20 years payable

    (i) Annually in arrears;

    (ii) Quarterly in arrears;

    (iii) Monthly in arrears;

    (iv) Continuously.

(b) Find the present values of the annuities described in (a) on the basis of an interest rate of 12% per annum convertible quarterly.

4.6 An annuity is payable in arrears for 15 years. The annuity is payable half-yearly for the first 5 years, quarterly for the next 5 years, and monthly for the final 5 years. The annual amount of the annuity is doubled after each 5-year period. On the basis of an interest rate of 8% per annum convertible quarterly for the first 4 years, 8% per annum convertible half-yearly for the next 8 years, and 8% per annum effective for the final 3 years, the present value of the annuity is £2,049. Find the initial annual amount of the annuity.

4.7 An annuity is payable for 20 years. The amount of the annuity in the tth year is £t^2. On the basis of an effective rate of interest of 5% per annum, find the present value of the annuity, assuming that it is payable
(a) Annually in advance,
(b) Quarterly in advance, the payments for each year being made in four equal installments,
(c) Half-yearly in arrears, the payments for each year being made in two equal installments,
(d) Continuously, the rate of payment being constant over each year.

4.8 An investor effects a contract under which he pays £50 to a savings account on 1 July 2006, and at 3-monthly intervals thereafter, the final payment being made on 1 October 2019. On 1 January 2020 the investor will be paid the accumulated amount of the account. Calculate how much the investor will receive if the account earns interest at the rate of
(a) 12% per annum effective,
(b) 12% per annum convertible half-yearly,
(c) 12% per annum convertible quarterly,
(d) 12% per annum convertible monthly.

4.9 On 1 November 2005 a man was in receipt of the following three annuities, all payable by the same insurance company:
(a) £200 per annum payable annually on 1 February each year, the final payment being on 1 February 2027,
(b) £320 per annum payable quarterly on 1 January, 1 April, 1 July, and 1 October each year, the final payment being on 1 January 2022,
(c) £180 per annum payable monthly on the first day of each month, the final payment being on 1 August 2024.
Immediately after receiving the monthly payment due on 1 November 2005, the man requested that these three annuities be combined into a single annuity payable half-yearly on 1 February and 1 August in each subsequent year, the final payment being made on 1 February 2027. The man's request was granted.

Find the amount of the revised annuity, given that it was calculated on the basis of an interest rate of 8% per annum effective, all months being regarded as of equal length.

4.10 **(a)** On 1 January and 1 July each year for the next 20 years a company will pay a premium of £200 into an investment account. In return the company will receive a level monthly annuity for 15 years, the first annuity payment being made on 1 January following payment of the last premium. Find (to the nearest pound) the amount of the monthly annuity payment, given that it is determined on the basis of an interest rate of

  (i)   12% per annum effective;

  (ii) 12% per annum convertible half-yearly;

  (iii) 12% per annum convertible monthly.

(b) Find the monthly annuity payment as in (a), except that the first payment of the annuity to the company is made 1 month after payment of the last premium.

CHAPTER 5

Loan Repayment Schedules

In this chapter we consider an important commercial application of compound interest and equations of value, namely *loans*. In particular, we discuss how to calculate regular repayment amounts and methods for calculating the loan outstanding at any time within the term of the loan. The chapter ends with a brief discussion of alternative definitions of the interest rates that can occur in commercial practice.

5.1 THE GENERAL LOAN SCHEDULE

Suppose that at time $t = 0$ an investor lends an amount L in return for a series of n payments, the rth payment, of amount x_r, being due at time r $(1 \leq r \leq n)$. Suppose further that the amount lent is calculated on the basis of an effective annual interest rate i_r for the rth year $(1 \leq r \leq n)$. In many situations i_r may not depend on r, but at this stage it is convenient to consider the more general case.

The amount lent is simply the present value, on the stated interest basis, of the repayments

$$
\begin{aligned}
L = {} & x_1(1 + i_1)^{-1} + x_2(1 + i_1)^{-1}(1 + i_2)^{-1} \\
& + x_3(1 + i_1)^{-1}(1 + i_2)^{-1}(1 + i_3)^{-1} + \cdots \\
& \cdots + x_n(1 + i_1)^{-1}(1 + i_2)^{-1}\cdots(1 + i_n)^{-1}
\end{aligned}
\tag{5.1.1}
$$

The investor may consider part of each payment as interest (for the latest period) on the outstanding loan and regard the balance of each payment as a capital repayment, which is used to reduce the amount of the loan outstanding. If any payment is insufficient to cover the interest on the outstanding loan, the shortfall in interest is added to the amount of the outstanding loan. In this situation the investor may draw up a schedule which shows the amount of interest contained in each payment and also the amount of the loan outstanding immediately after each payment has been received. It is desirable to consider this schedule in greater detail. The division of each payment into interest and capital is frequently necessary for taxation purposes. Also, in the event of default by the borrower, it may be necessary for the lender

CONTENTS

An Introduction to the Mathematics of Finance. http://dx.doi.org/10.1016/B978-0-08-098240-3.00005-9
© 2013 Institute and Faculty of Actuaries (RC000243). Published by Elsevier Ltd. All rights reserved.

to know the amount of loan outstanding at the time of default. In addition, a change in the terms of the loan would prompt a recalculation of the repayment rates based on the loan outstanding at that time.

Let $F_0 = L$ and, for $t = 1, 2, \ldots, n$, let F_t be the loan outstanding immediately *after* the payment due at time t has been made. The amount of loan repaid at time t is simply the amount by which the payment then made, x_t, exceeds the interest then due, $i_t F_{t-1}$. Also, the loan outstanding immediately after the tth payment equals the loan outstanding immediately after the previous payment *minus* the amount of loan repaid at time t. Hence,

$$F_t = F_{t-1} - (x_t - i_t F_{t-1}) \qquad\qquad 1 \le t \le n \qquad\qquad (5.1.2)$$

Note that this equation holds for $t = 1$, since we have defined $F_0 = L$. Therefore,

$$F_t = (1 + i_t)F_{t-1} - x_t \qquad\qquad t \ge 1 \qquad\qquad (5.1.3)$$

Hence,

$$\begin{aligned} F_1 &= (1 + i_1)F_0 - x_1 \\ &= (1 + i_1)L - x_1 \end{aligned}$$

Then

$$\begin{aligned} F_2 &= (1 + i_2)F_1 - x_2 \\ &= (1 + i_2)[(1 + i_1)L - x_1] - x_2 \\ &= (1 + i_1)(1 + i_2)L - (1 + i_2)x_1 - x_2 \end{aligned}$$

and so on.

More generally, it is easily seen that

$$\begin{aligned} F_t = {}&(1 + i_1)(1 + i_2)\cdots(1 + i_t)L - (1 + i_2)(1 + i_3)\cdots(1 + i_t)x_1 \\ &-(1 + i_3)(1 + i_4)\cdots(1 + i_t)x_2 - \cdots - (1 + i_t)x_{t-1} - x_t \end{aligned} \qquad (5.1.4)$$

The quantity F_t is therefore simply the amount lent *accumulated to time t* less the repayments received by time t *also accumulated to that time*, the accumulations being made on the appropriate varying interest basis. This is referred to as the *retrospective approach* for calculating the loan outstanding at time t, and refers to the fact that one looks *back* at previous cash flows in order to compute the loan outstanding.

An alternative expression for F_t is obtained by multiplying Eq. 5.1.1 by $(1 + i_1)(1 + i_2)\cdots(1 + i_t)$. This gives

$$\begin{aligned} L(1 + i_1)(1 + i_2)\cdots(1 + i_t) = {}&(1 + i_2)(1 + i_3)\cdots(1 + i_t)x_1 + \cdots + x_t \\ &+ (1 + i_{t+1})^{-1}x_{t+1} + \cdots \\ &\cdots + (1 + i_{t+1})^{-1}(1 + i_{t+2})^{-1}\cdots(1 + i_n)^{-1}x_n \end{aligned}$$

By combining this equation with Eq. 5.1.4, we immediately obtain

$$F_t = (1 + i_{t+1})^{-1}x_{t+1} + (1 + i_{t+1})^{-1}(1 + i_{t+2})^{-1}x_{t+2} + \cdots$$
$$+ (1 + i_{t+1})^{-1}(1 + i_{t+2})^{-1}\cdots(1 + i_n)^{-1}x_n$$

(5.1.5)

This shows that F_t is simply *the present value at time t of the outstanding repayments.*
This is referred to as the **prospective approach** for calculating the loan
outstanding and refers to the fact that one looks *forward* to future cash flows yet
to be made in order to calculate the loan outstanding. However, this remark
comes with a warning that the future payments required in the calculation are
those on the *original* terms of the loan. This is particularly important where the
loan outstanding is required in order to compute a new loan schedule
following changes to the term or interest rate, for example.

Equation 5.1.3 is

$$F_t = (1 + i_t)F_{t-1} - x_t$$

Similarly,

$$F_{t+1} = (1 + i_{t+1})F_t - x_{t+1}$$

If $i_t = i_{t+1}$, it follows by subtraction that

$$(F_t - F_{t+1}) = (1 + i)(F_{t-1} - F_t) + x_{t+1} - x_t$$

(5.1.6)

where i denotes the common value of i_t and i_{t+1}. Letting f_t denote the amount
of loan repaid at time t, we may write Eq. 5.1.6 as

$$f_{t+1} = (1 + i)f_t + x_{t+1} - x_t$$

(5.1.7)

It is important to realize that Eq. 5.1.7 holds *only* when the same rate of interest
i is applicable to *both* the tth year and the $(t + 1)$th year. In particular, when the
rate of interest is constant throughout the transaction and all the repayments
are of equal size, the amounts of successive loan repayments form a geometric
progression with common ratio $(1 + i)$.

5.2 THE LOAN SCHEDULE FOR A LEVEL ANNUITY

Consider the particular case where, on the basis of an interest rate of i per unit
time, a loan of amount $a_{\overline{n}|}$ is made at time $t = 0$ in return for n repayments, each
of amount 1, to be made at times 1, 2, . . . , n. The lender may construct
a schedule showing the division of each payment into capital and interest.

Immediately after the tth repayment has been made, there remain $(n - t)$
outstanding payments, and the prospective method (Eq. 5.1.5) shows that the
outstanding loan is simply $a_{\overline{n-t}|}$. In the notation of Section 5.1,

Table 5.2.1 Schedule for a Level Annuity (Amount of Loan $a_{\overline{n}|}$)

Payment	Interest Content of Payment	Capital Repaid	Loan Outstanding after Payment
1	$ia_{\overline{n}|} = 1 - v^n$	v^n	$a_{\overline{n}|} - v^n = a_{\overline{n-1}|}$
2	$ia_{\overline{n-1}|} = 1 - v^{n-1}$	v^{n-1}	$a_{\overline{n-1}|} - v^{n-1} = a_{\overline{n-2}|}$
$\vdots$	$\vdots$	$\vdots$	$\vdots$
t	$ia_{\overline{n-t+1}|} = 1 - v^{n-t+1}$	v^{n-1+1}	$a_{\overline{n-t+1}|} - v^{n-t+1} = a_{\overline{n-t}|}$
$\vdots$	$\vdots$	$\vdots$	$\vdots$
$n-1$	$ia_{\overline{2}|} = 1 - v^2$	v^2	$a_{\overline{2}|} - v^2 = a_{\overline{1}|}$
n	$ia_{\overline{1}|} = 1 - v$	v	$a_{\overline{1}|} - v = 0$

EXAMPLE 5.2.1

A loan of £50,000 is to be repaid over 5 years by a level annuity payable annually in arrears. The amount of the annual payment is calculated on the basis of an interest rate of 3% per annum effective. Immediately *after* the third repayment was made, the borrower requests that he be able to pay off the loan with a single lump sum. Use a repayment schedule to calculate the value of the lump sum required to repay the loan at this time.

Solution

We require the annual repayment amount, X. The equation of value at the time the loan was issued is $Xa_{\overline{5}|\,0.03} = 50{,}000$; this leads to $X = £10{,}917.75$.

The loan outstanding immediately *after* each repayment is given by the final column of the following table:

The first line here, for example, is calculated as in Table 5.2.1 with a constant factor $X = 10{,}917.75$.

> Interest content $= 0.03 \times 10{,}917.73 \times 4.5797 = 1{,}500$
> Capital repaid $= 10{,}91773 \times (1.03)^{-5} = 9{,}417.73$
> Loan outstanding $= 10{,}917.73 \times 3.7171 = 40{,}582.27$, etc.

Alternatively, one can obtain these and all other values from an iterative process as follows:

> Interest content $= 50{,}000 \times 0.03 = 1{,}500$
> Capital repaid $= 10{,}917.73 - 1{,}500 = 9{,}417.73$
> Loan outstanding $= 50{,}000 - 9{,}417.73 = 40{,}582.27$, etc.

The loan outstanding immediately after the third payment is then £20,890.74, and this is the value of the lump sum required to repay the loan at that time.

Payment	Interest Content	Capital Repaid	Loan Outstanding
1	1,500.00	9,417.73	40,582.27
2	1,217.47	9,700.26	30,882.01
3	926.46	9,991.27	20,890.74
4	626.72	10,291.01	10,599.74
5	317.99	10,599.74	0.00

EXAMPLE 5.2.2

Calculate the value of the lump sum required in Example 5.2.1 using (a) the retrospective method and (b) the prospective method.

Solution

Both methods require the annual repayment amount, $X = £10,917.75$. The loan outstanding is required at time $t = 3$.

(a) Under the *retrospective* approach, the cash flows are 50,000 at time 0 and $-X$ at times 1, 2 and 3. The accumulated values of these are

$$50,000(1.03)^3 - X(1.03)^2 - X(1.03)^1 - X = 20,890.68$$

(b) Under the *prospective* approach, despite the borrower wishing to repay the loan outstanding with a single payment, the original terms of the loan require payments of X at times $t = 4$ and 5. The present value of these are

$$Xv + Xv^2 = 20,890.78$$

We see that both methods result in the same loan outstanding with the slight difference arising from rounding errors.

$$F_t = a_{\overline{n-t}|} \qquad (5.2.1)$$

then the amount of loan repaid at time t is

$$f_t = F_{t-1} - F_t = a_{\overline{n-t+1}|} - a_{\overline{n-t}|}$$
$$= v^{n-t+1} \qquad (5.2.2)$$

The lender's schedule may be presented in the form of Table 5.2.1. More generally, if an amount L is lent in return for n repayments, each of amount $X = L/a_{\overline{n}|}$, the monetary amounts in the lender's schedule are simply those in the schedule of Table 5.2.1 multiplied by the constant factor X.

5.3 THE LOAN SCHEDULE FOR A pTHLY ANNUITY

No new principles are involved in the loan schedule for a $p$thly annuity, since this is simply a particular example of the general schedule discussed in the preceding sections. For a loan repayable by a level annuity payable $p$thly in arrears over n time units and based on an interest rate i per unit time, the schedule is best derived by working with an interest rate of $i^{(p)}/p$ per time interval of length $1/p$. Therefore, the interest due at time r/p ($r = 1, 2, \ldots, np$) is $i^{(p)}/p$ times the loan outstanding at time $(r - 1)/p$ (immediately after the repayment then due has been received).

For example, in relation to a loan of $a_{\overline{n}|}^{(p)}$ (at rate i), it is simple to show that the capital repaid in the rth annuity payment ($r = 1, 2, \ldots, np$) is $(1/p)v^{n - (r - 1)/p}$ and that the loan outstanding immediately after the rth payment has been

EXAMPLE 5.3.1

A loan of £10,000 is to be repaid over 10 years by a level annuity payable monthly in arrears. The amount of the monthly payment is calculated on the basis of an interest rate of 1% per month effective. Find

(a) The monthly repayment,

(b) The total capital repaid and interest paid in (i) the first year and (ii) the final year,

(c) After which monthly repayment the outstanding loan is first less than £5,000, and

(d) For which monthly repayment the capital repaid first exceeds the interest content.

Solution

We choose 1 month as our time unit and let $i = 0.01$.

(a) Since the period of the loan is 120 months, we need the value of $a_{\overline{120}|}$ at 1%. This may be obtained from one of the forms

$$a_{\overline{120}|} = a_{\overline{100}|} + v^{100} a_{\overline{20}|}$$

$$= \frac{1 - v^{120}}{i}$$

either of which give $a_{\overline{120}|} = 69.700\ 522$. The monthly repayment is £X, where

$$Xa_{\overline{120}|} = 10{,}000$$

from which it follows that $X = 143.47$. Note that the total payment in any year is $12X = 1{,}721.64$.

(b) **(i)** To find the capital repaid in the first year, we may use either of two approaches. The loan outstanding at the end of the first year (i.e., immediately after the 12th monthly payment) under the prospective method is simply the value then of the remaining repayments, i.e.,

$$Xa_{\overline{108}|} = 9{,}448.62 \qquad \text{at } 1\%$$

This means that the capital repaid in the first year is $(10{,}000 - 9{,}448.62)$, i.e., £551.38. Hence, the interest paid in the first year is $(1{,}721.64 - 551.38)$, i.e., £1,170.26.

Alternatively, note that the capital repaid in the first monthly repayment is $143.47 - (0.01 \times 10{,}000)$, i.e., 43.47. Since we are dealing with a level annuity, successive capital payments form a geometric progression with common ratio $(1 + i)$, i.e., 1.01. The total of the first 12 capital payments is therefore $43.47 s_{\overline{12}|} = £551.31$. (Because of rounding errors, this number differs slightly from the value of £551.38 found by the first method. In practice, to avoid such errors, one may adjust the capital repayments slightly in order that their total exactly equals the original amount lent.)

(ii) The capital repaid in the final year is simply the loan outstanding at the start of the final year. This is $143.47 a_{\overline{12}|} = £1{,}614.77$ under the prospective method. The interest paid in the final year is therefore $(1{,}721.64 - 1{,}614.77) = £106.87$.

(c) After the tth monthly repayment, the outstanding loan is $143.47\, a_{\overline{120-t}|}$. Consider the equation

$$143.47 a_{\overline{120-t}|} = 5{,}000$$

i.e.,

$$a_{\overline{120-t}|} = 34 \cdot 850 \qquad \text{at } 1\%$$

Since $a_{\overline{43}|} = 34.8100$ and $a_{\overline{44}|} = 35.4555$, the outstanding loan is first less than £5,000 when $120 - t = 43$ (so $t = 77$), i.e., after the 77th monthly repayment has been made.

(d) The tth capital payment is of amount $43.47(1.01)^{t-1}$. We need to know when this is first greater than one-half of the total monthly payment. We therefore seek the least integer t for which

$$43 \cdot 47(1 \cdot 01)^{t-1} > \frac{143 \cdot 47}{2}$$

i.e.,

$$t - 1 > \frac{\ln 1.6502}{\ln 1.01} = 50.34$$

Hence, the required value of t is 52.

EXAMPLE 5.3.2

An investor takes out a loan on the basis of an interest rate of 4% per annum, repayable annually in arrears for 20 years. The first repayment payment is £2,000, and subsequent payments increase by £100 each year. The investor draws up a schedule showing the division of each repayment into capital and interest.

Find the original amount of the loan. Derive expressions for the capital and interest content of the tth payment and for the loan outstanding after the tth payment has been made.

Solution

The original loan is

$$1,900a_{\overline{20}|} + 100(Ia)_{\overline{20}|} \quad \text{at } 4\% \quad = 38,337.12$$

Let f_t denote the capital repaid in the tth annuity payment. Then

$$f_1 = 2,000 - (0 \cdot 04 \times 38,337.12) = 466.515$$

To avoid significant rounding errors, we work with three decimal places. Since each payment is £100 greater than the previous payment, Eq. 5.1.7 implies that

$$f_{t+1} = 1.04 f_1 + 100$$

from which it follows easily by induction that, for $t > 1$,

$$f_t = 1.04^{t-1} f_1 + 100(1.04^{t-2} + 1.04^{t-3} + \cdots + 1)$$

$$= (1.04^{t-1} \times 466.515) + 100\frac{1.04^{t-1} - 1}{0.04}$$

$$= 2,966.515 \times 1.04^{t-1} - 2,500$$

Since the tth payment is $(1,900 + 100t)$, the interest content is

$$1,900 + 100t - [(2,966.515 \times 1.04^{t-1}) - 2,500]$$

$$= 4,400 + 100t - 2,966.515 \times 1.04^{t-1}$$

The loan repaid in the first t payments is

$$\sum_{r=1}^{t} f_r = \sum_{r=1}^{t} [(2,966.515 \times 1.04^{r-1}) - 2,500]$$

$$= 2,966.515 \frac{1.04^t - 1}{0.04} - 2,500t$$

$$= 74,162.875(1.04^t - 1) - 2,500t$$

The loan outstanding after t payments have been received is therefore

$$38,337.12 - [74,162,875(1.04^t - 1) - 2,500t]$$

$$= 112,500 + 2,500t - 74,162.875 \times 1.04^t$$

A check on this final expression is obtained by observing that the loan outstanding immediately after the repayment at time t is the value then of the remaining payments. Since the payment at time $t + r$ is £$(1,900 + 100t + 100r)$, the value at time t of the outstanding payments is

$$(1,900 + 100t) \, a_{\overline{20-t}|} + 100(Ia)_{\overline{20-t}|}$$

at 4%. The reader should verify that this equals the expression above for the outstanding loan at time t. (Both expressions are equal to $112,500 + 2,500t - 162,500 \times (1.04)^{t-20}$.)

received is $a^{(p)}_{\overline{n-r/p}|}$ (at rate i). This is simply the value of the outstanding payments from the prospective method.

5.4 CONSUMER CREDIT LEGISLATION

Annual Percentage Rate (APR)

Various countries around the world have enacted laws aimed at making people who borrow money or buy goods or services on credit more aware of the true

EXAMPLE 5.4.1

A loan of £1,000 is repayable by a level annuity payable half-yearly in arrears for 3 years and calculated on the basis of an interest rate of 15% per annum effective. Construct the lender's schedule showing the subdivision of each payment into capital and interest and the loan outstanding after each repayment.

Solution

At 15%, $i^{(2)}/2 = 0.072381$, so the interest due at the end of each half-year is 7.2381% of the loan outstanding at the start of the half-year. The amount of the annual repayment is $1,000/a_{\overline{3}|}^{(2)}$ at 15%, i.e., 422.68. The half-yearly payment is then £211.34. The schedule is easily drawn up as follows:

(1)	(2)	(3)	(4)
	Loan Outstanding at	Interest Due at End of nth Half-year (£)	Capital Repaid at End of nth Half-year (£)
n	Start of nth Half-year (£)	[0.072 381 × (2)]	[211.34 − (3)]
1	1000.00	72.38	138.96
2	861.04	62.32	149.02
3	712.02	51.54	159.80
4	552.22	39.97	171.37
5	380.85	27.57	183.77
6	197.08	14.26	197.08

EXAMPLE 5.4.2

A 25-year loan of £250,000 is made to an individual in order to purchase a house. The repayment terms are initially based on an interest rate of 4% per annum effective, and repayments are to be made monthly in arrears. The interest rate changes to 5% per annum effective immediately after the 12th payment. Calculate the revised monthly repayment amount after this change.

Solution

We begin by calculating the initial repayment amounts. Working in years, let the annual repayment by X such that

$$250,000 = X\, a_{\overline{25}|}^{(12)} = 15.9065X$$

leading to $X = £15,716.89$, and so the monthly repayments are £1,309.74.

When we use the retrospective approach, the capital outstanding immediately after the 12th payment (1 year) is X such that

$$250,000 \times (1+i)^1 - X\, s_{\overline{1}|}^{(12)} = 243,997.01$$

The reader should verify this amount using the prospective approach, resulting in $X\, a_{\overline{24}|}^{(12)}$.

The revised monthly repayments are then Y such that

$$243,997.01 = 12Y\, a s_{\overline{24}|}^{(12)} \text{ at } 5\%$$

leading to $Y = £1,440.81$.

cost of credit; in particular, the laws enable people to compare the true interest rates implicit in various lending schemes. Examples of laws of this type are the Consumer Credit Act 1974 (revised in 2006) in the UK and the Consumer Credit Protection Act 1968 (which contains the "truth in lending" provisions) in the USA.

Regulations made under powers introduced in the UK Consumer Credit Act 1974 lay down what items should be treated as entering into the total charge for credit and how the rate of charge for credit should be calculated. The rate is known as the *Annual Percentage Rate of Charge (APR)* and is defined in such a way as to be the effective annual rate of interest on the transaction, obtained by solving the appropriate equation of value for i, taking into account all the items entering into the total charge for credit. The APR is therefore closely associated with the internal rate of return for the loan that we will cover in Chapter 6. In all cases the APR is to be quoted to the *lower one-tenth of 1%*. For example, if the rate i is such that $0.155 \leq i < 0.156$, the quoted APR is 15.5%. The total charge for credit and the APR have to be disclosed in advertisements and in quotations for consumer credit agreements.

Regulation Z of the US Consumer Credit Protection Act 1968 requires the disclosure of the "finance charge" (defined as the excess of the total repayments over the amount lent) and the "annual percentage rate", which is the nominal rate of interest per annum convertible as often as the repayments are made (e.g., monthly or weekly). The value quoted must be accurate to one-quarter of 1%.

Flat Rates of Interest

In many situations in which a loan is to be repaid by level installments at regular intervals, it is occasionally common commercial practice to calculate the amount of each repayment installment by specifying a *flat rate* of interest for the transaction. The operation of flat rates of interest is as follows.

Consider a loan of L_0. Which is to be repayable over a certain period by n level installments. Suppose that the flat rate of interest for the transaction is F per specified time unit. (Note that the time unit used to specify F need *not* be the time interval between repayments; in practice the time unit used to specify F is generally a year.) The total *charge for credit* for the loan is defined to be

$$D = L_0 Fk \tag{5.4.1}$$

where k is the repayment period of the loan, measured in units of time used in the definition of F.

The total amount repaid is defined to be the amount of the loan plus the charge for credit, i.e., $(L_0 + D)$. Each installment is therefore of amount

$$X = \frac{L_0 + D}{n} \tag{5.4.2}$$

This together with Eq. 5.4.1 leads to the definition of the flat rate

$$F = \frac{Xn - L_0}{L_0 k} \tag{5.4.3}$$

which defines the flat rate as the *total interest paid per unit time, per unit borrowed*.

One can calculate the flat rate from the repayment installments and the amount of the loan using Eq. 5.4.3. For example, if the loan L_0 is repaid over 2 years by level monthly installments of X, the total paid is $24X$. This amount includes the total capital and interest paid, so the interest is $24X - L_0$. The flat rate is therefore

$$F = \frac{24X - L_0}{2L_0}$$

The flat rate is a simple calculation that ignores the details of gradual repayment of capital over the loan. For this reason it is only useful for comparing loans of equal term. Since the flat rates ignore the repayment of capital over the term of the loan, it will be considerably lower than the true effective rate of interest charged on the loan.

EXAMPLE 5.4.3

A loan of £4,000 is to be repaid by level monthly installments over 2 years. What is the amount of the monthly repayment if the flat rate of interest for the transaction is

(a) 10% per annum,
(b) 1% per month?

Assuming that the payments are in arrears, find the APR in each case.

Solution

(a) The period of the loan is 2 years, so the charge for credit for the loan is £4,000 × 0.1 × 2 = £800. The total repaid is then £4,800, and the amount of each monthly installment is £4,800/24 = £200.

The APR is obtained from the equation

$$12 \times 200 a_{\overline{2}|}^{(12)} = 4,000 \qquad \text{at rate } i$$

from which $i = 0.197\,47$, so the APR is 19.7% per annum.

(b) The period of the loan is 24 months, so the charge for credit for the loan is £4,000 × 0.01 × 24 = £960. The total repaid is then £4,960, and the monthly installment is then £4,960/24 = £206.67.
The APR is obtained from the equation

$$12 \times 206.67 \times a_{\overline{2}|}^{(12)} = 4,000$$

from which $i = 0.238\,39$, so the APR is 23.8% per annum.

SUMMARY

- The calculation of loan repayment amounts and schedules is an important application of equations of value.
- The level loan repayment can be calculated with knowledge of the interest rate, term, and capital borrowed by solving the equation of value for the entire loan.
- Each repayment amount covers the interest owed on the capital outstanding in the prior period and a contribution towards the capital repayment.
- Under the *retrospective method*, the loan outstanding is calculated from the accumulated value of all prior repayments and the original capital receipt.
- Under the *prospective method*, the loan outstanding is calculated from the present value of all future repayments under the original terms of the loan.
- The *Annual Percentage Rate of Charge* (*APR*) for a loan is defined as the effective annual rate of interest on the transaction, obtained by solving the appropriate equation of value for i, taking into account all the items entering into the total charge for credit. It is quoted to the lower one-tenth of 1%.
- The *flat rate* is the total interest paid per unit time, per unit borrowed.

EXERCISES

5.1 Suppose that a 15-year mortgage for $200,000 is obtained. The mortgage is a level-payment, fixed-rate, fully amortized mortgage. The mortgage rate is 7.0% per annum nominal, converted monthly, and the monthly mortgage payment is $1,797.66.

Institute and Faculty of Actuaries

    (a) Compute an amortization (i.e., repayment) schedule for the first 6 months.

    (b) What will the mortgage balance be at the end of the 15th year?

    (c) If an investor purchased this mortgage, what will the timing of the cash flow be assuming that the borrower does not default?

5.2 An actuarial student has taken out two loans.

    Loan A: A 5-year car loan for £10,000 repayable by equal monthly installments of capital and interest in arrears with a flat rate of interest of 10.715% per annum.

Institute and Faculty of Actuaries

    Loan B: A 5-year bank loan of £15,000 repayable by equal monthly installments of capital and interest in arrears with an effective annual interest rate of 12% for the first 2 years and 10% thereafter.

    The student has a monthly disposable income of £600 to pay the loan interest after all other living expenses have been paid.

    Freeloans is a company that offers loans at a constant effective interest rate for all terms between 3 years and 10 years. After 2 years, the student is approached by a representative of Freeloans who offers the student a 10-year loan on the capital outstanding, which is repayable by equal monthly installments of capital and interest in arrears. This new loan is used to pay off the original loans and will have repayments equal to half the original repayments.

(a) Calculate the final disposable income (surplus or deficit) each month after the loan payments have been made.

(b) Calculate the capital repaid in the first month of the third year assuming that the student carries on with the original arrangements.

(c) Estimate the capital repaid in the first month of the third year assuming that the student has taken out the new loan.

(d) Suggest, with reasons, a more appropriate strategy for the student.

Institute and Faculty of Actuaries

5.3 A government is holding an inquiry into the provision of loans by banks to consumers at high rates of interest. The loans are typically of short duration and to high-risk consumers. Repayments are collected in person by representatives of the bank making the loan.

Campaigners on behalf of the consumers and campaigners on behalf of the banks granting the loans are disputing one particular type of loan. The initial loans are for £2,000. Repayments are made at an annual rate of £2,400 payable monthly in advance for 2 years.

The Consumers' Association Case

The consumers' association asserts that, on this particular type of loan, consumers who make all their repayments pay interest at an annual effective rate of over 200%.

The Banks' Case

The banks state that, on the same loans, 40% of the consumers default on all their remaining payments after exactly 12 payments have been made. Furthermore, half of the consumers who have not defaulted after 12 payments default on all their remaining payments after exactly 18 payments have been made. The banks also argue that it costs 30% of each monthly repayment to collect the payment. These costs are still incurred even if the payment is not made by the consumer. Furthermore, with inflation of 2.5% per annum, the banks therefore assert that the real rate of interest that the lender obtains on the loan is less than 1.463% per annum effective.

(a) (i) Calculate the flat rate of interest paid by the consumer on the loan described in this exercise.

(ii) State why the flat rate of interest is not a good measure of the cost of borrowing to the consumer.

(b) Determine, for each of the preceding cases, whether the assertion is correct.

5.4 A loan of £3,000 is to be repaid by a level annuity-certain, payable annually in arrears for 25 years and calculated on the basis of an interest rate of 12% per annum.

   (a) Find
- **(i)** The annual repayment;
- **(ii)** The capital repayment and interest paid at the end of (1) the 10th year and (2) the final year;
- **(iii)** After which repayment the outstanding loan will first be less than £1,800; and
- **(iv)** For which repayment the capital content will first exceed the interest content.

   (b) Immediately after making the 15th repayment, the borrower requests that the term of the loan be extended by 6 years, the annual repayment being reduced appropriately. Assuming that the lender agrees to the request and carries out his calculations on the original interest basis, find the amount of the revised annual repayment.

5.5 A loan of £16,000 was issued to be repaid by a level annuity-certain payable annually in arrears over 10 years and calculated on the basis of an interest rate of 8% per annum. The terms of the loan provided that at any time the lender could alter the rate of interest, in which case the amount of the annual repayment would be revised appropriately.

   (a) Find the initial amount of the annual repayment.

   (b) Immediately after the fourth repayment was made, the annual rate of interest was increased to 10%. Find the revised amount of the level annual repayment.

   (c) Immediately after the seventh repayment was made, the annual rate of interest was reduced to 9%. There was no further change to the rate of interest. Find the final amount of the level annual repayment and the effective rate of interest paid by the borrower on the completed transaction.

5.6 A loan of £2,000 is repayable by a level annuity-certain, payable annually in arrears for 18 years. The amount of the annual repayment is calculated on the basis of an annual interest rate of 10% for the first 6 years and 9% thereafter.

   (a) Find (i) the amount of the annual repayment and (ii) the amount of capital contained in (1) the 4th repayment and (2) the 12th repayment.

   (b) Immediately after making the 12th repayment, the borrower makes an additional capital repayment of £100, the amount of the annual repayment being appropriately reduced. Assuming that the interest basis is unaltered, find the amount of the revised repayment.

5.7 An annuity is payable continuously between time 0 and time n (where n is not necessarily an integer). The rate of payment of the annuity at time t $(0 \leq t \leq n)$ is t per unit time.

On the basis of an effective interest rate of 5% per annum, the value of the annuity at time $t = 0$ is equal to one-half of the total amount which will be paid. Find n.

5.8 A loan of £9,880 was granted on 10 July 1998. The loan is repayable by a level annuity payable monthly in arrears (on the 10th of each month) for 25 years and calculated on the basis of an interest rate of 7% per annum effective. Find

  (a) The monthly repayment,

  (b) The loan outstanding immediately after the repayment on 10 March 2012,

  (c) The capital repaid on 10 October 2009,

  (d) (i) The total capital to be repaid, and (ii) the total amount of interest to be paid, in the monthly installments due between 10 April 2016 and 10 March 2017 (both dates inclusive),

  (e) The month when the capital to be repaid first exceeds one-half of the interest payment.

5.9 A loan of £19,750 was repayable by a level annuity payable monthly in arrears for 20 years and calculated on the basis of an interest rate of 9% per annum effective. The lender had the right to alter the conditions of the loan at any time and, immediately after the 87th monthly repayment had been made, the effective annual rate of interest was increased to 10%. The borrower was given the option of either increasing the amount of his level monthly repayment or extending the term of the loan (the monthly repayment remaining unchanged).

  (a) Show that, if the borrower had opted to pay a higher monthly installment, the monthly repayment would have been increased by £8.45.

  (b) Assume that the borrower elected to continue with the monthly repayment unchanged. Find the revised term of the loan and, to the nearest pound, the (reduced) amount of the final monthly repayment.

5.10 A loan of £11,820 was repayable by an annuity payable quarterly in arrears for 15 years. The repayment terms provided that at the end of each 5-year period the amount of the quarterly repayment would be increased by £40. The amount of the annuity was calculated on the basis of an effective rate of interest of 12% per annum.

  (a) Find the initial amount of the quarterly repayment.

  (b) On the basis of the lender's original schedule, find the amount of principal repaid in (i) the 3rd year and (ii) the 13th year.

(c) Immediately after paying the 33rd quarterly installment, the borrower requested that in future the repayments be of a fixed amount for the entire outstanding duration of the loan. The request was granted, and the revised quarterly repayment was calculated on the original interest basis. Find the amount of the revised quarterly repayment.

5.11 A loan is repayable over 10 years by a special decreasing annuity, calculated on the basis of an effective interest rate of 10% per annum. The annuity payment each year is divided into an interest payment (equal to 10% of the loan outstanding at the start of the year) and a capital payment, which is used to reduce the amount of the loan outstanding. The annuity decreases in such a way that, if income tax of 30% of the interest content of each annuity payment were to be deducted from each payment, the *net* amount of the payment (i.e., the capital payment plus the interest payment less tax) would be £5,000 each year. An investor who is not liable to tax will in fact receive the *gross* amount of each annuity payment (i.e., the payment without any deduction). What price should such an investor pay for the annuity to achieve an effective yield of 8% per annum?

CHAPTER 6

Project Appraisal and Investment Performance

This chapter is largely concerned with a number of applications of compound interest theory to the assessment of investments and business ventures. These matters are, of course, considered by accountants, economists, and others as well as by actuaries. Some writers use terminology and symbols which differ from those usually employed by actuaries, but there are no differences of principle. The chapter also introduces the concept of *real returns* that will be revisited in later chapters.

CONTENTS

6.1 NET CASH FLOWS

Suppose that an investor (who may be a private individual or a corporate body in all that follows) is considering the merits of an investment or business project. The investment or project will normally require an initial outlay and possibly other outlays in future, which will be followed by receipts (although in some cases the pattern of income and outgo is more complicated). The cash flows associated with the investment or business venture may be completely fixed (as in the case of a secure fixed-interest security maturing at a given date; see Chapters 7 and 8), or they may have to be estimated. The estimation of the cash inflows and outflows associated with a business project usually requires considerable experience and judgment, and all relevant factors (such as taxation) should be considered. It is often prudent to perform calculations on more than one set of assumptions, for example, on the basis of 'optimistic', 'average', and 'pessimistic' forecasts, respectively. More complicated techniques using statistical theory, for example, are available to deal with this kind of uncertainty but are beyond the scope of this book. Precision is not attainable in the estimation of cash flows for many business projects, and so extreme accuracy is out of place in many of the calculations that follow.

Recall from Section 3.2 that the net cash flow c_t at time t (measured in suitable time units) is

$$c_t = \text{cash inflow at time } t - \text{cash outflow at time } t \qquad (6.1.1)$$

An Introduction to the Mathematics of Finance. http://dx.doi.org/10.1016/B978-0-08-098240-3.00006-0

If any payments may be regarded as continuous, then $\rho(t)$, the net rate of cash flow per unit time at time t, is defined as

$$\rho(t) = \rho_1(t) - \rho_2(t) \tag{6.1.2}$$

where $\rho_1(t)$ and $\rho_2(t)$ denote the rates of inflow and outflow at time t, respectively.

EXAMPLE 6.1.1

A businessman is considering a particular project. He estimates that the venture will require an initial outlay of £20,000 and a further outlay of £10,000 after 1 year. There will be an estimated inflow of £3,000 per annum payable continuously for 10 years beginning in 3 years' time and a final inflow of £6,000 when the project ends in 13 years' time. Measuring time in years, describe the net cash flows associated with this venture and illustrate the position on a diagram that uses the standard convention for inflows and outflows.

Solution
We have the net cash flows:

$$c_0 = -20,000$$
$$c_1 = -10,000$$
$$c_{13} = +6,000$$
$$\rho(t) = +3,000 \qquad \text{for } 3 \le t \le 13$$

Figure 6.1.1 illustrates these cash flows.

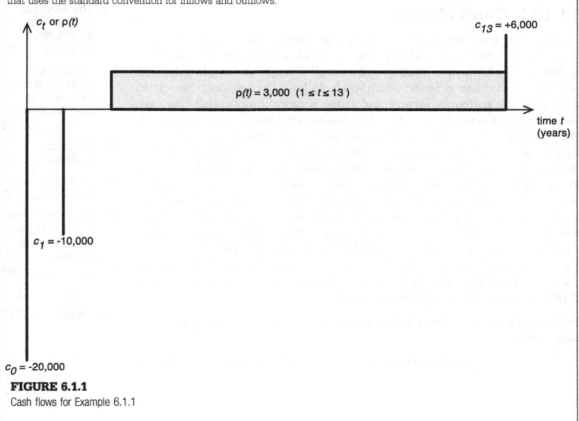

FIGURE 6.1.1
Cash flows for Example 6.1.1

6.2 NET PRESENT VALUES AND YIELDS

Having ascertained or estimated the net cash flows of the investment or project under scrutiny, the investor will wish to measure its profitability in relation to other possible investments or projects. In particular, he may wish to determine whether or not it is prudent to borrow money to finance the venture.

Assume for the moment that the investor may borrow or lend money at a fixed rate of interest i per unit time. The investor could accumulate the net cash flows connected with the project in a separate account in which interest is payable or credited at this fixed rate. By the time the project ends (at time T, say), the balance in this account will be

$$\sum c_t(1+i)^{T-t} + \int_0^T \rho(t)(1+i)^{T-t}dt \qquad (6.2.1)$$

where the summation extends over all t such that $c_t \neq 0$.

The present value at rate of interest i of the net cash flows is called the **net present value** at rate of interest i of the investment or business project, and is usually denoted by $NPV(i)$. Hence,

$$NPV(i) = \sum c_t(1+i)^{-t} + \int_0^T \rho(t)(1+i)^{-t}dt \qquad (6.2.2)$$

(Note that if the project continues indefinitely, the accumulation of Eq. determined by Eq. 6.2.1 is not defined, but the net present value may be defined by Eq. 6.2.2 with $T = \infty$.) If $\rho(t) = 0$, we obtain a simpler formula resulting from the discrete payments

$$NPV(i) = \sum c_t v^t \qquad (6.2.3)$$

where $v = (1+i)^{-1}$. Since the equation

$$NPV(i) = 0 \qquad (6.2.4)$$

is the *equation of value* for the project at the present time, the yield i_0 on the transaction is the solution of this equation, provided that a unique solution exists. Conditions under which the yield exists, and numerical methods for solving Eq. 6.2.4, were discussed in Chapter 3.

It may readily be shown that $NPV(i)$ is a smooth function of the rate of interest i and that $NPV(i) \rightarrow c_0$ as $i \rightarrow \infty$.

In economics and accountancy, the yield per annum is often referred to as the **internal rate of return (IRR)** or the **yield to redemption**. The latter term is frequently used when dealing with fixed-interest securities, for which the 'running' yield is also considered (see Chapter 7).

EXAMPLE 6.2.1

Find the net present value function $NPV(i)$ and the yield for the business venture described in Example 6.1.1. Plot the graph of $NPV(i)$ for $0 \leq i \leq 0.05$.

Solution

By Eq. 6.2.2,

$$NPV(i) = -20{,}000 - 10{,}000v + 3{,}000\left(\bar{a}_{\overline{13}|} - \bar{a}_{\overline{3}|}\right)$$
$$+ 6{,}000v^{13} \qquad \text{at rate } i$$

The graph of $NPV(i)$ is as in Figure 6.2.1 (for $0 \leq i \leq 0.05$). We see from the graph that the yield i_0 (which must exist since the cash flow changes sign only once) is more than 2%, and by interpolation between the rates of interest 2% and 2½%, we obtain $i_0 \approx 2.2\%$. (A more accurate value is 2.197%, but this degree of accuracy is not usually necessary.)

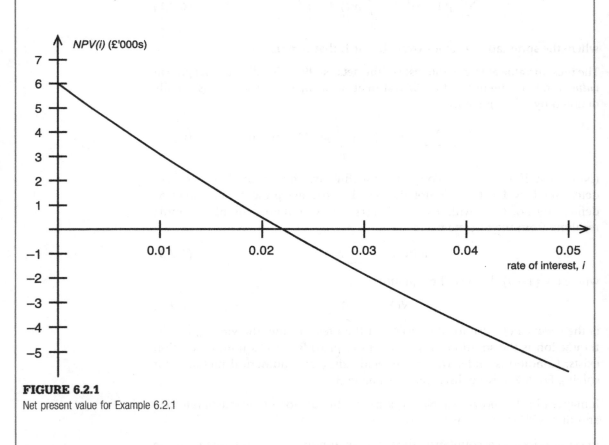

FIGURE 6.2.1
Net present value for Example 6.2.1

The practical interpretation of the net present value function $NPV(i)$ and the yield is as follows. Suppose that the investor may lend or borrow money at a fixed rate of interest i_1. Since, from Eq. 6.2.2, $NPV(i_1)$ is the present value at rate of interest i_1 of the net cash flows associated with the project, we conclude that the project will be profitable if and only if

$$NPV(i_1) > 0 \qquad (6.2.5)$$

Also, if the project ends at time T, then the profit (or, if negative, loss) at that time is

$$NPV(i_1)(1 + i_1)^T \qquad (6.2.6)$$

Let us now assume that, as is usually the case in practice, the yield i_0 exists and $NPV(i)$ changes from positive to negative when $i = i_0$. Under these conditions, it is clear that the project is profitable if and only if

$$i_1 < i_0 \qquad (6.2.7)$$

i.e., the yield exceeds the rate of interest at which the investor may lend or borrow money.

EXAMPLE 6.2.2

Assume that the businessman in Example 6.1.1 may borrow or lend money at 2% per annum. Determine whether or not the business venture of Example 6.2.1 is profitable, and find the profit or loss when the project ends in 13 years' time.

Solution

The net cash flow changes sign only once, so the yield i_0 exists, and it is clear that $NPV(i)$ changes sign from positive to negative at i_0. Since $i_0 \approx 2.2\%$ exceeds $i_1 = 2\%$, the project is profitable. By Eq. 6.2.6, the expected profit in 13 years' time is

$$NPV(0.02) \times (1.02)^{13} = 481 \times (1.02)^{13} = £622$$

6.3 THE COMPARISON OF TWO INVESTMENT PROJECTS

Suppose now that an investor is comparing the merits of two possible investments or business ventures, which we call projects A and B, respectively. We assume that the borrowing powers of the investor are not limited.

Let $NPV_A(i)$ and $NPV_B(i)$ denote the respective net present value functions, and let i_A and i_B denote the yields (which we shall assume to exist). It might be thought that the investor should always select the project with the higher yield,

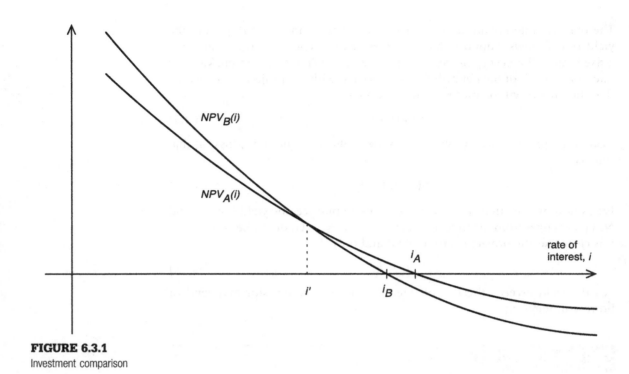

FIGURE 6.3.1
Investment comparison

but this is not always the best policy. A better criterion to use is *the profit at time T* (the date when the later of the two projects ends) or, equivalently, the *net present value*, calculated at the rate of interest i_1 at which the investor may lend or borrow money. The reason is that A is the more profitable venture if

$$NPV_A(i_1) > NPV_B(i_1) \tag{6.3.1}$$

The fact that $i_A > i_B$ may not imply that $NPV_A(i_1) > NPV_B(i_1)$ is illustrated in Figure 6.3.1. Although i_A is larger than i_B, the $NPV(i)$ functions 'cross over' at i'. It follows that $NPV_B(i_1) > NPV_A(i_1)$ for any $i_1 < i'$, where i' is the *cross-over rate*. There may even be more than one cross-over point, in which case the range of interest rates for which Project A is more profitable than Project B is more complicated. This behaviour reflects that the *NPVs* are sensitive to the profile of the timings of cash flows and the term of the investment.

Example 6.3.1 illustrates the fact that the choice of investment depends very much on the rate of interest i_1 at which the investor may lend or borrow money. If this rate of interest were 5.75%, say, then loan B would produce a loss to the investor, while loan A would give a profit.

EXAMPLE 6.3.1

An investor is considering whether to invest in either or both of the following instruments:

Investment A   For a purchase price of £10,000 the investor will receive £1,000 per annum payable quarterly in arrears for 15 years.

Investment B   For a purchase price of £11,000, the investor will receive an income of £605 per annum, payable annually in arrears for 18 years, and a return of his outlay at the end of this period.

The investor may lend or borrow money at 4% per annum. Would you advise him to invest in either loan, and, if so, which would be the more profitable?

Solution

We first consider Investment A:

$$NPV_A(i) = -10{,}000 + 1{,}000a_{\overline{15}|}^{(4)}$$

and the yield is found by solving the equation $NPV_A(i) = 0$, or $a_{\overline{15}|}^{(4)} = 10$, which gives $i_A \approx 5.88\%$.
For Investment B we have

$$NPV_B(i) = -11{,}000 + 605a_{\overline{18}|} + 11{,}000v^{18}$$

and the yield (i.e., the solution of $NPV_B(i) = 0$) is $i_B = 5.5\%$. The rate of interest at which the investor may lend or borrow money is 4% per annum, which is less than both i_A and i_B, so we compare $NPV_A(0.04)$ and $NPV_B(0.04)$.

Now $NPV_A(0.04) = £1{,}284$ and $NPV_B(0.04) = £2{,}089$, so it follows that, although the yield on Investment B is less than on Investment A, the investor will make a larger profit from loan B. We should therefore advise him that an investment in either loan would be profitable, but that, if only one of them is to be chosen, then loan B will give the higher profit.

6.4 DIFFERENT INTEREST RATES FOR LENDING AND BORROWING

So far we have assumed that the investor may borrow or lend money at the same rate of interest i_1. In practice, however, he will probably have to pay a higher rate of interest (j_1, say) on borrowings than the rate (j_2, say) he receives on investments. The difference $j_1 - j_2$ between these rates of interest depends on various factors, including the creditworthiness of the investor and the expense of raising a loan.

The concepts of net present value and yield are, in general, no longer meaningful in these circumstances. We must calculate the accumulation of the net cash flows from first principles, the rate of interest depending on whether or not the investor's account is in credit. In many practical problems, the balance in the investor's account (i.e., the accumulation of net cash flows) will be negative until a certain time t_1 and positive afterwards, except, perhaps, when the project ends. In order to determine this time t_1, we must solve one or more equations, as in Example 6.4.1.

EXAMPLE 6.4.1

A mining company is considering an opencast project. It is estimated that the opencast site will produce 10,000 tonnes of ore per annum continuously for 10 years, after which period there will be an outlay of £300,000 to restore the land. The purchase price of the mining rights will be £1,000,000 and mining operations will cost £200,000 per annum, payable continuously.

The company has insufficient funds to finance this venture, but can borrow the initial outlay of £1,000,000 from a bank, which will charge interest at 12% per annum effective; this loan is not for a fixed term, but may be reduced by repayments at any time. When the mining company has funds to invest, it will receive interest calculated at 10% per annum effective on its deposits.

On the assumption that the price of ore is such that this project will just break even, determine (to the nearest month) how long the mining company will take to repay its bank indebtedness and hence calculate this minimum ore price.

Solution

Let P be the break-even price per tonne of ore. The net cash flow of the opencast project to the mining company is as follows:

$$c_0 = -1,000,000$$
$$c_{10} = -300,000$$
$$\rho(t) = k = 10,000P - 200,000 \qquad \text{for } 0 \leq t \leq 10$$

Since the rates of interest for borrowing and lending (12% and 10% per annum, respectively) are different, we must find the

time t_1 years when the mining company will have repaid its bank indebtedness. This will occur when

$$1,000,000(1.12)^{t_1} = k\bar{s}_{\overline{t_1}|} \qquad \text{at 12\%}$$

which is equivalent to

$$1,00,000 = k\bar{a}_{\overline{t_1}|} \qquad \text{at 12\%} \qquad (1)$$

From time t_1 onwards, the net cash flow may be accumulated at 10% per annum interest and, since the balance at the end of the project is to be zero, we have

$$k\bar{s}_{\overline{10-t_1}|} - 300,000 = 0 \qquad \text{at 10\%} \qquad (2)$$

We now solve Eqs 1 and 2 for the two unknowns t_1 and k (from which we can find P). Eliminating k from these equations, we obtain

$$0.3\bar{a}_{\overline{t_1}|0.12} = \bar{s}_{\overline{10-t_1}|0.10}$$

This equation must be solved numerically for t_1. By trial and interpolation, we obtain $t_1 = 8.481$ (to three decimal places) and hence $k = 183,515$. Hence

$$P = \frac{200,000 + k}{10,000}$$
$$= £38.35$$

is the minimum price per tonne of ore which would make the project profitable.

In some cases the investor must finance his investment or business project by means of a fixed-term loan without an early repayment option. In these circumstances he cannot use a positive cash flow to repay the loan gradually, but must accumulate this money at the rate of interest applicable on lending, i.e., j_2. We reconsider the mining venture of Example 6.4.1 under these conditions in Example 6.4.2.

6.5 PAYBACK PERIODS

In many practical problems, the net cash flow changes sign only once, this change being from negative to positive. In these circumstances the balance in the investor's account will change from negative to positive at a unique time t_1,

EXAMPLE 6.4.2

Consider again the opencast ore project of Example 6.4.1, but now suppose that the bank loan of £1,000,000 is for a fixed term of 10 years with no early repayment options and that interest is payable continuously. What is the minimum price per tonne of ore which would make the project viable?

Solution

Let P' be the break-even price per tonne of ore under these conditions. Note that the annual rate of payment of loan interest is $1,000,000\,\delta_{0.12}$, i.e., 113,329. Hence, after paying loan interest, the mining company will have a continuous cash inflow at the rate of

$$k' = 10,000P' - 200,000 - 113,329$$
$$= 10,000P' - 313,329$$

per annum for 10 years, at which time it must repay the bank loan and the cost of restoring the land. Therefore, accumulating the cash inflow at 10% per annum, we must have

$$k'\,\bar{s}_{\overline{10}|} = 1,300,000 \qquad \text{at } 10\%$$

from which we obtain $k' = 77,744$, and so $P' = 39.11$. The minimum viable ore price is then £39.11 per tonne. As is clear by general reasoning, the new condition on the bank loan (i.e., that it cannot be repaid early) is disadvantageous to the mining company because it must accumulate cash at the (lower) rate of interest applicable to lending rather than pay off bank indebtedness (on which interest is chargeable at 12% per annum).

or it will always be negative, in which case the project is not viable. If this time t_1 exists, it is referred to as the *discounted payback period* (**DPP**). It is the smallest value of t such that $A(t) \geq 0$, where

$$A(t) = \sum_{s \leq t} c_s(1+j_1)^{t-s} + \int_0^t \rho(s)(1+j_1)^{t-s}ds \qquad (6.5.1)$$

Note that t_1 does not depend on j_2 but only on j_1, the rate of interest applicable to the investor's borrowings.

Suppose that the project ends at time T. If $A(T) < 0$ (or, equivalently, if $NPV(j_1) < 0$), the project has no discounted payback period and is not profitable. If the project is viable (i.e., there is a discounted payback period t_1), the *accumulated profit* when the project ends at time T is

$$P = A(t_1)(1+j_2)^{T-t_1} + \sum_{t > t_1} c_t(1+j_2)^{T-t}$$
$$+ \int_{t_1}^T \rho(t)(1+j_2)^{T-t}dt \qquad (6.5.2)$$

This follows since the net cash flow is accumulated at rate j_2 (the rate of interest applicable to the investor's deposits) after the discounted payback period has elapsed.

If interest is ignored in Eq. 6.5.1 (i.e., if we put $j_1 = 0$), the resulting period is called the *payback period*. As is shown in Examples 6.5.1 and 6.5.2 however, its use instead of the discounted payback period often leads to erroneous results and is therefore not to be recommended.

The discounted payback period is often employed when considering a single investment of C, say, in return for a series of payments each of R, say, payable annually in arrears for n years. The discounted payback period t_1 years is clearly the smallest integer t such that $A^*(t) \geq 0$, where

$$A^*(t) = -C(1+j_1)^t + Rs_{\overline{t}|} \qquad \text{at rate } j_1 \qquad (6.5.3)$$

i.e., the smallest integer t such that

$$Ra_{\overline{t}|} \geq C \qquad \text{at rate } j_1 \qquad (6.5.4)$$

The project is therefore viable if $t_1 \leq n$, in which case the accumulated profit after n years is clearly

$$P = A^*(t_1)(1+j_2)^{n-t_1} + Rs_{\overline{n-t_1}|} \qquad \text{at rate } j_2 \qquad (6.5.5)$$

EXAMPLE 6.5.1

An investment of £100,000 will produce an annuity of £10,500 annually in arrears for 25 years. Find the discounted payback period when the interest rate on borrowed money is 9% per annum. Find also the accumulated profit after 25 years if money may be invested at 7% per annum.

Solution

By Eq. 6.5.4, the discounted payback period is the smallest integer t such that

$$10,500a_{\overline{t}|} \geq 100,000 \qquad \text{at 9%}$$

From compound interest tables we see that the discounted payback period is 23 years, i.e.,

$$a_{\overline{23}|} = 9.5802 \text{ at } i = 9\%.$$

The accumulated profit after 25 years is (from Eqs 6.5.3 and 6.5.5)

$$P = \left[-100,000(1.09)^{23} + 10,500s_{\overline{23}|0.09} \right](1.07)^2$$
$$+ 10,500s_{\overline{2}|0.07} = £26,656$$

EXAMPLE 6.5.2

Find the payback period for the investment detailed in Example 6.5.1.

Solution

If interest is ignored, the payback period (calculated from Eq. 6.5.3 with $j_1 = 0$) is 10 years. This is much less than the true discounted period from Example 6.5.1.

Likewise, £$(25 \times 10,500 - 100,000)$, i.e., £162,500, is far too high an estimate of the final profit.

If the rates (or forces) of interest on borrowing and/or lending are assumed to vary with time, one may find the accumulation of the net cash flow by formulae given in Chapter 3. In practice, the determination of the net cash flow and its accumulation at any future time is done computationally. It is usual in many such calculations to consider the net cash flow and its accumulation on a yearly basis. The resulting analysis may be easily understood and interpreted by those responsible for making investment decisions.

6.6 THE EFFECTS OF INFLATION

Consider the simplest situation in which an investor can lend and borrow money at the same rate of interest i_1. In certain economic conditions, the investor may assume that some or all elements of the future cash flows should incorporate allowances for inflation (i.e., increases in prices and wages). The extent to which the various items in the cash flow are subject to inflation may differ. For example, wages may increase more rapidly than the prices of certain goods, or vice versa, and some items (such as the income from rent-controlled property) may not rise at all, even in highly inflationary conditions.

The case when *all* items of cash flow are subject to the same rate of escalation e per time unit is of special interest. In this case we find or estimate c_t^e and $\rho^e(t)$, the net cash flow, and the net rate of cash flow allowing for escalation at rate e per unit time, by the formulae

$$c_t^e = (1+e)^t c_t \tag{6.6.1}$$

$$\rho^e(t) = (1+e)^t \rho(t) \tag{6.6.2}$$

where c_t and $\rho(t)$ are estimates of the net cash flow and the net rate of cash flow, respectively, at time t without any allowance for inflation. It follows that, with allowance for inflation at rate e per unit time, the net present value of the investment or business project at rate of interest i is

$$NPV_e(i) = \sum c_t(1+e)^t(1+i)^{-t} + \int_0^\infty \rho(t)(1+e)^t(1+i)^{-t}dt$$

$$\tag{6.6.3}$$

$$= \sum c_t(1+j)^{-t} + \int_0^\infty \rho(t)(1+j)^{-t}dt$$

where

$$1+j = \frac{1+i}{1+e}$$

or

$$j = \frac{i-e}{1+e} \tag{6.6.4}$$

If e is not too large, one sometimes uses the approximation

$$j \approx i - e \qquad (6.6.5)$$

Combining Eqs 6.6.3 and 6.6.4, we have

$$NPV_e(i) = NPV_0\left(\frac{i-e}{1+e}\right) \qquad (6.6.6)$$

where NPV_0 is the net present value function with no allowance for inflation. It follows that, with inflation at rate e per unit time, the yield (or internal rate of return) i_0^e of the project is such that

$$\frac{i_0^e - e}{1+e} = i_0$$

EXAMPLE 6.6.1

A smallholder is considering increasing the size of his flock from 300 to 450 head. According to his calculations (which are all based on 2012 prices), this would give an additional annual profit, after allowing for replacement of sheep but not for financing costs, of £19.10 per extra head. The initial costs associated with increase of size of the flock are as follows:

Purchase price of 150 sheep	4,292
Fencing	1,850
Reseeding 40 acres	8,000
Total	£14,142

In addition, there is an annual cost of £32.70 per reseeded acre for fertilizer, i.e., £1,308 per annum. The fencing may be assumed to last for 20 years, and the resale value (after 20 years) of the extra sheep is taken as zero.

(a) Assuming that there will be no inflation and that the net profit each year will be received at the end of the year, find the internal rate of return for the project.

(b) What uniform annual rate of inflation will make the project viable, if the farmer may borrow and invest money at 10% per annum interest?

Solution
(a) With no inflation, the net income each year from the project is £$(19.10 \times 150 - 1{,}308) = £1{,}557$, so that the net cash flows associated with the project are

$$c_0 = -14{,}142$$
$$c_1 = c_2 = \cdots = c_{20} = 1{,}557$$

Hence,

$$NPV_0(i) = -14{,}142 + 1{,}557 a_{\overline{20}|}$$

from which it follows that the internal rate of return is $i_0 \approx 9.07\%$.

(b) Now suppose that the sheep farmer may borrow and lend money at the fixed rate of 10% per annum. Assuming that all prices and costs escalate at a compound annual rate e, we obtain the net present value function

$$NPV_e(i) = NPV_0\left(\frac{i-e}{1+e}\right)$$

and the internal rate of return per annum becomes (see Eq. 6.6.7)

$$i_0^e = 0.0907(1+e) + e$$

which is greater than 0.1 if $e > 0.0085$. It follows that, if the annual rate of inflation is greater than 0.85%, i_0^e will exceed 10% and the venture will be profitable.

where i_0 is the corresponding yield if there were no inflation (also see Section 7.9 where *real* yields on investments are discussed). This means that

$$i_0^e = i_0(1 + e) + e \qquad (6.6.7)$$

or, if e is small,

$$i_0^e \approx i_0 + e \qquad (6.6.8)$$

EXAMPLE 6.6.2

In response to a period of high interest rates, a building society can offer borrowers a 'low-start' repayment plan. Under these schemes, the borrower of an initial sum of L_0 will pay $P(1 + e)^{t-1}$ at the end of year t ($t = 1, 2, 3, \ldots, n$), where n is the term of the loan and e is a fixed rate of escalation. The payments are used to pay interest at rate i on the outstanding loan and, if any money remains, to reduce the loan. If the interest on the outstanding loan exceeds the year's payment, the excess is added to the capital outstanding. The building society makes no allowance for expenses or for taxation.

(a) Show that, if L_t denotes the amount of loan outstanding immediately after the repayment due at time t has been made, then

$$L_t = L_0(1 + i)^t - P(1 + i)^{t-1}\ddot{a}_{\overline{t}|j}$$

where

$$j = \frac{i - e}{1 + e}$$

Hence, show that

$$P = \frac{(1 + i)L_0}{\ddot{a}_{\overline{n}|j}}$$

(b) Given that $i = 0.12$ and $e = 0.1$, find, in respect of an initial loan of £200,000 repayable over 20 years, the payment P due at the end of the first year.

Solution
(a) The loan outstanding at the end of a given year can be obtained (from the retrospective method) as the

accumulation at rate i of the amount lent less the accumulation of the repayments made (see Chapter 5). Hence,

$$L_t = L_0(1 + i)^t - \sum_{r=1}^{t} P(1 + e)^{r-1}(1 + i)^{t-r}$$

$$= L_0(1 + i)^t - P(1 + i)^{t-1}\sum_{r=1}^{t}\left(\frac{1 + i}{1 + e}\right)^{-(r-1)}$$

$$= L_0(1 + i)^t - P(1 + i)^{t-1}\sum_{r=1}^{t}(1 + j)^{-(r-1)}$$

$$= L_0(1 + i)^t - P(1 + i)^{t-1}\ddot{a}_{\overline{t}|j}$$

as required.

Since the term of the loan is n years, we must have $L_n = 0$. Hence,

$$L_0(1 + i)^n - P(1 + i)^{n-1}\ddot{a}_{\overline{n}|j} = 0$$

so

$$P = \frac{L_0(1 + i)}{\ddot{a}_{\overline{n}|j}}$$

as required.

(b) Using the preceding results, we have

$$P = \frac{200,000 \times 1.12}{\ddot{a}_{\overline{20}|j}} \qquad \text{(where } j = 0.02/1.1)$$

$$= £13,219.56$$

These results are of considerable practical importance because projects that are apparently unprofitable when rates of interest are high may become highly profitable when even a modest allowance is made for inflation. It is, however, true that in many ventures the positive cash flow generated in the early years of the venture is insufficient to pay bank interest, so recourse must be had to further borrowing (unless the investor has adequate funds of his own). This in itself does not undermine the profitability of the project, but the investor would require the agreement of his lending institution before further loans could be obtained, and this might cause difficulties in practice.

Note that in Example 6.6.1 the interest on the initial loan of £14,142 will be £1,414.20 in the first year. Since this is less than £1,557 $(1 + e)$, the smallholder will not have to borrow more money at any time. This is not always the case, as shown in Example 6.6.2.

6.7 MEASUREMENT OF INVESTMENT FUND PERFORMANCE

It is frequently the case in practice that one wishes to measure the investment performance of a given fund (e.g., a pension fund or life assurance fund) over a period of several years. Suppose that the period in question is from time t_0 to time t_n (where time is measured in years). Suppose further that

$$t_0 < t_1 < t_2 < \cdots < t_n$$

so that the overall period $[t_0, t_n]$ may be divided into n specified subintervals $[t_0, t_1], [t_1, t_2], \ldots, [t_{n-1}, t_n]$. In practice these subintervals are often periods of length 1 year, but this is not always the case.

Using appropriate methods, we may calculate the annual yield obtained on the fund in each of these subintervals. For $r = 1, 2, \ldots, n$, let i_r denote the yield per annum on the fund over the period $[t_{r-1}, t_r]$.

Consider now an investment of 1 at time t_0. If, for $r = 1, 2, \ldots, n$, this investment had earned interest at an effective annual rate i_r over the period $[t_{r-1}, t_r]$, the accumulated amount of the investment at time t_n would have been

$$(1 + i_1)^{t_1 - t_0}(1 + i_2)^{t_2 - t_1} \cdots (1 + i_n)^{t_n - t_{n-1}}$$

The constant effective annual rate of interest which would have given the same accumulation over the period t_0 to t_n is i, where

$$(1 + i)^{t_n - t_0} = (1 + i_1)^{t_1 - t_0}(1 + i_2)^{t_2 - t_1} \cdots (1 + i_n)^{t_n - t_{n-1}}$$

from which it follows that

$$i = \left[(1 + i_1)^{t_1 - t_0}(1 + i_2)^{t_2 - t_1} \cdots (1 + i_n)^{t_n - t_{n-1}}\right]^{1/(t_n - t_0)} - 1 \qquad (6.7.1)$$

This value of i is generally known as the *linked internal rate of return* per annum on the fund for the period t_0 to t_n (with reference to the subdivisions defined by the intermediate times $t_1, t_2, \ldots, t_{n-1}$). In an obvious sense, the linked internal rate of return is an average of the yields obtained in each subinterval. Note, however, that this average takes no account of changes in the size of the fund from period to period. It also depends on the particular subdivision of the entire period.

To allow for changes in the size of the fund with time, we may also calculate the yield per annum on the fund over the entire period t_0 to t_n. This yield is generally referred to as the *money-weighted rate of return* per annum on the fund for the period. It does not depend upon any particular subdivision of the entire period and generally gives greater weight to the yields pertaining to the times when the fund is largest.

A third index, which is often used in practice, is the *time-weighted rate of return* per annum over a given period. The time-weighted rate of return does not depend upon any particular subdivision of the period, but does require that the precise times of all cash flows of *new money* into and out of the fund over the period are known. Suppose that the cash flows of *new money*, strictly within the period, occur at times $t_1, t_2, \ldots, t_{n-1}$, where $t_1 < \cdots < t_{n-1}$. Let the given period be $[t_0, t_n]$. There may also be cash flows of new money at times t_0 and t_n. For $0 \leq r \leq n$, let V_r denote the amount of the fund at time t_r, *after* receipt of all interest and capital gains due up to and including time t_r, but *before* payment to or withdrawal from the fund of any new money at that time. Let c_r ($0 \leq r \leq n$) denote the *net* inflow of new money to the fund at time t_r. Therefore, c_r is the amount of new investment in the fund at time t_r less the amount of money withdrawn from the fund at that time. Hence, $V_r + c_r$ is the amount of the fund at time t_r *after* new money has been received or paid at that time.

For $0 \leq r < n$, define

$$R_r = \frac{V_{r+1}}{V_r + c_r}. \qquad (6.7.2)$$

That is, R_r is the 'accumulation factor' for the fund over the time interval $[t_r, t_{r+1}]$. Let

$$R = R_0 R_1 R_2 \ldots R_{n-1} \qquad (6.7.3)$$

and define i by the equation

$$(1+i)^{t_n - t_0} = R,$$

i.e.,

$$i = R^{1/(t_n - t_0)} - 1. \qquad (6.7.4)$$

This value of i is the *time-weighted rate of return* per annum for the period $[t_0, t_n]$.

The money-weighted rate of return and the time-weighted rate of return are best understood in the context of an example.

EXAMPLE 6.7.1

A fund had a value of £150,000 on 1 July 2009. A net cash flow of £30,000 was received on 1 July 2010, and a further net cash flow of £40,000 was received on 1 July 2011. The fund had a value of £175,000 on 30 June 2010 and a value of £225,000 on 30 June 2011. The value of the fund on 1 January 2012 was £280,000.

(a) Calculate the time-weighted rate of return per annum earned on the fund between 1 July 2009 and 1 January 2012.

(b) Calculate the money-weighted rate of return per annum earned on the fund between 1 July 2009 and 1 January 2012.

Solution

(a) The time-weighted rate of return per annum is i such that

$$(1+i)^{2.5} = \frac{175}{150} \times \frac{225}{175+30} \times \frac{280}{225+40} = 1.352968$$

hence,
$$i = 12.85\%$$

(b) The money-weighted rate of return is i such that

$$150(1+i)^{2.5} + 30(1+i)^{1.5} + 40(1+i)^{0.5} = 280$$

This can be solved through trial and interpolation to find that $i = 12.58\%$ per annum.

EXAMPLE 6.7.2

A fund had assets totaling £600m on 1 January 2011. It received net income of £40m on 1 January 2012 and £100m on 1 July 2012. The value of the fund was £450m on 31 December 2011, £500m on 30 June 2012, and £800m on 31 December 2012.

(a) Calculate the time-weighted rate of return per annum for the period 1 January 2011 to 31 December 2012.

(b) Calculate the linked-internal rate of return over the same period, using subintervals of a calendar year.

Solution

(a) The time-weighted rate of return per annum is i such that

$$(1+i)^2 = \frac{450}{600} \times \frac{500}{450 \times 40} \times \frac{800}{500+100} = 1.020408$$

hence, $i = 1.015\%$ per annum.

(b) The first subinterval is the first year, and the money-weighted rate of return in this period is i_1. We have

$$1+i_1 = \frac{450}{600}, \text{and so } i_1 = -25\%$$

The second subinterval is the second year, and the money-weighted rate of return in this period is i_2. We have

$$490(1+i_2) + 100(1+i_2)^{0.5} = 800$$

This can be solved as a quadratic in $(1+i_2)^{0.5}$, leading to $i_2 = 39.188\%$ per annum.

The linked internal rate of return is then i, such that

$$(1+i)^2 = 0.75 \times 1.39188, \text{leading to } i$$
$$= 2.1719\% \text{ per annum.}$$

Note that the time-weighed rate of return in Example 6.7.1 is better for appraising the performance of a fund (and therefore fund manager), as it is not sensitive to cash flow amounts and the timing of payments, which are, of course, beyond the control of the fund manager. The money-weighted rate of return is sensitive to both amount and timing of cash flows, and this is a disadvantage of the method as a means of fund appraisal. A further disadvantage of the money-weighted rate of return is that the equation may not have a unique solution, or any solution at all. A significant disadvantage of both the time-weighted and money-weighted rates of return are that the calculations require information about all the cash flows of the fund during the period of concern. This disadvantage motivates the use of the linked internal rate of return with reference to subdivisions that coincide with standard reporting periods of the fund.

In some cases the linked internal rate of return (with reference to a particular subdivision), the money-weighted rate of return, and the time-weighted rate of return do not differ significantly. In many situations, however (especially if the amount of the fund varies greatly over the relevant period), these rates of return may differ considerably.

SUMMARY

- The *net present value* (*NPV*) of a project is calculated from the present values of the net cash flows. The project is profitable at a particular rate of interest if the NPV is positive. NPVs can be used to compare the profitability of different projects at a particular interest rate.
- The *internal rate of return* (*IRR*) is the value of i that solves the equation of value for a project. The project is profitable if the IRR is positive. IRRs can be used to compare the return per unit investment achieved by different projects.
- The *payback period* of a project is the time after which the cumulative cash flow of a project becomes positive. The time value of money is not considered in the calculation.
- The *discounted payback period* is the time after which the cumulative NPV of the cash flow of a project becomes positive.
- The *linked internal rate of return* of an investment fund over a period of time is obtained from the geometric average of returns from the fund within specified subintervals.
- The *money-weighted rate of return* of an investment fund over a period of time is obtained from solving the equation of value for the yield, i.
- The *time-weighted rate of return* is obtained from the returns between flows of new money into or out of the fund.

EXERCISES

 CFA Institute

6.1 Westcott-Smith is a privately held investment management company. Two other investment-counseling companies, which want to be acquired, have contacted Westcott-Smith about purchasing their business. Company A's price is £2 million. Company B's price is £3 million. After analysis, Westcott-Smith estimates that Company A's profitability is consistent with a perpetuity of £300,000 a year. Company B's prospects are consistent with a perpetuity of £435,000 a year. Westcott-Smith has a budget that limits acquisitions to a maximum purchase cost of £4 million. Its opportunity cost of capital relative to undertaking either project is 12% per annum.
   (a) Determine which company or companies (if any) Westcott-Smith should purchase according to the NPV measure.
   (b) Determine which company or companies (if any) Westcott-Smith should purchase according the IRR measure.
   (c) State which company or companies (if any) Westcott-Smith should purchase. Justify your answer.

 CFA Institute

6.2 John Wilson buys 150 shares of ABM on 1 January 2002 at a price of $156.30 per share. A dividend of $10 per share is paid on 1 January 2003. Assume that this dividend is not reinvested. Also on 1 January 2003, Wilson sells 100 shares at a price of $165 per share. On 1 January 2004, he collects a dividend of $15 per share (on 50 shares) and sells his remaining 50 shares at $170 per share.
   (a) Write the formula to calculate the money-weighted rate of return on Wilson's portfolio.
   (b) Using any method, compute the money-weighted rate of return.
   (c) Calculate the time-weighted rate of return on Wilson's portfolio.
   (d) Describe a set of circumstances for which the money-weighted rate of return is an appropriate return measure for Wilson's portfolio.
   (e) Describe a set of circumstances for which the time-weighted rate of return is an appropriate return measure for Wilson's portfolio.

CFA Institute

6.3 Projects 1 and 2 have similar outlays, although the patterns of future cash flows are different. The cash flows, as well as the NPV and IRR, for the two projects are as follows. For both projects, the required rate of return is 10%.

Year	Cash Flows					NPV	IRR (%)
	0	1	2	3	4		
Project 1	−50	20	20	20	20	13.40	21.86
Project 2	−50	0	0	0	100	18.30	18.92

The two projects are mutually exclusive. What is the appropriate investment decision?

    A. Invest in both projects.

    B. Invest in Project 1 because it has higher IRR.

    C. Invest in Project 2 because it has higher NPV.

6.4 Consider the following two projects. The cash flows as well as the NPV and IRR for the two projects are given. For both projects, the required rate of return is 10%.

	Cash Flows						
Year	0	1	2	3	4	NPV	IRR (%)
Project 1	−100	36	36	36	36	14.12	16.37
Project 2	−100	0	0	0	175	19.53	15.02

What discount rate would result in the same NPV for both projects?

    A. A rate between 0.00% and 10.00%.

    B. A rate between 10.00% and 15.02%.

    C. A rate between 15.02% and 16.37%.

6.5 A computer manufacturer is to develop a new chip to be produced from 1 January 2008 until 31 December 2020. Development begins on 1 January 2006. The cost of development comprises £9 million payable on 1 January 2006 and £12 million payable continuously during 2007. From 1 January 2008, the chip will be ready for production, and it is assumed that income will be received half-yearly in arrears at a rate of £5 million per annum.

    (a) Calculate the discounted payback period at an effective rate of interest of 9% per annum.

    (b) Without doing any further calculations, explain whether the discounted payback period would be greater than, less than, or equal to that given in part (a) if the effective interest rate were substantially greater than 9% per annum.

6.6 A piece of land is available for sale for £5,000,000. A property developer, who can lend and borrow money at a rate of 15% per annum, believes that she can build housing on the land and sell it for a profit. The total cost of development would be £7,000,000, which would be incurred continuously over the first 2 years after purchase of the land. The development would then be complete. The developer has three possible project strategies. She believes that she can sell the completed housing:

    • in 3 years' time for £16,500,000;

    • in 4 years' time for £18,000,000;

    • in 5 years' time for £20,500,000.

The developer also believes that she can obtain a rental income from the housing between the time that the development is completed and the time of sale. The rental income is payable quarterly in advance and is expected to be £500,000 in the first year of payment. Thereafter, the rental income is expected to increase by £50,000 per annum at the beginning of each year that the income is paid.

(a) Determine the optimum strategy if this is based upon using net present value as the decision criterion.

(b) Determine which strategy would be optimal if the discounted payback period was to be used as the decision criterion.

(c) If the housing is sold in six years' time, the developer believes that she can obtain an internal rate of return on the project of 17.5% per annum. Calculate the sale price that the developer believes that she can receive.

(d) Suggest reasons why the developer may not achieve an internal rate of return of 17.5% per annum even if she sells the housing for the sale price calculated in (c).

6.7 A fund had a value of £21,000 on 1 July 2003. A net cash flow of £5,000 was received on 1 July 2004, and a further net cash flow of £8,000 was received on 1 July 2005. Immediately before receipt of the first net cash flow, the fund had a value of £24,000, and immediately before receipt of the second net cash flow, the fund had a value of £32,000. The value of the fund on 1 July 2006 was £38,000.

(a) Calculate the annual effective money-weighted rate of return earned on the fund over the period 1 July 2003 to 1 July 2006.

(b) Calculate the annual effective time-weighted rate of return earned on the fund over the period 1 July 2003 to 1 July 2006.

(c) Explain why the values in (a) and (b) differ.

6.8 A company is considering two capital investment projects. Project A requires an immediate expenditure of £1,000,000 and will produce returns of £270,000 at the end of each of the next 8 years. Project B requires an immediate investment of £1,200,000 together with further expenditure of £20,000 at the end of each of the first 3 years, and will produce returns of £1,350,000 at the end of each of the sixth, seventh, and eighth years.

(a) Calculate (to the nearest 0.1%) the internal rate of return per annum for each project.

(b) Find the net present value of each project on the basis of an effective annual interest rate of 15%.

Comment briefly on your answers.

6.9 A businessman is considering two projects. Each project requires an immediate initial outlay and a further outlay in 1 year's time. For each project, the return will be a level income, payable annually for

.7 years and commencing in 5 years' time. The projects are described in the following table:

Project	Initial Outlay	Outlay after 1 Year	Annual Income
A	£160,000	£80,000	£60,000
B	£193,000	£80,000	£70,000

(a) Find the internal rate of return per annum for each project.
(b) As an alternative to undertaking Project B, the businessman may undertake Project A and, at the same time as making the initial investment for Project A, the businessman could also purchase, with a single premium of £33,000, a level annuity payable annually for 7 years and commencing in 5 years' time. Given that the amount of the annuity would be calculated on the basis of a fixed annual rate of interest, determine how large the rate must be in order that this combined transaction be more advantageous to the businessman than Project B.

6.10 A business venture requires an initial investment of £10,000 and a further investment of £3,000 in a year's time. The venture will produce an income of £500 in 2 years' time, £1,000 in 3 years' time, £1,500 in 4 years' time, and so on, the final income being £4,000 in 9 years' time.
(a) Find the internal rate of return for this project.
(b) An investor has no spare cash but may borrow money at any time at the fixed-interest rate of 5% per annum, all loans being repayable in whole or in part at any time at the borrower's discretion. The investor is considering borrowing to finance the venture described in this exercise. Should he undertake the venture and, if so, what will his profit be in 9 years' time on the completed transaction?

6.11 You have recently inherited a small island, which may be used for sheep rearing, goat breeding, or forestry, and have calculated that the cash flows associated with these three projects will be as follows:

Sheep rearing	Initial cost: £20,000
	Annual income: £1,100, payable annually in arrears for 20 years
	Sale price after 20 years: £20,000
Goat breeding	Initial cost: £20,000
	Annual income: £900, payable annually in arrears for 20 years
	Sale price after 20 years: £25,000
Forestry	Planting cost: £20,000
	Sales of mature trees after 20 years: £57,300

(a) Calculate the internal rates of return of each of these projects (to the nearest 0.1%).

(b) You do not have the capital for any of these ventures, but you have been offered a bank loan of £20,000 at 5% per annum interest payable annually in arrears. This loan is repayable in 20 years' time, there being no early repayment option. Should you require further loans, they will be granted by the bank on the same interest basis and will be repayable at the same date as the original loan. If you have any money to invest after paying bank interest, you may invest it to secure interest at 4% per annum effective.

Which of the three projects will give the largest profit in 20 years' time?

6.12 Two business projects, each of which takes 2 years to complete, produce the following income and expenditure:

Project A	Initial income of £1,000
	After 1 year, expenditure of £2,000
	After 2 years, income of £2,000
Project B	Initial expenditure of £4,000
	After 1 year, income of £7,000
	After 2 years, expenditure of £1,500

(a) For each project, find the rates of interest, if any, which make the present value of the income equal to the present value of the expenditure.

(b) For what range of positive interest rates does the net present value of Project A exceed the net present value of Project B?

(c) An investor, who has no spare cash, can borrow money at any time at a fixed annual rate of interest for any desired term. He may also lend money for any desired term at this same rate.

Which of these projects will be the more profitable, if the fixed annual rate of interest is (i) 20%, (ii) 25%? Calculate the accumulated profit after 2 years from each project in both cases.

6.13 (a) In return for an immediate outlay of £10,000 an investor will receive £6,000 in 1 year's time and £6,600 in 2 years' time. Find, to the nearest 0.1%, the internal rate of return on this investment.

(b) A person who has no spare cash may make the investment described in (a) by borrowing the initial outlay from a bank.

(i) If bank loans are granted on the basis of an interest rate of 16% per annum and may be partially repaid at any time, should the person make the investment and, if so, what profit will he have made at the end of the completed transaction?

(ii) Suppose instead that the bank requires that its loan be for a fixed 2-year period and that interest of 16% per annum be paid annually in arrears. If the person will be able to earn interest of 13% per annum on any spare funds, should he make the investment and, if so, what profit will he have made when the transaction is completed?

6.14 A chemical company has agreed to supply a quantity of a certain compound for each of the next 7 years. The company estimates that the cash flows associated with this project will be as follows:

Outlays	Initial outlay (cost of building plant): £500,000, payable immediately. Manufacturing costs: £400,000 per annum payable annually for 7 years, the first payment being made in 1 year's time. Disposal of waste and demolition of plant: £250,000 per annum payable annually for 3 years, the first payment being made in 8 years' time.
Income	Sales of compound: £600,000 per annum payable annually for 7 years, the first income being received in 1 year's time.

The company proposes to finance the project by borrowing the entire initial cost from a bank, which charges interest on loans at 15% per annum. Partial repayment of the loan will be allowed at any time, and the loan will be repaid by installments as soon as possible from the profits on the manufacture and sale of the compound.

(a) After how many years will the bank loan be paid off?

(b) After repayment of the bank loan, the company will accumulate its profits in a deposit account at the bank, which pays 12% per annum interest on deposits. The costs of disposal of chemical waste and demolition of the plant will be met from this account. Calculate the anticipated balance in the company's deposit account when the project ends in 10 years' time.

6.15 A businessman has decided to purchase a leasehold property for £80,000, with a further payment of £5,000 for repairs in 1 year's time. The income associated with letting the property will be £10,000 per annum, payable continuously for 20 years commencing in 2 years' time.

(a) **(i)** Given that the venture will be financed by bank loans on the basis of an effective annual interest rate of 7% and that the loans may be repaid continuously, find the discounted payback period for the project.

(ii) Given, further, that after the loans have been repaid the businessman will deposit all the available income in an account which will earn interest at 6% per annum effective, find the accumulated amount of the account in 22 years' time.

(b) Suppose that the bank loans may be repaid partially, but only at the end of each *complete* year, interest being paid annually in arrears, and that the businessman may still deposit money at any time for any term at an annual rate of interest of 6% effective. Find (i) the discounted payback period for the project and (ii) the accumulated amount in the businessman's account in 22 years' time.

6.16 Over the period 1 January 2004 to 1 January 2005, the unit prices of two accumulation funds, a 'property fund' and an 'equity fund', took the values given in the following table:

| Fund | Unit Prices (£) | | | | |
| | 2004 | | | | 2005 |
	1 Jan.	1 April	1 July	1 Oct.	1 Jan.
Property	1.24	1.31	1.48	1.58	1.64
Equity	1.21	0.92	1.03	1.31	1.55

(a) Find the time-weighted rate of return for each fund for the year 2004.

(b) An investor bought units in the property fund on 1 January, 1 April, 1 July, and 1 October 2004, and sold his units on 1 January 2005. Find his yield on the completed transaction if
  (i) He bought the same number of units on each date;
  (ii) He invested the same sum of money on each date.

(c) As in (b), except that the investments were made in the equity fund. Comment on your answers.
  (Ignore expenses and taxation. You should assume that investors may buy fractional parts of a unit.)

6.17 In a particular accumulation fund, income is retained and used to increase the value of the fund unit. The 'middle price' of the unit on 1 April in each of the years 1999 to 2005 is given in the following table:

Year	1999	2000	2001	2002	2003	2004	2005
Middle price of unit on 1 April (£)	1.86	2.11	2.55	2.49	2.88	3.18	3.52

(a) On the basis of the preceding prices and ignoring taxation and expenses, find
  (i) The time-weighted rate of return for the fund over the period 1 April 1999 to 1 April 2005;

(ii) The yield obtained by an investor who purchased 200 units on 1 April in each year from 1999 to 2004 inclusive and who sold his holding on 1 April 2005;

(iii) The yield obtained by a person who invested £500 in the fund on 1 April each year from 1999 to 2004 inclusive and who sold back his holding to the fund managers on 1 April 2005. (You should assume that investors may purchase fractional parts of a unit.)

(b) Suppose that, in order to allow for expenses, the fund's managers sell units at a price which is 2% above the published middle price and that they buy back units at a price which is 2% below the middle price. On this basis, find revised answers to (a) (ii) and (iii).

CHAPTER 7

The Valuation of Securities

One of the most important areas of practical application of compound interest theory is in the valuation of stock market securities and the determination of their yields. Accordingly, we shall consider this subject in some detail over the next two chapters. In the present chapter we begin by introducing many of the securities in question. Formulae for calculating the yields obtainable on fixed-interest securities are then discussed and *Makeham's formula* is introduced. This chapter also introduces the complications of incorporating taxation into calculations. In particular, *income tax* is considered here, while *capital gains tax* is considered in Chapter 8. This chapter is necessarily long.

Note that the terms *security* and *loan* are used interchangeably. This reflects that such securities are issued as a source of loan capital.

7.1 FIXED-INTEREST SECURITIES

A government, local authority, private company, or other body may raise money by floating a loan on a stock exchange. The terms of the issue are set out by the borrower, and investors may be invited to subscribe to the loan at a given price (called the *issue price*); or the issue may be by tender, in which case investors are invited to nominate the price that they are prepared to pay, and the loan is then issued to the highest bidders, subject to certain rules of allocation. In either case the loan may be underwritten by a financial institution, which thereby agrees to purchase, at a certain price, any of the issue that is not subscribed to by other investors. Where the interest payments are known in monetary terms in advance (i.e., as a percentage of the nominal value, and not linked to an index), the loans are referred to as *fixed-interest securities*.

Fixed-interest securities normally include in their title the rate of interest payable, e.g., 8% Treasury Stock 2021 or 5% Treasury Gilt 2018, both British government stock. The annual interest payable to each holder, which is often, but not invariably, payable half-yearly, is found by multiplying the *nominal amount* of his holding N by the rate of interest per annum D, which is generally called the *coupon rate*. For example, in the case of 8% Treasury Stock 2021, $D = 0.08$; for 5% Treasury Gilt 2018, $D = 0.05$.

An Introduction to the Mathematics of Finance. http://dx.doi.org/10.1016/B978-0-08-098240-3.00007-2
© 2013 Institute and Faculty of Actuaries (RC000243). Published by Elsevier Ltd. All rights reserved.
For End-of-chapter Questions: © 2013. CFA Institute, Reproduced and republished with permission from the CFA Institute. All rights reserved.

At this stage it is important to introduce the concept of *income tax*. This is the tax levied by governments on financial income of all entities within their jurisdiction. It is distinct from *capital gains tax*, which we consider in Chapter 8. If an investor is liable to income tax, at rate t_1 on the interest payments from a fixed-interest security, his annual income *after* tax will be

$$(1 - t_1)DN$$

The money payable at redemption is calculated by multiplying the nominal amount held N by the *redemption price R* per unit nominal (which is often quoted "per cent" in practice). If $R = 1$, the stock is said to be redeemable *at par*; if $R > 1$, the stock is said to be *redeemable above par* or *at a premium*; and if $R < 1$, the stock is said to be *redeemable below par* or *at a discount*. Some securities have varying coupon rates D or varying redemption prices R; these will be discussed later. The redemption payment is a return of the amount initially lent and is not subject to income tax; it may, however, be subject to capital gains tax.

The *redemption date* is the date on which the redemption money is due to be paid. Some bonds have variable redemption dates, in which case the redemption date may be chosen by the borrower (or perhaps the lender) as any interest date within a certain period, or any interest date on or after a given date. In the latter case the stock is said to have no final redemption date, or to be *undated*. Some banks allow the interest and redemption proceeds to be bought and sold separately, effectively creating bonds with no coupon and bonds redeemable at zero.

We consider as illustrations the following British government stocks.

(a) *8% Treasury 2015*: This stock, which was issued in 1995, bears interest at 8% per annum, payable half-yearly on 7 June and 7 December. The stock is redeemable at par on 7 December 2015.

(b) $7\frac{3}{4}$*% Treasury Loan 2012—2015*: This stock was issued in 1972 and bears interest at $7\frac{3}{4}$%, payable half-yearly on 26 January and 26 July. It was redeemable at par on any interest date between 26 January 2012 and 26 July 2015 (inclusive) at the option of the government. In fact, the government opted to redeem it on 26 January 2012, the earliest possible date, reflecting the low prevailing borrowing rates at that time. In general, since the precise redemption date is not predetermined at issue, but may be chosen between certain limits by the borrower (i.e., issuer), this stock is said to have *optional redemption dates* (see Section 7.7).

(c) $3\frac{1}{2}$*%War Loan*: This stock was issued in 1932 as a conversion of an earlier stock, issued during the 1914—1918 war. Interest is at $3\frac{1}{2}$% per annum, payable half-yearly on 1 June and 1 December. This stock is redeemable at par on any interest date the government chooses, there being no final redemption date. The stock may therefore be considered as having optional redemption dates, the second of them being infinity, and may be valued by methods discussed later.

In all three examples, $R = 1$ and the frequency of interest payments per annum (which we denote by p) is 2. In fact, all British government stocks are redeemable at par (although, as in example (c) above, there may be no final redemption date), and all pay interest half-yearly. This is with the exception of $2\frac{1}{2}$% Consols, $2\frac{1}{2}$% Annuities, and $2\frac{3}{4}$% Annuities, which pay interest quarterly on 5 January, 5 April, 5 July, and 5 October. Note that these undated issues were made in the 1800s and can be considered as remnants of very early, non-standard issues.

The issue price and subsequent market prices of any stock are usually quoted in terms of a certain nominal amount, e.g., £100 or £1 nominal. A loan may therefore be considered to consist of, say, £10,000,000 nominal divided into 100,000 bonds each of £100 nominal. The statement that an investor owns a bond of £100 nominal does not, in general, imply that his holding of the stock in question is worth £100 under the prevailing market conditions at that time. If it is worth £105, say, it is said to be *above par* or *at a premium*; if it is worth £90, say, it is said to be *below par* or *at a discount*; and if it is worth £100, the stock is said to be *at par*. We use the symbol P to denote the price per unit nominal and the symbol A for the price of nominal amount N of the stock. Therefore,

$$A = NP \tag{7.1.1}$$

In practice, stocks are generally quoted "per cent", i.e., per £100 nominal. We also use the symbol C, where

$$C = NR \tag{7.1.2}$$

to denote the *cash* received on redemption in respect of a nominal amount N of the stock.

The coupon rate, redemption price, and term to redemption of a fixed-interest security serve to define the cash payments promised to a tax-free investor in return for the purchase price. If the investor is subject to taxation, appropriate deductions from the cash flow must, of course, be made. The value of fixed-interest stocks at a given rate of interest and the determination of their yields at a given price may be found as for any other investment or business project (see Chapter 6), but, in view of the practical importance of fixed-interest securities and the special terminology used, we discuss them in detail in this chapter.

In practice, a wide range of actively traded securities can be classified as fixed-interest securities and analysed using the techniques detailed later in this chapter and elsewhere in this book. Particular examples include those discussed in the following sections.

Fixed-interest Government Bonds

Fixed-interest government bonds (also known as *gilt-edged stocks* or *gilts*, if issued by the British government) are long-term securities, which exist in large volumes. They may be considered to offer perfect security against default by the borrower if issued by governments of developed countries in their domestic currency. In most developed countries, bonds issued by the government form the largest, most important, and most liquid part of the bond market. They can be bought and sold with relatively little expense and dealt with in large quantities with little or no impact on the price. Since February 1986 almost all fixed-interest securities have been dealt with on the basis that the purchaser pays and the seller receives not only the quoted market price but also *accrued interest* (see Eq. 7.8.3) calculated from the last interest date. In recent years *index-linked* government securities have been issued in the UK. These securities, which carry interest and redemption payments linked to the Retail Prices Index, will be discussed in Section 7.9.

Government Bills

Government bills are short-dated securities issued by governments to fund their short-term spending requirements. They are issued at discount and redeemed at par with no coupon, and mostly denominated in the domestic currency, although issues can be made in other currencies. Government bills are absolutely secure and highly marketable; as a result, they are often used as a benchmark risk-free, short-term investment.

Securities Issued by Local Authorities, Public Boards, and Nationalized Industries

Stocks such as securities issued by local authorities, public boards, and nationalized industries may, in general, be considered to have little risk of default, although the historic example of the Mersey Docks and Harbour Board (which went bankrupt in 1970) shows that caution should be exercised in considering the security of some of these loans.

Loans Issued by Overseas Governments

A UK investor, for example, considering securities such as loans issued by overseas governments must consider questions of currency appreciation or depreciation (unless the loan is in sterling), exchange control, taxation agreements, and, in some cases, the possibility of default.

Loans Issued by Overseas Provinces, Municipalities, Etc.

Stocks such as loans issued by overseas provinces, municipalities, etc., should, of course, be subjected to closer scrutiny than those of the country itself. In the USA, for example, many local stocks went into default in the 1930s.

Debentures

Debentures are issued by various companies and are provided with some form of security. Typically, they are long-dated. Even a very secure debenture will be less marketable than a government stock of similar term and have greater dealing costs. Accordingly, the yield required by investors will be higher than for a comparable government stock. In the United States such stocks are known as *bonds* and are classified according to their credit ratings; yields are, of course, higher for the lower-rated stocks. Yield calculations allowing for the possibility of default may be carried out, but generally we shall ignore this possibility in the valuation of such securities.

Unsecured Loan Stocks

Unsecured loan stocks are issued by various companies and differ from debentures in that they are *unsecured*. In the event of the issuing company winding up (i.e., dissolving), holders of unsecured loan stocks rank alongside other unsecured creditors. Yields will be higher than on comparable debentures by the same company, to reflect the higher risk of default.

Eurobonds

Eurobonds are a form of unsecured medium- or long-term borrowing made by issuing bonds which pay regular interest payments and a final capital repayment at par. Eurobonds are issued and traded internationally and are often *not* denominated in a currency native to the country of the issuer. Large companies, governments, and supra-national organisations issue them. Yields very much depend on the riskworthiness of the issuer and issue size, which determines the marketability. Typically, however, the yield will be slightly lower than for conventional unsecured loan stocks of the same issuer. Issuers are free to add novel features to their issues, and this enables them to appeal to different investors.

Certificates of Deposit

Certificates of deposit are certificates stating that some money has been deposited, usually issued by banks and building societies. They are short-dated securities, and terms to redemption are usually in the range of 28 days to 6 months. Interest is payable on maturity. The degree of security and marketability will depend on the issuing bank.

The valuation of these various types of fixed-interest securities is the topic of much of this chapter. There are also stocks with a coupon rate that varies according to changes in a standard rate of interest, such as the rate on Treasury Bills. These stocks, and the index-linked securities mentioned previously, are not fixed-interest securities in the strict sense, but have more in common with *ordinary shares*, which we briefly discuss in the following section.

7.2 RELATED ASSETS

Ordinary Shares

In addition to dealing in fixed-interest securities, the stock market deals in *ordinary shares* or *equities* (known as *common stocks* in the USA). Ordinary shares or equities are securities, issued by commercial undertakings and other bodies, which entitle their holders to receive all the net profits of the company after interest on loans and fixed-interest stocks has been paid. The cash paid out each year is called the *dividend*, the remaining profits (if any) being retained as reserves or to finance the company's activities. Dividend payments are not a legal obligation of the company, but are paid at the discretion of the directors. The holders of ordinary shares are proprietors, not creditors, of the company, and they usually have voting rights at the company's Annual General Meeting, although non-voting shares can be issued in the UK (but not the USA). We shall not be much concerned with equities in this book, although in Example 7.2.1 we consider the *discounted cash flow* valuation of ordinary shares.

Ordinary shares are the principal way in which companies in many countries are financed, and offer investors high potential return for high risk, particularly risk of capital losses. The expected overall future return on ordinary shares ought to be higher than for most other classes of security to compensate for the greater risk of default and for the variability of returns. Furthermore, ordinary shares are the lowest ranking form of finance issued by companies. In addition to the dividends, the return on ordinary shares can also arise from an increase in the market price of the shares. The initial running yield on ordinary shares is low, but dividends should increase with inflation and real growth in a company's earnings.

Marketability of ordinary shares varies according to the size of the company, but will be better than for the loan capital of the same company if the bulk of the company's capital is in the form of ordinary shares and the loan capital is fragmented into several different issues. Investors buy and sell ordinary shares more frequently than they trade in loan capital, perhaps because the residual nature of ordinary shares makes them more sensitive to changes in investors' views about a company.

Preference Shares

Reference should also be made to *preference shares*, which, despite their name, are fixed-interest securities issued by commercial companies. Interest (or, as they are usually called, dividend) payments are met only after the interest due on any bank loans and debentures has been paid, but before anything is paid to the ordinary shareholders. If there are insufficient divisible profits in any year to meet the dividends for the preference shareholders, the dividend payments are reduced or, in extreme cases, passed. In the case of *cumulative preference shares*, all arrears of dividend are carried forward until they are paid off (although the arrears themselves do not earn interest).

For investors, the expected return on preference shares is likely to be lower than on ordinary shares because the risk of holding preference shares is lower. Preference shares rank higher on the wind up of the issuing company, and the level of income payments is more certain. Marketability of preference shares is likely to be similar to loan capital marketability.

If a security is bought *ex dividend* (x.d.), the seller, not the buyer, will receive the next interest or dividend payment. If it is bought *cum dividend*, then the buyer will receive the next interest or dividend payment. As previously mentioned, most fixed-interest stocks are traded on the basis that the purchaser makes a payment to the seller for accrued interest, in proportion to the number of days between the date of purchase and the date of the last interest payment.

EXAMPLE 7.2.1

A pension fund, which is not subject to taxation, has a portfolio of UK ordinary shares with a current market value of £14,700,000. The current rate of dividend payment from the portfolio is £620,000 per annum. The pension fund wishes to value its holding, at an interest rate of 6% per annum effective, on the assumption that (a) both dividend income from the shares and their market value will increase continuously at the rate of 2% per annum, and (b) the shares will be sold in 30 years' time. What value should the pension fund place on the shares?

Solution

The value to be placed on the holding is from the continuous receipt of dividends and the sale proceeds:

$$620,000 \int_0^{30} (1.02v)^t \, dt + 14,700,000(1.02v)^{30} \qquad \text{at } 6\%$$

$$= 620,000 \left[\frac{1 - (1.02v)^{30}}{-\ln(1.02v)} \right] + 14,700,000(1.02v)^{30}$$

$$= (620,000 \times 17.798) + 4,636,000 = £15,671,000$$

This method for valuing share prices is often called the *discounted cash flow model*, for obvious reasons.

Note that the discounted cash flow method used to value ordinary shares in Example 7.2.1 gives an answer rather different from the market value of £14,700,000 and hence may be used only under certain circumstances, a discussion of which is beyond the scope of this book. The difference between the two values lies in the fact that the market is making different assumptions concerning the future. The method also ignores the variability of share prices and dividends, and is therefore more suitable for valuing a large, diverse portfolio than a single holding. More sophisticated methods for valuing equities, using statistical techniques, have been developed, but we do not discuss them here. We do, however, briefly consider *stochastic models* in Chapter 12.

Property Investment

For the purposes of analysis, property (offices, shops, industrial properties) may be considered another example of fixed-interest security, although property is a real investment, and, as such, rents and capital values might be expected to increase broadly with inflation in the long term, which makes the returns from property similar in nature to those from ordinary shares. We consider inflation-linked investments in Section 7.9.

The return from investing in property comes from rental income and from capital gains, which may be realized on sale. Neither rental income nor capital values are guaranteed, and there can be considerable fluctuations in capital values in particular, in real and monetary terms. Again, these properties suggest similarities with ordinary shares.

Rental terms are specified in lease agreements. Typically, it is agreed that rents are reviewed at specific intervals such as every 3 or 5 years. The rent is changed, at a review time, to be more or less equal to the market rent on similar properties at the time of the review. Some leases have clauses that specify upward-only adjustments of rents.

The following characteristics are particular to property investments, which are fairly unique compared with other examples of investments considered in this chapter:

(a) Large unit sizes lead to less flexibility than investment in shares,
(b) Each property is unique, so can be difficult and expensive to value,
(c) The actual value obtainable on sale is uncertain, i.e., values in property markets can fluctuate just as stock markets can,
(d) Buying and selling expenses are higher than for shares and bonds,
(e) Net rental income may be reduced by maintenance expenses,
 (f) There may be periods when the property is unoccupied and no income is received.

Marketability is poor because each property is unique, and because buying and selling incur high costs.

The *running yield*, as defined in Section 7.3, from property investments will normally be higher than that for ordinary shares. The reasons for this are that dividends usually increase annually (whereas rents are often reviewed less often); property is much less marketable; expenses associated with property investment are much higher; and large, indivisible units of property are much less flexible.

7.3 PRICES AND YIELDS

We now return to the discussion of fixed-interest stocks. As in other compound interest problems, one of two questions may be asked:

1. What price A, or P per unit nominal, should be paid by an investor to secure a net yield of i per annum?
2. Given that the investor pays a price A, or P per unit nominal, what net yield per annum will he obtain?

To answer question 1, we set A equal to the present value, at rate of interest i per annum, of the interest and capital payments, less any taxes payable by the investor. That is,

$$A = \begin{pmatrix} \text{present value, at rate} \\ \text{of interest } i \text{ per annum,} \\ \text{of } net \text{ interest payments} \end{pmatrix} + \begin{pmatrix} \text{present value, at rate} \\ \text{of interest } i \text{ per annum,} \\ \text{of } net \text{ capital payments} \end{pmatrix}$$

(7.3.1)

The price per unit nominal is, of course, $P = A/N$, where N is the nominal amount of stock to which the payments relate.

To answer question 2, we set A in Eq. 7.3.1 equal to the purchase price and solve the resulting equation of value for the net yield i. If the investor is not subject to taxation, the yield i is referred to as a *gross yield*. The yield quoted in the press for a fixed-interest security is often the gross nominal yield per annum, convertible

---

EXAMPLE 7.3.1

A certain *debenture* was redeemable at par on 1 October 2012. The stock bore interest at 6% per annum, payable half-yearly on 1 April and 1 October.

(a) What price per cent should have been offered for this stock on 1 August 1990 to secure a yield of 5% per annum for a tax-free investor?

(b) What yield per annum did this stock offer to a tax-free investor who bought it at 117% on 1 August 1990?

Solution

(a) In this example we have $R = 1$, $N = 100$, $C = 100$, $D = 0.06$, and $p = 2$. The price A which should be offered on 1 August 1990 to secure a yield of 5% per annum is, by Eq. 7.3.1,

$A = $ present value at 5% of interest payments
$\quad$ + present value at 5% of capital payment
$\quad = v^{1/6}[3 + 6a^{(2)}_{\overline{22}|} + 100v^{22}]$ $\qquad$ at 5%
$\quad = 116.19$

(b) We now solve the equation of value

$$117 = v^{1/6}\left[3 + 6a^{(2)}_{\overline{22}|} + 100v^{22}\right]$$

for the rate of interest i. By part (a), when $i = 5\%$, the right side of this equation is 116.19, so the yield is below 5% per annum. Further trials and interpolation give $i \approx 4.94\%$ per annum.

half-yearly. If the investor sells his holding before redemption, or if he is subject to taxation, his actual yield will in general be different from that quoted.

The yield on a security is sometimes referred to as the *yield to redemption* or the *redemption yield* to distinguish it from the *flat* (or *running*) *yield*, which is defined as D/P, the ratio of the coupon rate to the price per unit nominal of the stock.

Note that in Example 7.3.1, one could also work with a period of half a year; the corresponding equation of value would then be

$$117 = v^{1/3}\left(3 + 3a_{\overline{44}|} + 100v^{44}\right) \qquad \text{at rate } i'$$

which has approximate solution $i' \approx 0.0244$, so the effective yield per annum is

$$i \approx (1.0244)^2 - 1 \qquad \text{or} \qquad 4.94\%, \text{ as before}$$

When one has to solve an equation of value by interpolation (to find a yield), it is convenient to have a rough idea of the order of magnitude of the required solution. In most situations, upper and lower bounds may be found quite simply, as the following discussion indicates.

Consider a loan which will be redeemed after n years at a redemption price of R per unit nominal. Suppose that the loan bears interest, payable annually in arrears at a coupon rate of D per annum, and that an investor who is liable to income tax at rate t_1 buys the loan at a price P per unit nominal. What can be said about the magnitude of i, the investor's net annual yield?

In return for a payment of P, the investor receives net interest each year of $D(1 - t_1)$ and redemption proceeds of R. His net yield i is therefore that interest rate for which

$$P = D(1 - t_1)a_{\overline{n}|} + Rv^n \tag{7.3.2}$$

If $R = P$, then clearly

$$i = \frac{D(1 - t_1)}{P}$$

If $R > P$, there is a gain on redemption, and therefore

$$i > \frac{D(1 - t_1)}{P}$$

In this case the gain on redemption is $(R - P)$. If the investor were to receive this gain in equal installments each year over the n years rather than as a lump sum after n years, he would clearly be in a more advantageous position. In this case each year he would receive $D(1 - t_1) + (R - P)/n$ as income (and P as redemption proceeds), so his net annual yield would be $[D(1 - t_1) + (R - P)/n]/P$. This overstates i, so

$$\frac{D(1 - t_1)}{P} < i < \frac{D(1 - t_1) + (R - P)/n}{P}$$

If $R < P$, there is a loss on redemption, and hence,

$$i < \frac{D(1 - t_1)}{P}$$

The loss on redemption is $(P - R)$. If the investor had to bear this loss in equal installments each year over the n years rather than as a lump sum after n years, he would clearly be in a less advantageous position. In this case each year he would receive $D(1 - t_1) - (P - R)/n$ as income (and P as redemption proceeds), so his net annual yield would be $[D(1 - t_1) - (P - R)/n]/P$. This understates i, so

$$\frac{D(1 - t_1)}{P} > i > \frac{D(1 - t_1) - (P - R)/n}{P} = \frac{D(1 - t_1) + (R - P)/n}{P}$$

Therefore, in all cases i lies between $D(1 - t_1)/P$ and $[D(1 - t_1) + (R - P)/n]/P$. For most practical purposes, these bounds are sufficient to indicate suitable values to be used for interpolation.

EXAMPLE 7.3.2

A stock bears interest at $7\frac{1}{2}\%$ per annum, payable annually in arrears, and is redeemable at par in 20 years' time. Assuming that any interest now due will not be received by a purchaser, find the net yield per annum to an investor, liable to income tax at $33\frac{1}{3}\%$, who buys a quantity of this stock at 80%.

Solution

Note that, since the net annual interest payment is £7.5 × $(1-1/3)$ = £5 on an outlay of £80 (i.e., $6\frac{1}{4}\%$) and the stock is redeemed for £100, the net yield will certainly exceed $6\frac{1}{4}\%$ per annum. The gain on redemption is £20 per £100 nominal. If this gain were paid in equal annual installments (each of amount £1), an outlay of £80 would provide net income each year of £6, or $7\frac{1}{2}\%$. This would be a more advantageous investment than is actually available. The net annual yield is therefore less than $7\frac{1}{2}\%$.

To price the stock, note that we have coupon rate $D = 0.075$, price paid per unit nominal $P = 0.8$, redemption price per unit nominal $R = 1$, rate of income tax $t_1 = \frac{1}{3}$, and term to redemption $n = 20$. The equation of value is

$$P = D(1 - t_1)a_{\overline{n}|} + Rv^n \qquad \text{at rate } i$$

i.e.,

$$0.8 = 0.05a_{\overline{20}|} + v^{20}$$

The preceding remarks indicate that i lies between 0.0625 and 0.075. When $i = 0.065$ the right side of the last equation equals 0.8347, and when $i = 0.07$, the value is 0.7881. By interpolation, we estimate i as 0.0687 or 6.87%. (In fact, to four decimal places, the net annual yield is 6.8686%.)

7.4 PERPETUITIES

Certain loans, especially those made in connection with the purchase or lease of land, may continue indefinitely or for so long (e.g., 999 years) as to be considered of indefinite duration for practical purposes. Such loans are referred to as *perpetuities* or *perpetual loans*.

Let us assume that the next interest payment is due at time t years from the present and that interest is at rate D per annum per unit nominal, payable p times per annum. The price P per unit nominal to give a net yield of i per annum, or i, the net yield per annum to an investor who pays P per unit nominal, and is liable to income tax at rate t_1, may be found from the equation

$$P = D(1 - t_1)v^t \ddot{a}_{\overline{\infty}|}^{(p)} \qquad \text{at rate } i$$

i.e.,

$$P = \frac{D(1 - t_1)v^t}{d^{(p)}} \qquad \text{at rate } i \qquad (7.4.1)$$

EXAMPLE 7.4.1

The interest on a $3\frac{1}{2}$% War Loan is payable on 1 June and 1 December each year. Considering this stock to be a perpetuity, find (a) the effective yield per annum and (b) the nominal yield per annum, convertible half-yearly, to a tax-free investor on 22 August 2012, when the price was 34.875%.

Solution

In the notation of Eq. 7.4.1, we have $P = 0.348\ 75$, $D = 0.035$, $p = 2$, and $t = 101/365 = 0.276\ 71$. We therefore solve the equation

$$0.348\ 75 = 0.035 \left[v^{0.27671} / d^{(2)} \right]$$

for i. This gives $i = 10.53\%$, and hence, $i^{(2)} = 2\left[(1+i)^{1/2} - 1\right] \approx 10.27\%$.

7.5 MAKEHAM'S FORMULA

Consider a loan, of nominal amount N, which is to be repaid after n years at a price of R per unit nominal, and let $C = NR$. The quantity C is the *cash* payable on redemption. Let the coupon rate (i.e., the annual interest per unit *nominal*) be D, and assume that interest is payable $p$thly in arrears. Each interest payment is then of amount $DN/p = gC/p$, where

$$g = \frac{DN}{C} \qquad (7.5.1)$$

$$= \frac{D}{R} \qquad (7.5.2)$$

Note that g is the annual rate of interest *per unit of redemption price*.

Consider an investor, liable to income tax at rate t_1, who wishes to purchase the loan at a price to provide an effective net yield of i per annum. Let the price he should pay be A. (We assume that n is an integer multiple of $1/p$ and that any interest now due will not be received by the purchaser.) The price is simply the present value (at rate i) of the redemption proceeds and the future net interest payments. Therefore,

$$A = NRv^n + (1 - t_1)DNa^{(p)}_{\overline{n}|} \qquad \text{at rate } i$$

$$= Cv^n + (1 - t_1)gCa^{(p)}_{\overline{n}|}$$

$$= Cv^n + (1 - t_1)gC\frac{1 - v^n}{i^{(p)}}$$

$$= Cv^n + \frac{g(1 - t_1)}{i^{(p)}}(C - Cv^n)$$

Hence,

$$A = K + \frac{g(1 - t_1)}{i^{(p)}}(C - K) \qquad\qquad (7.5.3)$$

where $K = Cv^n$ (at rate i) is the present value of the capital repayment, and $[g(1 - t_1)/i^{(p)}](C - K)$ is the present value of the net interest payments. Note that $g(1 - t_1)$ is the net rate of annual interest payment *per unit redemption price* or per unit "indebtedness". Equation 7.5.3 is valid only when

(a) g, t_1, and R are constant throughout the term of the loan, and
(b) n is an integer multiple of $1/p$.

See Section 7.8 for the necessary modifications when these conditions are not satisfied.

Equation 7.5.3 is known as **Makeham's formula,** and it is important to realize that it is simply an alternative statement of the equation of value; its derivation contains no new concepts.

EXAMPLE 7.5.1

A 10-year loan is to be issued bearing interest at a rate of 3% per annum payable quarterly in arrears. The loan is redeemable at 110% of par. Calculate the price of this loan if a yield of 4% per annum is required for an investor subject to 40% income tax.

Solution

Using the standard notation, $i = 0.04$, $i^{(4)} = 0.039414$, $g = 0.03/1.10$, $C = 110$, $K = 110v^{10}$, $t_1 = 0.40$, and Makeham's formula 7.5.3 gives the price for £100 nominal as

$$A = 110v^{10} + \frac{0.03}{1.10}(1 - 0.4)\frac{(110 - 110v^{10})}{0.039414} = £89.13$$

An alternative approach is to evaluate the present value of the cash flows directly, leading to

$$A = 3(1 - 0.4)a^{(4)}_{\overline{10}|} + 110v^{10} = £89.13$$

Makeham's formula remains true when the loan is repayable by installments, provided that the coupon rate D, the rate of income tax t_1, and the redemption price R per unit nominal remain constant. To show this, we consider a loan of nominal amount $N = N_1 + N_2 + \cdots + N_m$ where N_j is the nominal amount to be redeemed at time n_j, for $j = 1, 2, \ldots, m$.

The cash received on repayment of part of the loan at time n_j is $C_j = RN_j$. Equation 7.5.3 implies that the value of the capital and net interest payments associated with the jth "tranche" of the loan is

$$A_j = K_j + \frac{g(1 - t_1)}{i^{(p)}}(C_j - K_j)$$

where

$$K_j = C_j v^{n_j}$$

The value of the entire loan is clearly

$$A = \sum_{j=1}^{m} A_j$$

$$= \sum_{j=1}^{m}\left[K_j + \frac{g(1 - t_1)}{i^{(p)}}(C_j - K_j)\right]$$

$$= K + \frac{g(1 - t_1)}{i^{(p)}}(C - K)$$

where

$$K = \sum_{j=1}^{m} K_j = \sum_{j=1}^{m} C_j v^{n_j}$$

is the value of the capital payments and

$$C = \sum_{j=1}^{m} C_j = \sum_{j=1}^{m} RN_j = R\sum_{j=1}^{m} N_j = RN$$

as before. Equation 7.5.3 is therefore still valid in this situation.

The present value, or price, per unit nominal is, of course, $P = A/N$, provided that one purchases the *entire* loan (see Section 7.6). The attractiveness of Makeham's formula lies in the fact that it enables the value of the (net) interest payments and the total value of the security to be obtained quickly from the value of the capital K, even when the stock is redeemable by installments.

Makeham's formula may also be established by general reasoning as follows.

Consider a second loan of the same total nominal amount N as the loan just described. Suppose that, as before, interest is payable $p$thly in arrears and that a nominal amount N_j of this second loan will be redeemed at time n_j $(1 \le j \le m)$ at a price of R per unit nominal. (The capital repayments for this second loan are therefore identical to those of the original loan.) Suppose, however, that for this second loan the *net* annual rate of interest *per unit of redemption price* is $i^{(p)}$. The total "indebtedness" (i.e., capital to be repaid) for either loan is $C = NR$. Since, by hypothesis, the net annual rate of interest payment per unit indebtedness for the second loan is $i^{(p)}$, the value at rate i of this loan is clearly C. Let K be the value of the capital payments of this second loan. (Of course, K is also the value of the capital payments of the first loan.) Then the value of the net interest payments for the second loan must be $(C - K)$. The difference between the two loans lies simply in the rate of payment of net interest. The net annual rate of interest per unit redemption price is $g(1 - t_1)$ for the original loan and $i^{(p)}$ for the second loan. *By proportion*, therefore, the value of the net interest payments for the *first* loan is obviously $g(1 - t_1)/i^{(p)}$ times the value of the net interest payments for the second loan, i.e.,

$$\frac{g(1 - t_1)}{i^{(p)}}(C - K)$$

The value of the first loan, being the value of the capital plus the value of the net interest, is then

$$K + \frac{g(1 - t_1)}{i^{(p)}}(C - K)$$

as given by Eq. 7.5.3. A clear grasp of the preceding "proportional" argument can simplify the solution of many problems but is, of course, not essential.

Note that Makeham's formula may be used either to value the security at a given rate of interest i or to find the yield when the price A is given. It is clear that A decreases as i increases, since A is the present value of a series of positive cash flows. Equivalently, as A increases, the corresponding yield i decreases. It also follows directly from Makeham's formula that

$$
\left.
\begin{array}{lll}
\text{(a)} & A = C & \text{if} \quad g(1 - t_1) = i^{(p)} \\[4pt]
\text{(b)} & A > C & \text{if} \quad g(1 - t_1) > i^{(p)} \\[4pt]
\text{(c)} & A < C & \text{if} \quad g(1 - t_1) < i^{(p)}
\end{array}
\right\}
\qquad (7.5.4)
$$

These results will prove useful in Sections 7.6 and 7.7.

The reader who chooses to value the interest payments from first principles in Example 7.5.2 will rapidly be convinced of the value of Makeham's formula. This is left as an exercise.

EXAMPLE 7.5.2

A loan of £75,000 is to be issued bearing interest at the rate of 8% per annum payable quarterly in arrears. The loan will be repaid at par in 15 equal annual installments, the first installment being repaid 5 years after the issue date. Find the price to be paid on the issue date by a purchaser of the whole loan who wishes to realize a yield of (a) 10% per annum effective and (b) 10% per annum convertible half-yearly. (Ignore taxation.)

Solution

The capital repayments are each of amount £5,000. The first repayment is after 5 years and the final repayment is after 19 years.

(a) Choose 1 year as the basic unit of time. The required yield per unit time is 10% so $i = 0.10$. Using the previous notation, we have $C = 75,000$ (since redemption is at par). The value of the capital repayment is

$$K = 5,000(a_{\overline{19}|} - a_{\overline{4}|}) \text{ at } 10\% \quad = 25,975.27$$

Note that, since redemption is at par, $g = D/R = 0.08$ and interest is paid quarterly (i.e., four times per time unit), so $p = 4$. From Makeham's formula, we obtain the required price as

$$25,975.27 + \frac{0.08}{0.10^{(4)}}(75,000 - 25,975.27) = £66,636.60$$

Since $66,636.60/75,000 = 0.8885$, this price may be quoted as £88.85%.

(b) Choose 6 months as the basic unit of time. The required yield per unit time is 5%, and so $i = 0.05$. Note now that interest is paid *twice* per time unit, so in the previous notation, $p = 2$. Also, per time unit, the amount of interest payable is 4% of the outstanding loan, so now we have $g = 0.04$. The capital repayments occur at times 10, 12, 14, ..., 38, so

$$K = \frac{5,000}{a_{\overline{2}|}}(a_{\overline{40}|} - a_{\overline{10}|}) \quad \text{at } 5\% \quad = 25,377.27$$

Hence, the value of the entire loan is

$$25,377.27 + \frac{0.04}{0.05^{(2)}}(75,000 - 25,377.27)$$
$$= £65,565.63 \text{ or } £87.42\%$$

Note that this price is lower than that in (a).
An alternative approach to (b) is to use the annual effective rate $i = 10.25\%$, which is implied by $i^{(2)} = 10\%$ and work in units of 1 year, as in (a).

EXAMPLE 7.5.3

In relation to the loan described in Example 7.5.2, find the price to be paid on the issue date by a purchaser of the entire loan who is liable to income tax at the rate of 40% and wishes to realize a net yield of 7% per annum effective.

Solution

The capital payments have value

$$K = 5,000(a_{\overline{19}|} - a_{\overline{4}|}) \text{ at } 7\% \quad = 34,741.92$$

Hence, the price to provide a net yield of 7% per annum effective is

$$34,741.92 + \frac{0.08(1 - 0.4)}{0.07^{(4)}}(75,000 - 34,741.92)$$
$$= £63,061.89 \text{ or } £84.08\%$$

EXAMPLE 7.5.4

A loan of nominal amount £80,000 is redeemable at 105% in four equal installments at the end of 5, 10, 15, and 20 years. The loan bears interest at the rate of 10% per annum payable half-yearly. An investor, liable to income tax at the rate of 30%, purchased the entire loan on the issue date at a price to obtain a net yield of 8% per annum effective. What price did he pay?

Solution

Note that the total indebtedness C is $80{,}000 \times 1.05$, i.e., £84,000. Each year the total interest payable is 10% of the outstanding nominal loan, so that the interest payable each year is g times the outstanding indebtedness, where $g = 0.1/1.05$.

Choose 1 year as the unit of time. Then $i = 0.08$, and at the issue date, the capital payments have present value

$$K = 20{,}000 \times 1.05(v^5 + v^{10} + v^{15} + v^{20}) \qquad \text{at } 8\%$$

$$= 21{,}000\frac{a_{\overline{20}|}}{s_{\overline{5}|}} \quad \text{at } 8\% \;= 35{,}144.90$$

Using the value of g described previously, we obtain the price paid by the investor as

$$35{,}144.90 + \frac{0.1}{1.05}\frac{(1-0.3)}{0.08^{(2)}}(84{,}000 - 35{,}144.90) = 76{,}656.07$$

Note that the price "per cent" is the price per £100 *nominal* amount of loan, i.e., $(76{,}656.07/80{,}000) \times 100 = £95.82$.

EXAMPLE 7.5.5

A loan of nominal amount £1,200,000 is to be issued bearing interest of 11% per annum payable half-yearly. At the end of each year, part of the loan will be redeemed at 105%. The nominal amount redeemed at the end of the first year will be £10,000, and each year thereafter the nominal amount redeemed will increase by £10,000 until the loan is finally repaid. The issue price of the loan is £98.80. Find the net effective annual yield to an investor, liable to income tax at 40%, who purchases the entire loan on the issue date.

Solution

The term of the loan is n years, where

$$1{,}200{,}000 = 10{,}000(1 + 2 + \cdots + n)$$
$$= 5{,}000n(n + 1)$$

from which it follows that $n = 15$.

Note that, since the redemption price is 105%, the total indebtedness is $C = 1{,}200{,}000 \times 1.05 = 1{,}260{,}000$ and

$$g = \frac{0.11}{1.05} \qquad (1)$$

Our unit of time is 1 year and $p = 2$.

At rate of interest i, the capital repayments have value

$$K = 10{,}000 \times 1.05 \times (Ia)_{\overline{15}|}$$

so that

$$K = 10{,}500(Ia)_{\overline{15}|} \quad \text{at rate } i \qquad (2)$$

The value of the loan to provide the investor with a net yield of i per annum is then (from Eq. 1)

$$A = K + \frac{0.11}{1.05}\frac{(1-0.4)}{i^{(2)}}(1{,}260{,}000 - K) \qquad (3)$$

where K is given by Eq. 2.

Since the issue price is £98.80 per £100 *nominal*, the price paid by the investor was $0.988 \times 1{,}200{,}000$, i.e., £1,185,600. We require the value of i such that A (as given by Eqs 2 and 3) equals this figure.

Note that each £98.80 invested generates net income of £6.60 per annum and is repaid as £105. The net yield will then be somewhat greater than $6.60/98.80 = 0.0668$ or 6.68%. As a first step, therefore, we value the loan at 7%. It is left for the

Continued

EXAMPLE 7.5.5   *(cont'd)*

reader to verify that, when $i = 0.07$, $A = 1,206,860$ or 100.57%. The net yield is therefore greater than 7% per annum. The reader should confirm that, when $i = 0.08$, $A = 1,127,286$ or 93.94%. By linear interpolation, we estimate the net yield as

$$0.07 + \frac{(1,206,860 - 1,185,600)}{(1,206,860 - 1,127,286)} \times 0.01 = 0.0727 \text{ or } 7.27\%.$$

An alternative approach to solving Eq. 1 for i would be to start with $i = 10\%$ and reduce in 1% increments to determine that the value lies between 7% and 8%. Interpolation could then be used between these values.

(The actual answer is 7.597%.)

EXAMPLE 7.5.6

Ten years ago a loan was issued bearing interest payable annually in arrears at the rate of 8% per annum. The terms of issue provided that the loan would be repaid by a level annuity of £1,000 over 25 years.

An annuity payment has just been made, and an investor is considering the purchase of the remaining installments. The investor will be liable to income tax at the rate of 40% on the interest content (according to the original loan schedule) of each payment. What price should the investor pay to obtain a net yield of 10% per annum effective?

Solution

We consider four alternative solutions. Each is correct, but the relative lengths of the solutions should be noted. Our last solution makes use of a technique known as the *indirect valuation of the capital*, which is very useful in certain problems.

(a) (Solution from first principles)

The amount of the original loan was $1,000a_{\overline{25}|}$ at 8%. The rth annuity payment consists of a capital payment of $1,000v_{0.08}^{26-r}$ and an interest payment of $1,000(1 - v_{0.08}^{26-r})$.

Fifteen payments remain to be made, the tth of which is the $(10 + t)$th overall payment. Hence, the tth remaining payment consists of capital $1,000v_{0.08}^{16-t}$ and interest $1,000(1 - v_{0.08}^{16-t})$.

After tax, the investor receives only 60% of the interest content of each payment. Accordingly, to obtain a net yield of 10% per annum, he should pay

$$\sum_{t=1}^{15} (1.1)^{-t} \left[1,000 v_{0.08}^{16-t} + (1 - 0.4) \times 1,000 \left(1 - v_{0.08}^{16-t} \right) \right]$$

$$= \sum_{t=1}^{15} (1.1)^{-t} \left(600 + v_{0.08}^{16} \times 400 \times v_{0.08}^{-t} \right)$$

$$= 600 a_{\overline{15}|0.1} + 1.08^{-15} \times 400 \frac{1 - \left(\frac{1.08}{1.1} \right)^{15}}{1.1 - 1.08}$$

$$= £6,080.64$$

(b) (Solution using Makeham's formula)

Note that the redemption price and rate of payment of interest are constant, so Makeham's formula may be applied. The outstanding loan C is simply the value (on the *original* interest basis) of the remaining installments. Therefore, $C = 1,000a_{\overline{15}|}$ at 8% = £8,559.48.

The value at 10% of the remaining capital payments is

$$K = \sum_{t=1}^{15} (1.1)^{-t} \times 1,000 \times v_{0.08}^{16-t}$$

$$= 1,000 v_{0.08}^{15} \frac{1 - \left(\frac{1.08}{1.1} \right)^{15}}{1.1 - 1.08} = 3,792.48$$

Hence, from Makeham's formula (with $g = 0.08$), the price to be paid for a net yield of 10% is as before

Continued

EXAMPLE 7.5.6   (cont'd)

$$3{,}792.48 + \frac{0.08(1 - 0.4)}{0.1}(8{,}559.48 - 3{,}792.48) = £6{,}080.64$$

(c) (Solution using the value of the capital, but not requiring Makeham's formula)

The value of the entire outstanding loan (with no taxation) is

$$A = 1{,}000a_{\overline{15}|} \text{ at } 10\% = 7{,}606.08$$

The value of the capital payments K has been found to be 3,792.48. The value of the *gross* interest payments I is simply the value of the entire loan less the value of the capital payments. Therefore,

$$I = A - K = 3{,}813.60$$

The value of the net interest payments is $0.6I$, so the price to be paid by the investor is as before

$$K + 0.6I = 3792.48 + (0.6 \times 3813.60) = 6{,}080.64$$

(d) (Solution using indirect valuation of the capital)

Ignoring tax, we may value the loan as simply the value of the remaining installments. Therefore, to an investor who is *not* liable to tax, the value of the loan to give a yield of 10% per annum is

$$A = 1{,}000a_{\overline{15}|} \text{ at } 10\% = 7{,}606.08 \qquad (1)$$

as in (c). However, if K denotes the value at 10% of the remaining capital payments, by Makeham's formula, the value of the loan to an investor who is not liable to tax is

$$K + \frac{0.08}{0.1}(8{,}559.48 - K) \qquad (2)$$

since £8,559.48 (i.e., $1{,}000a_{\overline{15}|0.08}$) is the outstanding loan. By equating these two expressions for the price, we obtain

$$7{,}606.08 = K + \frac{0.08}{0.1}(8{,}559.48 - K)$$

from which it follows that $K = 3{,}792.48$. Note now that the value of the *capital* payments to an investor who pays no tax is the same as the value of the capital payments to an investor who is liable only to *income* tax. (The reader should study the preceding argument carefully. It is important to realize that, instead of finding the value of K and then using this value to find the price, we have found the price [ignoring tax] directly and then used Makeham's formula backwards to obtain the value of K.)

Having found the value of K, we may use Makeham's formula to find the price to be paid by the investor who is liable to income tax as

$$3{,}792.48 + \frac{0.08(1 - 0.4)}{0.1}(8{,}559.48 - 3{,}792.48)$$
$$= £6{,}080.64$$

The arithmetical simplicity of this solution should be noted. The indirect valuation avoids the need to determine K as the sum of a geometric series.

Note that the indirect determination of K could be used in conjunction with the solution (c).

7.6 THE EFFECT OF THE TERM TO REDEMPTION ON THE YIELD

Consider first a loan of nominal amount N, which has interest payable $p$thly at the annual rate of D per unit nominal. Suppose that the loan is redeemable after n years at a price of R per unit nominal. An investor, liable to income tax at rate t_1, wishes to purchase the loan at a price to obtain a net effective annual yield of i.

As before, let $g = D/R$ and $C = NR$, so that $gC = DN$. The price to be paid by the investor is

$$A(n, i) = \begin{cases} (1 - t_1)DNa_{\overline{n}|}^{(p)} + Cv^n \\ (1 - t_1)gCa_{\overline{n}|}^{(p)} + C[1 - i^{(p)}a_{\overline{n}|}^{(p)}] \\ C + [(1 - t_1)g - i^{(p)}]Ca_{\overline{n}|}^{(p)} \end{cases} \quad \text{at rate } i \quad (7.6.1)$$

If the net annual rate of interest per unit indebtedness were $i^{(p)}$, the value of the loan would be C. In fact, the net annual rate of interest per unit indebtedness is $(1 - t_1)g$, and the second term on the right side of the last equation is the value of net interest in excess of the rate $i^{(p)}$.

The following are immediate consequences of Eq. 7.6.1.

$$\left.\begin{array}{l} \text{(a) If } i^{(p)} = (1-t_1)g, \text{then, for any value of } n, A(n, i) = C. \\ \text{(b) If } i^{(p)} < (1-t_1)g, \text{then, regarded as a function of } n, A(n,i) \text{ is} \\ \quad \text{an increasing function (i.e. if } n_2 > n_1, \text{then } A(n_2, i) > A(n_1, i)). \\ \text{(c) If } i^{(p)} > (1-t_1)g, \text{then, regarded as a function of } n, A(n, i) \text{ is} \\ \quad \text{a decreasing function.} \\ \text{(d) For any fixed } n, A(n, i) \text{ is a decreasing function of } i. \end{array}\right\} \quad (7.6.2)$$

These simple observations are consistent with Eq. 7.5.4. One very important application of the preceding is to be found in relation to loans which have optional redemption dates (see Section 7.7). It is, however, convenient to first discuss some related matters.

Consider two loans, each of which is as described in the first paragraph of this section except that the first loan is redeemable after n_1 years and the second loan after n_2 years, where $n_1 < n_2$. Suppose that an investor, liable to income tax at a fixed rate, wishes to purchase one of the loans for a price B. Then

$$\left.\begin{array}{l} \text{(a) If } B < C \text{ (where } C = NR), \text{the investor will obtain a higher} \\ \quad \text{net yield by purchasing the first loan (i.e., the loan that is} \\ \quad \text{repaid earlier),} \\ \text{(b) If } B > C, \text{the investor will obtain a higher net yield by} \\ \quad \text{purchasing the second loan (i.e., the loan which is} \\ \quad \text{repaid later),} \\ \text{(c) If } B = C, \text{the net yield will be the same for either loan.} \end{array}\right\} \quad (7.6.3)$$

This is easily seen as follows.

If $B = C$, then remark 7.6.2(a) shows that for either loan, the net annual yield is i, where $i^{(p)} = (1 - t_1)g$. This establishes 7.6.3(c).

Suppose now that $B < C$ and let i_1 be the net annual yield obtainable from the first loan. Clearly, $i_1^{(p)} > (1 - t_1)g$. Let B^* be the price which should be paid to provide a net annual yield of i_1 from the *second* loan. Then

$$B^* = A(n_2, i_1) \qquad \text{(by definition of } B^*)$$
$$< A(n_1, i_1) \qquad \text{(by 7.7.2(c), since } i_1^{(p)} > (1 - t_1)g \text{ and } n_2 > n_1)$$
$$= B \qquad \text{(by definition of } i_1)$$

Therefore, a purchaser of the second loan who wishes to obtain a net annual yield of i_1 should pay *less* than B. If, in fact, he pays B, then obviously [or by 7.6.2(d)] his net yield will be less than i_1. This establishes 7.6.3(a).

The justification of 7.6.3(b) follows similarly.

The preceding results are intuitively obvious. If $B < C$, the purchaser will receive a capital gain when either loan is redeemed. From the investor's viewpoint, the sooner he receives this capital gain the better. He will therefore obtain the greater yield on the loan that is redeemed first. However, if $B > C$, there will be a capital loss when either loan is redeemed. The investor will wish to defer this loss as long as possible. He will therefore obtain the greater yield on the loan that is redeemed later.

EXAMPLE 7.6.1

Consider the loan described in Example 7.5.4, and suppose that an investor, who is subject to income tax at 30%, purchases one bond of £100 nominal on the issue date for £95.82. Find the net yield per annum he will obtain, assuming redemption after (a) 5, (b) 10, (c) 15, and (d) 20 years.

Solution

Let the net yield per annum be i_r if redemption of the bond occurs at time $5r$ ($r = 1, 2, 3, 4$). The equation of value for finding i_r is

$$95.82 = K + \frac{0.1 \times 0.7}{1.05 i_r^{(2)}}(105 - K)$$

where $K = 105 v^{5r}$ at rate i_r. The solutions (expressed as percentages, correct to three decimal places) are as follows:

$$i_1 = 9.066\%, \qquad i_2 = 8.109\%, \qquad i_3 = 7.803\%,$$
$$i_4 = 7.660\%$$

7.7 OPTIONAL REDEMPTION DATES

Sometimes a security is issued without a fixed redemption date. In such cases the terms of issue may provide that the *borrower* (i.e., issuer) can redeem the security *at his option* at any interest date on or after some specified date. Alternatively, the issue terms may allow the borrower to redeem the security at his option at any interest date on or between two specified dates (or possibly on any one of a series of dates between two specified dates).

An example of the first situation is provided by $2\frac{1}{2}$% Consols. When this stock was issued, the UK government reserved the right to repay the stock at par on 5 April 1923 or at any later interest date, so it may now be repaid at any interest date the government chooses. An illustration of a range of redemption dates is given by 12% Exchequer 2013–2017. This stock was issued in 1978 and the UK government reserved the right to redeem it at par on any 12 June or 12 December between 12 December 2013 and 12 December 2017 inclusive at the government's choice. (The government has since redeemed this stock on 12 December 2013, reflecting the considerably lower borrowing rates available on the market at that time.)

The latest possible redemption date is called the **final redemption date** of the stock, and if there is no such date (as in the case of $2\frac{1}{2}$% Consols), then the stock is said to be **undated**. It is also possible for a loan to be redeemable between two specified interest dates, or on or after a specified interest date, at the option of the lender, but this arrangement is less common than when the borrower chooses the redemption date.

An investor who wishes to purchase a loan with redemption dates at the option of the borrower cannot, at the time of purchase, know how the market will move in the future and hence when the borrower will repay the loan. He therefore cannot know the net yield that he will obtain. However, by using Eq. 7.6.2 or Eq. 7.6.3, he can determine either

1. The maximum price to be paid, if his net yield is to be *at least* some specified value; or
2. The minimum net yield he will obtain, if the price is some specified value.

We shall now show how he may do so. Suppose that the outstanding term of the loan, n years, may be chosen by the borrower subject to the restriction that $n_1 \leq n \leq n_2$. (We assume that n_1 is an integer multiple of $1/p$ and that n_2 is an integer multiple of $1/p$ or is infinite.) Using the notation of Section 7.6, we let $A(n, i)$ be the price to provide a net annual yield of i, if the loan is redeemed at time n.

Suppose that the investor wishes to achieve a net annual yield of at least i. It follows from Eq. 7.6.2 that

(a) If $i^{(p)} < (1 - t_1)g$, then $A(n_1, i) < A(n, i)$ for *any* value of n such that $n_1 < n \leq n_2$. In this case, therefore, the investor should value the loan on the assumption that redemption will take place at the *earliest* possible date. If this does, in fact, occur, his net annual yield will be i. If redemption occurs at a later date, the net annual yield will exceed i.

(b) If $i^{(p)} > (1 - t_1)g$, then $A(n_2, i) < A(n, i)$ for *any* value of n such that $n_1 \leq n < n_2$. In this case, therefore, the investor should value the loan on the assumption that redemption will occur at the *latest* possible date. (If $n_2 = \infty$, the loan should be valued as a perpetuity.) This strategy will ensure that his net annual yield will be at least i.

(c) If $i^{(p)} = (1 - t_1)g$, the net annual yield will be i, irrespective of the actual redemption date chosen.

Suppose, alternatively, that the price of the loan is given. What can be said about the yield that the investor will obtain? As before, let P and R be the purchase price per unit nominal and the redemption price per unit nominal, respectively. The minimum net annual yield is obtained by solving an appropriate equation of value. The following are immediate consequences of Eq. 7.6.3.

(a) If $P < R$, the investor should determine the net annual yield on the assumption that the loan will be repaid at the *latest* possible date. If this does, in fact, occur, his net annual yield will be that calculated. If redemption takes place at an earlier date, the net annual yield will be greater than that calculated.

(b) If $P > R$, the investor should determine the net annual yield on the assumption that the loan will be repaid at the *earliest* possible date. The actual yield obtained will then be at least the value calculated on this basis.

(c) If $P = R$, the net annual yield is i, where $i^{(p)} = g(1 - t_1)$, irrespective of the actual redemption date chosen.

EXAMPLE 7.7.1

A fixed-interest security bears interest at 7% per annum payable half-yearly and is to be redeemed at 105% at an interest date n years from the present, where $10 \leq n \leq 15$ and n is to be determined by the borrower. What price per cent should be paid for this stock by

(a) A tax-free investor who requires a net yield of at least 5% per annum?

(b) An investor subject to income tax at 40% who also requires a net yield of at least 5% per annum?

Solution

(a) We have

$$g = \frac{0.07}{1.05} = 0.06667, \quad t_1 = 0, \quad i^{(p)} = 0.05^{(2)}$$

$$= 0.04939$$

so

$$g(1 - t_1) > i^{(p)}$$

and there is a capital loss. It is then prudent to assume redemption as early as possible, i.e., at time 10 years, to calculate the price for the *minimum* yield. By Makeham's formula, the required price per unit nominal is

$$K + \frac{g}{i^{(p)}}(C - K) = 1.05v^{10} + \frac{0.06667}{0.04939}$$

$$\times (1.05 - 1.05v^{10}) \text{ at } 5\% = 1.1918$$

Therefore, the *maximum* price is £119.18.

(If the investor pays this price and the loan is redeemed after 15 years, his net annual yield will be 5.42%.)

(b) We now have

$$t_1 = 0.4 \quad \text{and} \quad g(1 - t_1) = 0.04$$

so

$$g(1 - t_1) < i^{(2)}$$

and there is a capital gain. It is then prudent to assume redemption after 15 years (i.e., as late as possible) to calculate the price for the *minimum* yield. By Makeham's formula, the required price per unit nominal is

$$1.05v^{15} + \frac{0.04}{0.04939}(1.05 - 1.05v^{15}) \quad \text{at } 5\%$$

$$= 0.9464$$

Therefore, the *maximum* price is £94.64.

(If the investor pays this price and the loan is redeemed after 10 years, his net annual yield will be 5.35%.)

EXAMPLE 7.7.2

A loan bearing interest of 12% per annum payable quarterly is to be issued at a price of £92. The entire loan will be redeemed at par on any interest date chosen by the borrower between 20 and 25 years (inclusive) from the issue date. Find the minimum net annual yield that will be obtained by an investor, liable to income tax at 30%, who purchases part of the loan.

Solution

Since the issue price is less than the redemption price, our preceding remarks indicate that the investor should value the loan on the assumption that it will be redeemed at the latest possible date. Net interest each year is £8.40 per £100 nominal, so that the appropriate equation of value is

$$92 = 8.40a_{\overline{25}|}^{(4)} + 100v^{25} \qquad \text{at rate } i$$

The yield is clearly somewhat greater than 9%. By interpolation, we find that $i = 0.095\,46$; hence, the net annual yield will be at least 9.546%. (If the loan is, in fact, repaid after 20 years, the net annual yield will be 9.612%.)

Note that if the stock is being valued between two interest dates (or between the issue date and the first interest date), the preceding rules concerning prices and yields may not be correct, as the assumption that n_1 is an integer multiple of $1/p$ and n_2 is an integer multiple of $1/p$, or is infinite, does not hold. In this case the term which gives the greatest price (or least yield) should be determined by further analysis or by numerical calculations.

7.8 VALUATION BETWEEN TWO INTEREST DATES: MORE COMPLICATED EXAMPLES

In the preceding sections we generally assumed that a fixed-interest security was being valued at the issue date or just after receipt of an interest payment. If one is valuing a stock at an interest date immediately *before* receipt of the interest payment, the value to an investor who pays income tax at rate t_1 is clearly

$$A^* = A + (1 - t_1)DN/p = A + (1 - t_1)gC/p \qquad (7.8.1)$$

where A denotes the corresponding "ex dividend" value, which may be found by Makeham's formula or otherwise.

When valuing between two interest dates, or between the date of issue and the first interest date, we may use one of the following methods, (a) and (b). Let the valuation date be a fraction m of a year $(0 < m < 1/p)$ after the preceding interest date or the date of issue, and hence a fraction m' $(= 1/p - m)$ of a year from the next interest date.

Method (a): Let A denote the value of the security, at rate of interest i per annum, just *after* the last interest payment was made, or at the issue date if there has been no such payment; A may be found by Makeham's formula or by other methods. The value a fraction m of a year later is then

$$A' = (1 + i)^m A \qquad (7.8.2)$$

Method (b): Instead of working from the preceding interest date (or issue date), let us work from the next interest date. Let A^* be the value of the stock, at rate of interest i per annum, just *before* the next interest payment is made; A^* may be found by Eq. 7.8.1. It follows that the present value of this stock is

$$A' = (1+i)^{-m'}A^* \qquad (7.8.3)$$

Our first example illustrates the use of both these methods. In the subsequent examples, we consider how to deal with the valuation of securities when the redemption price, coupon rate, or rate of income tax is not constant. In some cases it is simplest to proceed from first principles, while in others Makeham's formula may be used to value the loan in sections, or in other ways. Our final example, the solution of which may at first seem difficult, illustrates how a clear grasp of the principles underlying Makeham's formula enables an indirect valuation of the capital to shorten the solution considerably.

EXAMPLE 7.8.1

The middle market price (i.e., the average of the buying and selling prices) quoted on the London Stock Exchange on 15 December 2011 for a 6% stock redeemable at par at the end of September 2016 was £66. If this stock paid interest half-yearly on 1 April and 1 October, find the gross yield per annum obtainable on a purchase of this stock at this price.

Solution

Method (a): The value per £100 nominal of this stock on 1 October 2011 *after* the payment of the interest then due is, by Makeham's formula,

$$A = K + \frac{0.06}{i^{(2)}}(100 - K)$$

where $K = 100v^5$ at rate i (the gross annual yield). Hence, the value on 15 December 2011 is

$$A' = (1+i)^{75/365}A$$

and we now solve the equation $A' = 66$ to give $i \approx 0.1776$ so that the gross annual yield is 17.76%.

Method (b): The value per £100 nominal just after receipt of the interest payment on 1 April 2012 is, by Makeham's formula,

$$A = K + \frac{0.06}{i^{(2)}}(100 - K)$$

where $K = 100v^{4.5}$ at rate i, the gross annual yield. The value just *before* receipt of this interest payment is $(A + 3)$ and the value on 15 December 2011 is

$$A' = (1+i)^{-107/365}(A+3)$$

We then find the gross yield per annum by solving $A' = 66$. This gives $i \approx 0.1776$ as before.

In more complicated problems involving varying redemption prices, it is often helpful to use one of the following approaches:

Method (a): Consider a fixed redemption price, value using Makeham's formula, and then adjust the answer to take "excess" or "shortfall" in the redemption money into account.

Method (b): Split the loan into sections, for each of which the redemption price is constant, and value each section by Makeham's formula.

Method (b) is generally more suitable when the redemption price R has only two or three values, whereas method (a) may be more suitable when R takes many values.

Similar methods apply when the coupon rate D or the rate of income tax t_1 varies. One may use first principles or either of the following methods.

Method (a): Consider the net rate of interest to be constant, value by Makeham's formula, and adjust the answer to allow for the actual net interest rates.

Method (b): Split the loan into sections, assume that for each the net rate of interest is constant, and value each section by Makeham's formula. Note, however, that adjustments will generally have to be made, since the net interest rate for some sections may not be constant (see Example 7.8.5.).

In certain, rather complicated, examples, the redemption price and the net rate of interest may vary. Such problems should be tackled by a combination of the methods described previously.

EXAMPLE 7.8.2

An issue of 1,000 debentures each of £100 nominal bears interest at $1\frac{1}{2}$% per annum, payable yearly, and is redeemable as follows:

250 bonds after 30 years at £110%
250 bonds after 40 years at £120%
250 bonds after 50 years at £130%
250 bonds after 60 years at £140%

Find the price payable for the issue by an investor requiring a yield of $2\frac{1}{2}$% per annum. (Ignore taxation.)

Solution

Method (a): If the bonds were redeemed at 60%, a yield of 2.5% per annum would be obtained, and their total value would be £60,000 (since they each carry an interest payment of £1.50 per annum on an outlay of £60). The value of the "extra" redemption money is

$$250 \times (50v^{30} + 60v^{40} + 70v^{50} + 80v^{60}) \qquad \text{at } 2\frac{1}{2}\%$$
$$= £21,183$$

The total price is then $60,000 + 21,183 = £81,183$.

Method (b): Makeham's formula can be used for each of the four distinct components:

Component 1: $R = 1.1$, $g = 0.015/1.1$, $n = 30$, leading to $A_1 = 83.84$
Component 2: $R = 1.2$, $g = 0.015/1.2$, $n = 40$, leading to $A_2 = 82.35$
Component 3: $R = 1.3$, $g = 0.015/1.3$, $n = 50$, leading to $A_3 = 80.37$
Component 4: $R = 1.4$, $g = 0.015/1.4$, $n = 60$, leading to $A_4 = 78.18$

The total price is then $250 \times (A_1 + A_2 + A_3 + A_4) = £81,183$.

EXAMPLE 7.8.3

A loan stock of £1,000,000 nominal is redeemable at 110% by ten equal annual installments of capital, the first due in 1 year's time. Interest is payable half-yearly in arrears at the rate of $(4.5 + t)$% per annum in year t ($t = 1, 2, \ldots, 10$). What price should be paid for the stock to yield 8% per annum convertible half-yearly? (Ignore taxation.)

Solution

Work in half-yearly periods so that the effective rate is $i^{(2)}/2 = 4$%. The value of the capital repayments is

$$K = 110,000(v^2 + v^4 + \cdots + v^{20}) \qquad \text{at } 4\%$$

$$= \frac{110,000 a_{\overline{20}|}}{s_{\overline{2}|}} \qquad \text{at } 4\% \qquad = 732,810$$

Assume firstly that interest is constant at 5.5% per annum, or 2.75% per period. By Makeham's formula, the value of the loan is

$$K + \frac{0.0275}{1.1 \times 0.04}(1,100,000 - K) = 962,300$$

To this must be added the value of the extra interest payments, which is

$$\frac{1}{2}\Big[9,000\big(v^3 + v^4\big) + 2 \times 8,000\big(v^5 + v^6\big)$$

$$+ 3 \times 7,000\big(v^7 + v^8\big) + \cdots + 9 \times 1,000\big(v^{19} + v^{20}\big)\Big]$$

$$= \frac{1,000}{2}\big(v^3 + v^4\big)\Big[9\big(1 + v^{16}\big) + 16\big(v^2 + v^{14}\big)$$

$$+ 21\big(v^4 + v^{12}\big) + 24\big(v^6 + v^{10}\big) + 25v^8\Big] = 106,680$$

Hence, the total value of the loan is £1,068,980.

EXAMPLE 7.8.4

A loan of nominal amount £10,000 is issued in bonds of £100 nominal bearing interest at 9% per annum payable annually in arrears. Five bonds will be redeemed at the end of each of the first 10 years and 10 bonds at the end of each of the next 5 years. During the first 10 years redemption will be at £112.50 per bond; thereafter, it will be at £120 per bond.

An investor has been offered the entire issue of bonds for £10,190. At present, the investor pays tax on income at the rate of 40%, but a date has been fixed after which the rate will be 30%. Given that these terms offer the investor a yield of 7% per annum net effective, find how many interest payments will be subject to the higher rate of tax.

Solution

Assume first that the redemption price is constant at 112.5%. The value of the capital repayments is

$$K = 562.5 a_{\overline{10}|} + 1,125 v^{10} a_{\overline{5}|} \qquad \text{at } 7\% \qquad = 6,295.63$$

and the value of the issue, assuming income tax to be at 30%, is

$$K + \frac{0.08 \times 0.7}{0.07}(11,250 - K) = 10,259$$

The addition for the higher redemption price in the last 5 years is

$$1,000 \times 0.075(a_{\overline{15}|} - a_{\overline{10}|}) \qquad \text{at } 7\% \qquad = 156$$

Hence, the total value of the loan, assuming that income tax is constant at 30%, is £10,415. From this must be deducted the value of the additional income tax on the first n interest payments, where n is such that the value of the loan is reduced to £10,190. We draw up the following schedule:

(1)	(2)	(3)	(4)	(5)
	Nominal Amount Outstanding at Beginning	Gross Interest at Time	Value of 10% of Gross Interest at	Cumulative Total
Year (t)	of Year t	t 0.09 × (2)	Time t $0.1 \times (3) \times v_{0.07}^t$	of (4)
1	10,000	900	84.1	84.1
2	9,500	855	74.7	158.8
3	9,000	810	66.1	224.9
4	8,500	765	58.4	

Since the actual price is £225 less than £10,415, we have $n = 3$; i.e., the 40% rate of income tax applies to the interest payments made at the end of the first 3 years, but not to those made later.

EXAMPLE 7.8.5

A loan of £3,000,000 nominal bears half-yearly interest at $3\frac{1}{2}\%$ per annum for the first 10 years and at 4% per annum thereafter. It is redeemable at par by equal annual installments of capital over the next 30 years. Find the net yield per annum to a financial institution, subject to income tax at the rate of 37.5%, which bought the loan at par.

Solution

We split the loan into two sections: the first section, of nominal amount £1,000,000, consists of that part of the loan which is redeemable in the first 10 years, and the second section, of nominal amount £2,000,000, is the remainder of the loan. Makeham's formula may be applied to value each section, but an adjustment must be made to allow for the fact that interest on the second section is at $3\frac{1}{2}\%$, not 4%, during the first 10 years. Let the unknown yield be i; we have (in an obvious notation)

$$K_1 = 100,000a_{\overline{10}|} \qquad \text{at rate } i$$
$$K_2 = 100,000(a_{\overline{30}|} - a_{\overline{10}|}) \qquad \text{at rate } i$$

The value of the first section is

$$A_1 = K_1 + \frac{0.035 \times 0.625}{i^{(2)}}\left(1,000,000 - K_1\right)$$

We now value the second section. If interest on this section were always at 4%, its value would be

$$A_2 = K_2 + \frac{0.04 \times 0.625}{i^{(2)}}\left(2,000,000 - K_2\right)$$

During the first 10 years, interest is actually at $3\frac{1}{2}\%$ per annum, so we must deduct the value of the "excess" net interest, i.e.,

$$2,000,000 \times 0.005 \times 0.625 \times a_{\overline{10}|}^{(2)} \qquad \text{at rate } i$$

We therefore solve the equation

$$3,000,000 = A_1 + A_2 - 6,250a_{\overline{10}|}^{(2)}$$

A rough solution is 3.75% × 0.625 ≈ 2.3%, and interpolation can be used to determine that $i = 2.32\%$.

EXAMPLE 7.8.6

A loan of £1,000,000 is to be repaid over 40 years by a level annuity payable monthly in arrears. The amount of the annuity is determined on the basis of an interest rate of 8% per annum effective. An investor, liable to income tax at the rate of 40% on the interest content (according to the original schedule) of each annuity payment, wishes to purchase the entire loan on the issue date at a price to obtain a net yield of 6% per annum convertible half-yearly. What price should he pay?

Solution

Note first that the amount of the repayment annuity per annum is £X, where

$$Xa_{\overline{40}|}^{(12)} = 1,000,000 \qquad \text{at 8\%}$$

This gives $X = 80,933.82$.

The required net yield is 6% per annum convertible half-yearly. Choose a half-year as the unit of time and note that the required net yield is 3% per unit time. Per half-year, the annuity is of amount $X/2 = 40,466.91$. Since the annuity is payable six times per half-year, the value of the entire loan (ignoring tax) is

$$A = 40,466.91a_{\overline{80}|}^{(6)} \qquad \text{at 3\%} \qquad = 1,237,320$$
$$(1)$$

(Note that $0.03^{(6)} = 0.029\,632$.)

Let K be the value (at 3% per half-year) of the capital payments. Interest is paid each month of amount $0.08^{(12)}/12$ times the outstanding loan. If, in fact, interest were payable each month of amount $0.03^{(6)}/6$ times the outstanding loan, the value of the loan would be £1,000,000. Hence, using Makeham's formula "backwards", we have

Continued

EXAMPLE 7.8.6   *(cont'd)*

$$1{,}237{,}320 = K + \frac{0.08^{(12)}/12}{0.03^{(6)}/6}(1{,}000{,}000 - K)$$

from which we immediately obtain

$$K = 216{,}214 \qquad (2)$$

Clearly, $(A - K)$ is the value of the gross interest payments. Allowing for tax, the investor should pay

$$K + (1 - 0.4)(A - K) = A - 0.4(A - K) = 0.6A + 0.4K$$

From Eqs 1 and 2, this is easily calculated to be £828,878 (or, say, £82.89%).

Note that the value of K may, of course, be obtained directly as

$$K = \left[\frac{80{,}933.82}{12} - 1{,}000{,}000 \times \frac{0.08^{(12)}}{12}\right]$$
$$\times \sum_{t=1}^{480} (1.08)^{(t-1)/12}(1.03)^{-t/6}$$

However, the evaluation of this expression is somewhat longer, and more liable to error, than the indirect method described previously.

7.9 REAL RETURNS AND INDEX-LINKED STOCKS

Inflation may be defined as a fall in the purchasing power of money. It is usually measured with reference to an index representing the cost of certain goods and (perhaps) services. In the UK the index used most frequently is the Retail Prices Index (RPI), which is calculated monthly by the Office for National Statistics. It is the successor to various other official indices dating back many years.

Real investment returns, as opposed to the *money* returns we have so far considered, take into account changes in the value of money, as measured by the RPI or another such index. It is possible for all calculations relating to discounted cash flow, yields on investments, etc., to be carried out using units of real purchasing power rather than units of ordinary currency, provided, of course, that one defines the index being used.

It is a simple matter to write down the equation of value in "real" terms for any transaction. Suppose that a transaction involves cash flows $c_1, c_2, \ldots, c_n$, the rth cash flow occurring at time t_r. (Note that the cash flows are *monetary* amounts.) If the appropriate index has value $Q(t)$ at time t, the cash flow c_r at time t_r will purchase $c_r/Q(t_r)$ "units" of the index. By the real yield on the transaction, we mean the yield calculated on the basis that the investor's receipts and outlays are measured in units of index (rather than monetary units). The real internal rate of return (or yield) on the transaction, measured in relation to the index Q, is then that value of i for which

$$\sum_{r=1}^{n} \frac{c_r}{Q(t_r)}(1 + i)^{-t_r} = 0 \qquad (7.9.1)$$

This equation is, of course, equivalent to

$$\sum_{r=1}^{n} c_r \frac{Q(t_k)}{Q(t_r)}(1+i)^{-t_r} = 0 \text{ for any } t_k. \qquad (7.9.2)$$

EXAMPLE 7.9.1

On 16 January 2009 a bank lent £25,000 to a businessman. The loan was repayable 3 years later, and interest was payable annually in arrears at 5% per annum. Ignoring taxation and assuming that the RPI for any month relates to the middle of that month, find the real annual rate of return, or yield, on this transaction. Values of the RPI for the relevant months are as follows:

Calendar Year	2009	2010	2011	2012
Value of RPI for January	210.1	217.9	229.0	238.0

Solution
In money terms, the yield is, of course, 5% per annum. To obtain the annual yield in real terms, we work in units of purchasing power, so, for this example, the equation of value 7.9.1 becomes

$$\frac{25,000}{Q(0)} = 1,250\left[\frac{v}{Q(1)}+\frac{v^2}{Q(2)}+\frac{v^3}{Q(3)}\right] + 25,000\frac{v^3}{Q(3)}$$

where $v = 1/(1+i)$ and $Q(t)$ is the RPI at time t years, measured from mid-January 2009. We then have

$$1 = 0.05\left(\frac{210.1}{217.9}\right)v + 0.05\left(\frac{210.1}{229.0}\right)v^2 + 1.05\left(\frac{210.1}{238.0}\right)v^3$$

i.e.,

$$1 = 0.048\ 210v + 0.045\ 874v^2 + 0.926\ 912v^3$$

When $i = 0$, the right side of the last equation is greater than 1, so the real rate of return is positive. When $i = 0.01$, the right side is 0.992 354. By linear interpolation, $i \approx 0.0073$, so the real rate of return is approximately equal to 0.73% per annum.

EXAMPLE 7.9.2

An investment trust bought 1,000 shares at £135 each on 1 July 2005. The trust received dividends on its holding on 30 June each year that it held the shares. The rate of dividend per share was as given in the following table:

30 June in Year	2005	2006	2007	2008	2009	2010
RPI	192.2	198.5	207.8	216.8	213.4	224.1
Rate of Dividend Per Share (£)	—	7.9	8.4	8.8	9.4	10.1

On 1 July 2010, the investment trust sold its entire holding of the shares at a price of £151 per share. Using the RPI values shown in the table, calculate the real rate of return per annum effective achieved by the trust on its investment.

Solution
In real terms, the equation of value for this investment is

$$135,000 = 7,900\frac{192.2}{198.5}v + 8,400\frac{192.2}{207.8}v^2 + 8,800\frac{192.2}{216.8}v^3$$
$$+ 9,400\frac{192.2}{213.4}v^4 + \left(10,100+151,000\right)\frac{192.2}{224.1}v^5$$

measured from 30 June 2005. By linear interpolation, $i = 0.0522\%$, and so the real yield is approximately equal to 5.2% per annum.

In money terms, the equation of value is

$$35,000 = 7,900v + 8,400v^2 + 8,800v^3 + 9,400v^4$$
$$+ \left(10,100+151,000\right)v^5$$

This can be solved to give a rate of yield approximately equal to 8.5% per annum.

Equation 7.9.2 may be considered as the equation of value for the transaction, measured in units of purchasing power at a particular time t_k. Similar equations may, of course, be developed for continuous cash flows.

In the remainder of this chapter, we shall, unless it is otherwise stated, calculate real yields with reference to the RPI.

Note that the RPI for any month relates to a specific applicable date in that month (usually the second or third Tuesday), and is normally published on the Friday of the second or third week of the following month. For example, the RPI for October 2012 was 245.6, and this value was published on 13 November 2012. It is important to note that on 13 January 1987, the RPI reverted to the base of 100. Care should be taken to ensure that any data used for months prior to this has been converted to the new base.

For many purposes, it is necessary to postulate a value of the RPI for a date between the applicable dates in consecutive months. Such a value may be estimated by interpolation. It is generally better to interpolate between values of $\ln Q(t)$ rather than between values of $Q(t)$, since $\ln Q(t)$ is usually more nearly linear.

In conditions of relatively stable money, such as existed in Britain for about 100 years before 1914, real and money rates of return are roughly equal. In modern times, rates of inflation have been relatively high, and the real return on UK government stocks, even to a tax-free investor, has fluctuated between positive and negative values. The real rate of return on fixed-interest investments, after allowing for taxation, has often been negative.

Theoretically, at least, the idea of index-linking the outlays and receipts on any particular transaction is quite straightforward. In its simplest form, an index-linked investment has a series of cash flows of "nominal" amount $c_1, c_2, \ldots,$ c_n, the rth cash flow being due at time t_r. The actual *monetary* amount of the rth cash flow is defined to be

$$c_r \frac{R(t_r)}{R(t_0)} \qquad (7.9.3)$$

where R is some specified index and t_0 is a prescribed base date. (It is, of course, essential that, for $r = 0, 1, \ldots, n$, the value of $R(t_r)$ is known at time t_r.) The real rate of return per annum on such an investment, measured in relation to an index Q, is (see Eq. 7.9.1) that value of i for which

$$\sum_{r=1}^{n} c_r \frac{R(t_r)}{R(t_0)} \frac{1}{Q(t_r)} (1+i)^{-t_r} = 0 \qquad (7.9.4)$$

Note that if Q, the index used to measure the real rate of return, were to be identically equal to R, the index used to determine the amount of each cash flow, then Eq. 7.9.4 would simplify to

$$\sum_{r=1}^{n} c_r (1+i)^{-t_r} = 0$$

so the real rate of return would be the same as the money rate of return on the corresponding "unindexed" transaction (with cash flows of monetary amounts $c_1, c_2, \ldots, c_n$). In this special case, therefore, the real rate of return does *not* depend on the movement of the index. In general, however, the indices Q and R are distinct, and the real rate of return on any index-linked transaction does depend on the movement of the relevant indices.

In reality this simplest form of index-linking, in which the same index is used both to determine the cash flows and to measure the real rate of return, is not feasible, since there does not exist a suitable index with values which form a continuous function and are *immediately* known at *all* times. There are clearly essential requirements, at least for the index relative to which the real rate of return is calculated. We have already remarked that the RPI, for example, is known accurately on only 1 day in each month, its value on that day being known (somewhat misleadingly) as the value "for the month". Moreover, the value for any 1 month is not known until around the middle of the following month. Therefore, although real returns are, in the UK, nearly always measured in relation to the RPI (interpolation being used as necessary to estimate the index values $Q(t_r)$ in Eqs 7.9.1 and 7.9.4), the index R used to determine cash flows for an index-linked transaction is generally a step function, constant over any calendar month and based on the RPI with a time lag. The use of a time lag overcomes difficulties which would otherwise arise since the index value for a given month is not published during that month.

An important example of an index-linked investment in the UK is an index-linked government stock. The UK government issued its first index-linked stock in March 1981, and this particular stock (2% I-L Treasury Stock 1996) carried a coupon of 2% per annum payable half-yearly on 16 March and 16 September. The final interest payment on the particular stock was 16 September 1996, on which date the stock was redeemed at par. Both interest and redemption price were indexed by the RPI with a time lag of 8 months.

A more recent example is the $1\frac{1}{4}$% I-L Treasury Gilt 2055, which carries coupon payments of $1\frac{1}{4}$% payable half-yearly on 22 May and 22 November each year. The gilt was issued on 23 September 2005 and will be redeemed on 22 November 2055. Both the coupons and redemption are indexed by RPI with a lag of 3 months. The $1\frac{1}{4}$% I-L Treasury Gilt 2055 was the first index-linked stock to be issued with a lag of 3 months and this is now standard practice (replacing the 8-month lag).

All index-linked stocks have essentially the same method of indexation. Each stock is redeemable on a single date at an indexed par. The indexation is simply described, if we recall our previous definition of the value of the RPI applicable at any time and let

$R(t)$ = value of RPI applicable three calendar months before time t

$$(7.9.5)$$

(Although note that for securities issued prior to 2005, the required lag is 8 months.) For an index-linked stock issued at time t_0 with annual coupon rate D (e.g., 0.02 or 0.025) payable half-yearly, the monetary amount of an interest payment due at any particular time t_r is

$$\frac{D}{2} 100 \frac{R(t_r)}{R(t_0)}$$

per £100 nominal stock. Similarly, the redemption money per £100 nominal is

$$100 \frac{R(t_n)}{R(t_0)}$$

where t_n denotes the time of redemption.

The use of a 3-month time lag for indexation means that an investor who purchases stock at any time within the 3 months prior to the interest payment will, at the time of purchase, always know the monetary amount of the first interest payment he will receive. For example, an investor who purchased $1\frac{1}{4}$% I-L Treasury Gilt 2017 (issued 8 February 2006) ex dividend on 1 September 2011 would have received his first interest on 22 November 2011, on which date the amount of interest paid per £100 nominal was

$$\frac{0.0125}{2} 100 \frac{\text{RPI value for August 2011}}{\text{RPI value for November 2005}} = \frac{0.0125}{2} 100 \frac{236.1}{193.6}$$

On 1 September 2011 the value of this expression was known to be £0.76 (to two decimal places, rounded down).

EXAMPLE 7.9.3

On 21 March 2011, an investor, who is not liable to taxation, purchased a quantity of $1\frac{7}{8}$% I-L Treasury Gilt 2021 at a price of £106.375 nominal. At the time of purchase, using the latest available RPI value, he estimated his real yield on the transaction on the assumption that he would hold the stock until redemption and that the RPI would grow continuously at a constant rate of 5% per annum. Find the estimate made by the investor of his real yield.

Note that the stock was issued in July 2007, and interest payments are made on each 22 May and 22 November. The values of the RPI for April 2007 and February 2011 are 205.4 and 231.3, respectively.

Solution

This example is typical of the kind of calculation that must be carried out in practice. Although with hindsight we see that neither of the underlying assumptions has been realized, we are required to obtain the investor's original estimates. Accordingly, our solution assumes that the calculation is being made at the time of purchase.

The first interest payment will be received on 22 May 2011. Interest is payable half-yearly. The final interest payment will be made on 22 November 2021, which is the redemption date of the stock. It is convenient to label successive calendar months by an integer variable as follows: February 2011 as month 0, March 2011 as month 1, April 2011 as month 2, etc.

Continued

EXAMPLE 7.9.3   *(cont'd)*

We measure time in *half-years* from the date of purchase and let i be the real yield per half-year. The period from 21 March 2011 (the purchase date) to 22 May 2011 consists of 62 days, so the first interest payment will be received at time f where $f = 62/182.5$. Over the entire transaction, the investor will receive 22 interest payments, the kth payment being at time $k - 1 + f$, measured in units of half-years.

Let r be the assumed compound rate of growth of the RPI *per half-year*. (We shall later assume that $1 + r = 1.05^{1/2}$.) Then

$$Q(t) = Q(0)(1+r)^t$$

where $Q(t)$ is the RPI value at exact time t. The RPI value for month 0 is 231.1. We assume that, if $l \geq 0$, the RPI value for month l will be $231.1 \times (1 + r)^{l/6}$. As we do not know the precise day for the official calculation of the index each month, this is a reasonable hypothesis.

We work per £100 nominal of stock. Recalling the 3-month time lag for indexation, we note that the first interest payment will be of amount

$$\frac{0.01875}{2} \times 100 \times \frac{\text{RPI February 2011}}{\text{RPI April 2007}} = 0.9375 \times \frac{231.3}{205.4}$$
$$= 1.0557$$

The amount of this payment is known on the purchase date. The payment will be made at time f when the value of the RPI will be $Q(f) = Q(0) (1 + r)^f$.

For $j \geq 2$, the investor's jth interest payment will be received in month $(6j - 3)$, and will be of amount (i.e., in each May and November)

$$\frac{0.01875}{2} \times 100 \times \frac{\text{RPI month } (6j - 6)}{\text{RPI April 2007}}$$

Using the assumed behavior of the RPI index, we estimate this to be

$$\frac{0.01875}{2} \times 100 \times \frac{231.3(1+r)^{(6j-6)/6}}{205.4}$$
$$= \frac{0.01875}{2} \times 100 \times \frac{231.3}{205.4}(1+r)^{j-1} = 1.0557(1+r)^{j-1}$$

This payment will be received at time $j - 1 + f$ (measured in half-years) when the value of the RPI will be $Q(0)(1+r)^{j-1+f}$.

The redemption proceeds will be paid at time $(21 + f)$ with the final (22th) interest payment. The estimated redemption proceeds will then be

$$100 \times \frac{231.3}{205.4}(1+r)^{22-1} = 112.6095(1+r)^{22-1}$$

Combining the preceding results, we use Eq. 7.9.1 to determine the real yield per half-year. It is that value of i for which

$$0 = \frac{-106.375}{Q(0)} + \frac{1.0557}{Q(0)(1+r)^f}(1+i)^{-f}$$
$$+ 1.0557 \sum_{j=2}^{22} (1+r)^{j-1}\frac{1}{Q(0)(1+r)^{j-1+f}}(1+i)^{-(j-1+f)}$$
$$+ 112.6095(1+r)^{22-1}\frac{1}{Q(0)(1+r)^{21+f}}(1+i)^{-(21+f)}$$

Since the factor $Q(0)$ cancels throughout, this equation can be expressed as

$$106.375 = ((1+i)(1+r))^{-f}$$
$$\left\{ 1.0557 \sum_{j=1}^{22} (1+i)^{-(j-1)} + 112.6095(1+i)^{-21} \right\}$$

or

$$106.375 = ((1+i)(1+r))^{-f}$$
$$\left\{ 1.0557 \ddot{a}_{\overline{22}|\,i} + 112.6095(1+i)^{-21} \right\}$$

If $1 + r = 1.05^{1/2}$ and $f = 62/182.5$, we can solve the expression to give $i = 1.23\%$ per half-year. The investor therefore estimated an annual real yield of 2.47%.

The calculation of observed real returns for index-linked stocks is carried out quite simply using Eq. 7.9.1 (or 7.9.2). In order to estimate prospective future real yields, we must make explicit assumptions about the future movement of the RPI. In any given situation, it may be desirable to estimate future real returns on the basis of alternative assumptions relating to future values of the RPI. This approach is illustrated in Example 7.9.3.

SUMMARY

- Many practical examples of *fixed-interest securities* and related assets are amenable to analysis using compound-interest theory. Examples include, but are not limited to, *gilts, debentures, ordinary shares,* and *property.*
- The *price* of a fixed-interest security can be obtained from the present value of its proceeds at a particular interest rate.
- The *yield* on fixed-interest security of particular price can be obtained from solving the equation of value for i.
- *Makeham's formula* is a particular representation of the equation of value for a fixed-interest security. Its use often simplifies price and yield calculations in practical situations. Furthermore, Makeham's formula can be used to determine optimal redemption times for securities with optional redemption dates.
- The *real yield* on a fixed-interest security can be obtained by incorporating an appropriate inflation index into the equation of value. *Index-linked securities* can also be analysed in this way.

EXERCISES

7.1 For each of the following issues, indicate whether the price of the issue should be at the nominal value, above the nominal value, or below the nominal value. You should assume that each issue is redeemed at par.

	Issue	Annual Coupon Rate (%)	Annual Yield Required by Market (%)
(a)	A	5.25	7.25
(b)	B	6.625	7.15
(c)	C	0	6.2
(d)	D	5.875	5
(e)	E	4.5	4.5

7.2 What is the value of a 5-year 7.4% coupon bond selling to yield 5.6% per annum (nominal converted semi-annually) assuming the coupon payments are made semi-annually and the bond is redeemed at par?

CFA Institute

7.3   What is the value of a zero-coupon bond that matures in 20 years, has a maturity of $1 million, and is selling to yield 7.6% per annum (nominal converted semi-annually)?

7.4   An investor, who is liable to income tax at 20%, but is not liable to capital gains tax, wishes to earn a net effective rate of return of 5% per annum. A bond bearing coupons payable half-yearly in arrears at a rate 6.25% per annum is available. The bond will be redeemed at par on a coupon date between 10 and 15 years after the date of issue, inclusive. The date of redemption is at the option of the borrower. Calculate the maximum price that the investor is willing to pay for the bond.

7.5 **(a)**   Describe the characteristics of an index-linked government bond.

    (b)   On 1 July 2002, the government of a country issued an index-linked bond of term 7 years. Coupons are paid half-yearly in arrears on 1 January and 1 July each year. The annual nominal coupon is 2%. Interest and capital payments are indexed by reference to the value of an inflation index with a time lag of 8 months.
You are given the following values of the inflation index.

Date	Inflation Index
November 2001	110.0
May 2002	112.3
November 2002	113.2
May 2003	113.8

The inflation index is assumed to increase continuously at the rate of 2½% per annum effective from its value in May 2003.

An investor, paying tax at the rate of 20% on coupons only, purchased the stock on 1 July 2003, just after a coupon payment had been made. Calculate the price to this investor such that a real net yield of 3% per annum convertible half-yearly is obtained and assuming that the investor holds the bond to maturity.

7.6   An investor bought a number of shares at 78 pence each on 31 December 2005. She received dividends on her holding on 31 December 2006, 2007, and 2008. The rate of dividend per share is given in the following table:

Date	Rate of Dividend Per Share	Retail Price Index
31.12.2005	—	147.7
31.12.2006	4.1 pence	153.4
31.12.2007	4.6 pence	158.6
31.12.2008	5.1 pence	165.1

On 31 December 2008, she sold her shares at a price of 93 pence per share. Calculate, using the retail price index values shown in the table, the effective annual real rate of return achieved by the investor.

7.7 A bond pays coupons in perpetuity on 1 June and 1 December each year. The annual coupon rate is 3.5% per annum. An investor purchases a quantity of this bond on 20 August 2009. Calculate the price per £100 nominal to provide the investor with an effective rate of return per annum of 10%.

Institute and Faculty of Actuaries

7.8 For a certain perpetuity, interest is payable on 1 June and 1 December each year. The amount of each interest payment is £1.75 per £100 nominal. Find the effective annual yield obtained by an investor, not liable to taxation, who purchased a quantity of this stock on 14 August 2012, when the market price was 35.125%.

7.9 A fixed-interest loan bears interest of 10% per annum payable half-yearly in arrears. The loan will be redeemed at 105% by five installments, of equal nominal amount, on 31 July in each of the years 2036 to 2040 inclusive. An investor, who is liable to income tax at the rate of 40%, purchased the entire loan on 19 June 2006, at a price to obtain a net yield of 9% per annum effective. Find the price per cent paid, assuming that the purchase was made
(a) "Ex dividend",
(b) "Cum dividend".
in relation to the interest payment due on 31 July 2006.

7.10 A loan of nominal amount £300,000 in bonds of nominal amount £100 is to be repaid by 30 annual drawings, each of 100 bonds, the first drawing being 1 year after the issue date. Interest will be payable quarterly in arrears at the rate of 8% per annum. Redemption will be at par for the first 15 drawings and at 120% thereafter. An investor, who will be liable to income tax at the rate of 40%, purchases the entire loan on the issue date at a price to obtain a yield per annum of 7% net effective. What price does the investor pay for each bond?

7.11 On 7 April 2006, the government of a certain country issued two index-linked stocks, with terms of 20 years and 30 years, respectively. For each stock, interest is payable half-yearly in arrears, and the annual coupon rate is 3%. Both interest and capital payments are indexed by reference to the country's "cost of living index" with a time-lag of 8 months. The index value for August 2005 was 187.52, and at the issue date of the stocks, the latest known value of the index was 192.10 (i.e. at February 2006). The issue price of each stock was such that, if the cost of living index was to increase continuously at the rate of 6% per annum effective, a pur-chaser of either stock would obtain a real yield on his investment of 3% per annum convertible half-yearly. (This real yield is measured in relation to the cost of living index.)

(a) Show that the issue prices of the stocks were equal and find the common issue price.

(b) Show that, if the cost of living index was to increase continuously at the rate of 4% per annum effective, the 20-year stock would provide a greater real yield than the 30-year stock but that, if the cost of living index was to increase continuously at the rate of 8% per annum effective, the opposite would be true.

7.12 On 31 December 2001, a loan was issued to be repaid over 12 years by a level annuity payable quarterly in arrears on the last days of March, June, September, and December. The amount of the annuity was calculated on the basis of an interest rate of 12% per annum convertible quarterly. The total interest paid in 2005, according to the original schedule, was £6,374.41. On 31 December 2005, an investor, liable to income tax at the rate of 40% on the interest content (according to the original schedule) of each annuity installment, purchased the annuity installments due after that date.

Find the purchase price, if the net yield obtained by the investor was

(a) 8% per annum convertible quarterly;

(b) 8% per annum effective.

7.13 A loan of nominal amount £30,000 is to be redeemed at par in three installments, each of nominal amount £10,000, at the end of 8, 16, and 24 years. Interest will be payable annually in arrears at the rate of 6% per annum for the first 8 years, 4% per annum for the next 8 years, and 2% per annum for the final 8 years.

An investor, who will be liable to income tax on the interest payments at the rate of 30% for 12 years and at the rate of 50% thereafter, pays £26,000 to purchase the entire loan on the issue date. Calculate the net effective annual yield obtained by the investor.

7.14 A loan of £100,000 is to be repaid over 15 years by a level annuity payable monthly in arrears. The amount of the annuity is calculated on the basis of an interest rate of 16% per annum convertible quarterly. An investor, liable to income tax at the rate of 80% on the interest content (according to the original schedule) of each annuity installment, wishes to purchase the entire loan on the issue date and calculates that, in order to obtain a yield on his investment of 5% per annum net effective, he should pay £86,467 for the loan. In the event, however, he has to pay £88,000 for this investment.

Find (a) the amount of each monthly annuity repayment and (b) the net effective annual yield that the investor will obtain.

7.15 A loan of nominal amount £8,000 is to be issued bearing interest of 10% per annum payable quarterly in arrears. At the end of the second, fourth, sixth, and eighth years, a nominal amount of £2,000 of the loan is to be redeemed at a premium, which is to be proportional to the time elapsed from the issue date.

An investor, who will be liable to income tax on the interest payments at the rate of 40% for 5 years and at the rate of 50% thereafter, calculates that, in order to obtain a yield of 7% per annum net effective on his investment, he should offer a price of £7,880.55 for the entire loan. Find the redemption prices of the loan.

7.16 A loan of £100,000 will be repayable by a level annuity, payable annually in arrears for 15 years. The amount of the annuity is calculated on the basis of an interest rate of 8% per annum effective. An investor, who is liable to income tax at the rate of 40% on the interest content (according to the original schedule) of each annuity payment, wishes to purchase the entire loan on the issue date.

Find the price he should pay in order to achieve a net effective annual yield of (a) 7% and (b) 8%.

CHAPTER 8

Capital Gains Tax

Capital gains tax, which was introduced in the UK by the Finance Act 1965, is
a tax levied on the difference between the sale or redemption price of a stock (or
other asset) and the purchase price, if lower. In contrast to income tax, this tax is
normally payable once only in respect of each disposal, at the date of sale or
redemption. Certain assets, including some fixed-interest securities, may be
exempt from capital gains tax; in addition, exemption from this tax may be
granted after an asset has been held for a certain period. In some cases capital
losses may be offset against capital gains on other assets (see Section 8.5).
Unless there is some form of "indexation", capital gains tax may be criticized as
unfair in that in times of inflation it does not take into account the falling value
of money: it taxes "paper" gains, as well as real ones. A system of indexation of
capital gains was first introduced in the UK in 1981 and subsequently modified.
After assets have been held for a year, the calculation of any capital gains tax
liability takes into account movements in the Retail Prices Index.

In this chapter we will be concerned with the effect of capital gains tax on the
prices and yields of fixed-interest securities. In general, we shall assume that the
stock in question will be held to redemption. If a stock is sold before the final
maturity date, the capital gains tax liability will, in general, be different, since it
will be calculated with reference to the sale proceeds rather than the corresponding redemption money. (See Examples 8.3.2 and 8.3.3 for illustrations of
these calculations.) For simplicity, we shall ignore the possibility of indexation
in our calculations. In theory, it is relatively simple to allow for indexation; in
practice, however, changes in legislation may make some capital gains tax
calculations extremely complicated.

8.1 VALUING A LOAN WITH ALLOWANCE FOR CAPITAL GAINS TAX

We begin with a simple example demonstrating how capital gains tax can be
included in a yield calculation.

CONTENTS

161

An Introduction to the Mathematics of Finance. http://dx.doi.org/10.1016/B978-0-08-098240-3.00008-4

EXAMPLE 8.1.1

A loan of nominal amount £100,000 will be repaid at 110% after 15 years. The loan bears interest of 9% per annum payable annually in arrears. On the issue date the loan is purchased for £80,000 by an investor who is liable to income tax at the rate of 40% and to capital gains tax at the rate of 30%. Find the investor's net effective annual yield for the transaction.

Solution

It is important to understand the operation of capital gains tax. Since the redemption price is 110%, the purchase price of £80,000 acquires redemption proceeds of £110,000. There is then a capital gain of £30,000 on redemption. The capital gains tax payable is therefore £9,000 (i.e., 30% of £30,000),

so that after tax the investor retains £101,000 of the redemption proceeds.

Interest payable each year is £9,000 before tax. The net interest received by the investor is then £5,400 per annum (after tax at 40%). The investor's net yield per annum is therefore that rate of interest for which

$$80,000 = 5,400a_{\overline{15}|} + 101,000v^{15}$$

The reader should verify that the solution for the rate of interest is 0.07737. The net annual yield is then 7.74% per annum.

Example 8.1.1 is straightforward, as it is evident that there is a capital gain at redemption. Typically, however, this may not be clear from the outset, and a more sophisticated approach is required, in general.

Consider a loan which bears interest at a constant rate and is redeemable by installments. Suppose that at a particular time the total nominal amount of loan outstanding is $N = (N_1 + N_2 + \cdots + N_r)$ and that a nominal amount N_j will be redeemed after a further n_j years ($1 \leq j \leq r$), where $n_1, n_2, \ldots, n_r$ are integer multiples of $1/p$. Suppose further that the redemption price per unit nominal is constant and equal to R.

What price should be paid by an investor who wishes to purchase the entire outstanding loan to obtain a net yield of i per annum, allowing for his tax liability? Assume that interest is payable p times per annum at the rate g per annum on the capital outstanding, measured in cash, not nominal, terms. Let $C = RN$. If the investor has no tax liability, the price to be paid is, of course,

$$A = K + \frac{g}{i^{(p)}}(C - K) \tag{8.1.1}$$

or

$$A = K + I$$

where K and I denote the present value of the gross capital and interest payments, respectively.

If the investor is liable to income tax at rate t_1, but is not liable to capital gains tax, the value of the tax payable must be deducted from the interest income and the price to be paid is then A', where

$$A' = K + \frac{g(1 - t_1)}{i^{(p)}}(C - K) \qquad (8.1.2)$$

This is Makeham's formula, as discussed in Chapter 7.

Finally, consider the price to be paid by an investor who is liable to capital gains tax at the rate t_2 in addition to income tax at the rate t_1. If

$$g(1 - t_1) \geq i^{(p)} \qquad (8.1.3)$$

then $A' \geq C$, so the price paid allowing for only income tax is not less than the total redemption monies receivable. If this price is paid, there will be a capital *loss* on redemption. In this case, therefore, the price to be paid is simply A' (as given by Eq. 8.1.2), and, in fact, no capital gains tax is payable. (We are assuming that it is not permissible to offset the capital loss against any other capital gain; see Example 8.5.1.)

If, however,

$$g(1 - t_1) < i^{(p)} \qquad (8.1.4)$$

then A', the price paid to allow for only income tax, is less than the total redemption monies; this will obviously be true also for the price allowing in addition for capital gains tax. In this case, let A'' be the price to be paid allowing for both income tax and capital gains tax. At time n_j, when a nominal amount N_j is repaid, the redemption money received is $N_j R$. If the price paid for the entire outstanding loan is A'', under current UK practice, the tranche redeemed at time n_j is considered (for tax purposes) to have cost $(N_j/N) A''$ (i.e., for capital gains tax calculations, the total purchase price paid is divided among the different tranches in proportion to the *nominal* amounts redeemed). The capital gain at time n_j is therefore deemed to be $N_j R - (N_j/N)A''$, and the total value of the capital gains tax payable is

$$\sum_{j=1}^{r} t_2 \left(N_j R - \frac{N_j}{N} A'' \right) v^{n_j} = t_2 \frac{NR - A''}{NR} \sum_{j=1}^{r} N_j R v^{n_j}$$

$$\qquad (8.1.5)$$

$$= t_2 \frac{C - A''}{C} K$$

$$= t_2 \frac{R - P''}{R} K \qquad (8.1.6)$$

where P'' is the price per unit nominal.

It is essential to understand the derivation of the last expression. It may be recalled simply by the following argument, which depends crucially on the fact that a constant proportion of each loan repayment is absorbed by capital gains tax. The total capital repayment (i.e., the redemption monies actually paid) is C, and this has present value K. The total capital gains tax payable is $t_2(C - A'')$. By proportion, therefore, the present value of the capital gains tax is $t_2(C - A'')(K/C)$, as given by Eq. 8.1.5.

The price A'' is the present value of the net proceeds, after allowance for all tax liability. This implies that

$$A'' = K + \frac{g(1 - t_1)}{i^{(p)}}(C - K) - t_2 \frac{C - A''}{C}K \qquad (8.1.7)$$

from which we immediately obtain

$$A'' = \frac{(1 - t_2)K + (1 - t_1)(g/i^{(p)})(C - K)}{1 - t_2 K/C} \qquad (8.1.8)$$

If the principles underlying Eqs 8.1.5 and 8.1.6 are fully understood, the valuation of a stock with allowance for capital gains tax often involves little more work than when this form of taxation is ignored.

For future reference, it is convenient to summarize the preceding results as follows.

If $A''(i)$ denotes the price to provide a net annual yield of i, and K and I are the present values (at rate of interest i) of the *gross* capital and interest payments, respectively, we have

$$A''(i) = \begin{cases} K + (1 - t_1)I & \text{if } i^{(p)} \leq g(1 - t_1) \\[2ex] \dfrac{(1 - t_2)K + (1 - t_1)I}{1 - t_2 K/C} & \text{if } i^{(p)} > g(1 - t_1) \end{cases} \qquad (8.1.9)$$

where $I = (g/i^{(p)})(C - K)$. Note that the condition $i^{(p)} > g(1 - t_1)$ is equivalent to

$$i > \left[1 + \frac{g(1 - t_1)}{p}\right]^p - 1$$

As i increases, $A''(i)$ decreases. When $i^{(p)} = g(1 - t_1)$, each of the expressions on the right side of Eq. 8.1.9 is equal to C.

Equation 8.1.9 is Makeham's formula modified for capital gains tax.

EXAMPLE 8.1.2

A loan is to be issued bearing coupons payable quarterly in arrears at a rate of 5% per annum. Capital is to be redeemed at 103% 20 years after the date of issue. An investor who is liable to income tax at 20% and capital gains tax of 25% wishes to purchase the entire loan at the date of issue. Calculate the price which the investor should pay to ensure a net effective yield of at least 4% per annum.

Solution

We first need to check whether a capital gain is made by the investor. The required yield is $i = 4\%$, and so $i^{(4)} = 0.039414$. The net return generated from the loan is $g(1-t_1) = (0.05/1.03) \times 0.8 = 0.038835 < i^{(4)}$, and so there is a capital gain on the contract from Eq. 8.1.3.

The price of the stock from Eq. 8.1.9 is then

$$A = \frac{(1-0.25) \times 103v^{20} + 0.038835 \times (103 - 103v^{20})}{1 - 0.25v^{20}}$$

$$= 102.09\%$$

Alternatively, we can calculate this from first principles by forming the equation of value

$$A = (1-0.2) \times 5a_{\overline{20}|}^{(4)} + 103v^{20} - 0.25 \times (103 - A)v^{20}$$

where the third term on the right side is the reduction in the capital gain from the tax that is due. Rearranging this expression for A leads to the same $A = 102.09\%$.

EXAMPLE 8.1.3

A loan of nominal amount £500,000 was issued bearing interest of 8% per annum payable quarterly in arrears. The loan will be repaid at 105% by 20 annual installments, each of nominal amount £25,000, the first repayment being 10 years after the issue date. An investor, liable to both income tax and capital gains tax, purchased the entire loan on the issue date at a price to obtain a net effective annual yield of 6%. Find the price paid, given that his rates of taxation for income and capital gains are

(a) 40% and 30%, respectively;
(b) 20% and 30%, respectively.

Solution

Note that $C = 500,000 \times 1.05 = 525,000$. Also,

$$K = 25,000 \times 1.05(a_{\overline{29}|} - a_{\overline{9}|}) \text{ at } 6\% = 178,211$$

and

$$I = \frac{0.08}{1.05} \frac{1}{0.06^{(4)}}(C - K) = 450,158$$

The price payable if there had been no tax liability is

$$A = K + I = 628,368$$

(a) $t_1 = 0.4$ and $t_2 = 0.3$.
In this case the price allowing for income tax only is

$$A - t_1(A - K) = K + (1 - t_1)I = 448,306$$

which is less than 525,000 (i.e., C). Therefore, capital gains tax is payable, and the price actually paid by the investor is given by Eq. 8.1.7 as

$$A'' = 448,306 - t_2\frac{C - A''}{C}K$$

Therefore,

$$A'' = 448,306 - 0.3\left(1 - \frac{A''}{525,000}\right)178,211$$

so that

$$A'' = \frac{448,306 - (0.3 \times 178,211)}{1 - (0.3 \times 178,211/525,000)} = 439,610$$

The price paid was therefore £439,610 or 87.922%.

(b) $t_1 = 0.2$ and $t_2 = 0.3$.
In this case the price allowing for income tax only is

$$178,211 + (1 - 0.2) \times 450,158 = 538,337$$

Since this exceeds £525,000, the price paid by the investor was £538,337, and, in fact, he has no liability for capital gains tax.

Note that the above could also be approached by first comparing $i^{(p)}$ to $g(1-t_1)$.

8.2 CAPITAL GAINS TAX WHEN THE REDEMPTION PRICE OR THE RATE OF TAX IS NOT CONSTANT

When (in relation to a loan repayable by installments) either the redemption price or the rate of capital gains tax is not constant, it is important to fall back on first principles in order to value the capital gains tax. As we have already remarked, it is vital to appreciate that for taxation purposes the "price" deemed to have been paid for a part of the loan redeemed at any one time is calculated from the total price paid in proportion to the *nominal* amount redeemed.

EXAMPLE 8.2.1

Fifteen years ago a loan of nominal amount £300,000 was issued bearing interest of 8% per annum payable annually in arrears. The loan was to be repaid over 30 years, a nominal amount of £10,000 being repayable at the end of each year. The redemption price is par for the first 10 years, 105% for the next 10 years, and 110% for the final 10 years.

The 15th capital repayment has just been made, and an investor wishes to purchase the entire outstanding loan. The investor is not liable for income tax, but will be liable for capital gains tax (CGT) at the rate of 35% for the next 10 years and at the rate of 30% thereafter. What price should the investor pay if he wishes to realize a net effective annual yield of 10% on his investment?

Solution
The loan has an outstanding term of 15 years, during which three distinct situations will pertain:

(a) First 5 years:	redemption price 105%, CGT rate 35%;
(b) Second 5 years:	redemption price 110%, CGT rate 35%;
(c) Final 5 years:	redemption price 110%, CGT rate 30%.

Ignoring the premiums on redemption, the investor may value the future capital repayments, as $(K_1 + K_2 + K_3)$, where $K_1 = 10,000a_{\overline{5}|}$, $K_2 = 10,000_5|a_{\overline{5}|}$,   and $K_3 = 10,000_{10}|a_{\overline{5}|}$ (at 10%).   Therefore, $K_1 = 37,907.87$, $K_2 = 23,537.80$, $K_3 = 14,615.12$, and $K_1 + K_2 + K_3 = 76,060.79$.

The nominal amount of loan outstanding at the time of purchase is £150,000. Hence, the value of the loan without any allowance for tax (but with allowance for the redemption premiums) is

$$76,060.79 + \frac{0.08}{0.10}(150,000 - 76,060.79) + 0.05 \times 37,907.87$$
$$+ 0.1 \times (23,537.80 + 14,615.12) = 140,922.84$$

We must now value the capital gains tax. Let A'' be the price paid by the investor. Since equal nominal amounts are redeemed each year, each outstanding tranche is considered as having cost $A''/15$, so that the capital gains tax has present value

$$0.35 \sum_{t=1}^{5}\left[(10,000 \times 1.05) - \frac{A''}{15}\right]v^t$$

$$+ 0.35 \sum_{t=6}^{10}\left[(10,000 \times 1.1) - \frac{A''}{15}\right]v^t$$

$$+ 0.3 \sum_{t=11}^{15}\left[(10,000 \times 1.1) - \frac{A''}{15}\right]v^t$$

or

$$(0.35 \times 1.05 \times K_1) + (0.35 \times 1.1 \times K_2) + (0.3 \times 1.1 \times K_3)$$
$$-\frac{A''}{15}(0.35a_{\overline{10}|} + 0.3_{10}|a_{\overline{5}|})$$

which equals

$$27,816.18 - 0.172\,603\ A''$$

Hence, A'' is given by the equation

$$A'' = 140,922.84 - (27,816.18 - 0.172\,603A'')$$

from which it follows that $A'' = 136,701.80$. Note that since the outstanding nominal loan is £150,000, this price could be quoted as 91.13%.

Suppose, as before, that when the loan is purchased, the total nominal amount outstanding is $N = (N_1 + N_2 + \cdots + N_r)$. Suppose further that a nominal amount N_j will be redeemed after n_j years, at which time the redemption price per unit nominal will be R_j and the rate of capital gains tax t_j^*.

If the price paid for the entire loan is A'', the tranche redeemed at time n_j for an amount N_jR_j is considered as having cost $(N_j/N)A''$. This means that the total capital gains tax payable has present value

$$\sum_{j=1}^{r} t_j^* \left(N_jR_j - \frac{N_j}{N}A'' \right) v^{n_j} \tag{8.2.1}$$

$$= \sum_{j=1}^{r} t_j^* \left(\frac{R_j - P''}{R_j} \right) K_j \tag{8.2.2}$$

where $K_j = N_jR_jv^{n_j}$ is the value of the capital repayment for the jth tranche and P'' is the price per unit nominal. In practice, either of these summations is evaluated quite simply by grouping together those terms for which the tax rate and the redemption price are constant (see Example 8.2.1).

If the value obtained for P'' by using the preceding expression for the value of the capital gains tax payable is such that (for certain values of j) $R_j < P''$, there will be no tax liability in respect of these tranches. In this case P'' must be recalculated by omitting the appropriate terms from Eqs 8.2.1 and 8.2.2. If these capital losses can be offset against capital gains on other tranches, more detailed calculations are required (see Section 8.5).

8.3 FINDING THE YIELD WHEN THERE IS CAPITAL GAINS TAX

An investor who is liable for capital gains tax may wish to determine the net yield on a particular transaction in which he has purchased a loan at a given price.

One possible approach is to determine the price on two different net yield bases and then estimate the actual yield by interpolation. This approach, which is obviously comparable with that described in previous chapters, is not always the quickest method. Since the purchase price is known, so too is the amount of the capital gains tax, and the net receipts for the investment are therefore known. In this situation one may more easily write down an equation of value that will provide a simpler basis for interpolation, as illustrated by Example 8.3.1.

EXAMPLE 8.3.1

A loan of £1,000 bears interest of 6% per annum payable yearly and will be redeemed at par after 10 years. An investor, liable to income tax and capital gains tax at the rates of 40% and 30%, respectively, buys the loan for £800. What is his net effective annual yield?

Solution

Note that the net income each year of £36 is 4.5% of the purchase price. Since there is a gain on redemption, the net yield is clearly greater than 4.5%.

The gain on redemption is £200, so that the capital gains tax payable will be £60, and the net redemption proceeds will be £940. The net effective yield per annum is then that value of i for which

$$800 = 36a_{\overline{10}|} + 940v^{10}$$

If the net gain on redemption (i.e., £140) were to be paid in equal installments over the 10-year duration of the loan rather than as a lump sum, the net receipts each year would be £50 (i.e., £36 + £14). Since £50 is 6.25% of £800, the net yield actually achieved is less than 6.25%. When $i = 0.055$, the right side of the preceding equation takes the value 821.66, and when $i = 0.06$, the value is 789.85. By interpolation, we estimate the net yield as

$$i = 0.055 + \frac{821.66 - 800}{821.66 - 789.85}(0.06 - 0.055) = 0.0584$$

The net yield is then 5.84% per annum.

EXAMPLE 8.3.2

Assume that in respect of the loan described in Example 8.1.3 the investor's tax rates for income and capital gains were 40% and 30%, respectively, and that he purchased the entire loan on the issue date at a price to obtain a net yield of 6% per annum effective. Assume further that 8 years after purchasing the loan (immediately after receiving the interest payment then due), the investor sold the entire loan to another investor who was liable to the same rates of income tax and capital gains tax. The price paid by the second investor was such as to provide him with a net effective yield of 6% per annum, and the original purchaser paid capital gains tax on the proceeds of the sale. Find the net effective annual yield obtained on the completed transaction by the first investor.

Solution

The price paid by the first investor was £439,610 (see the solution to Example 8.1.3). At the time he sells the loan, it has an outstanding term of 21 years and the first payment of capital will occur after 2 years. The second investor values the gross capital payments as

$$K = (25,000 \times 1.05)(a_{\overline{21}|} - a_{\overline{1}|}) = 284,043$$

and the gross income payments as

$$I = \frac{0.08}{1.05}\frac{1}{0.06^{(4)}}(C - K) \qquad \text{(where } C = 525,000\text{)}$$
$$= 312,778$$

For the second investor the value of the gross receipts less income tax (only) is

$$K + (1 - 0.4)I = 471,710$$

which is less than £525,000. Hence, capital gains tax will be payable by the second investor, and the price he actually paid is A'', where

$$A'' = 471,710 - \frac{0.3 \times (525,000 - A'')}{525,000}284,043$$

from which it follows that $A'' = 461,384$.

The capital gain realized by the sale for the first investor is therefore £21,774 (i.e., £461,384 − £439,610). The capital gains tax payable is therefore £6,532.2, and the net sale proceeds are £454,851.8. The net yield per annum for the first investor is then the value of i for which

$$439,610 = (1 - 0.4) \times 40,000a_{\overline{8}|}^{(4)} + 454,851.8v^8$$

The reader should verify that $i = 0.0593$ or 5.93%.

Note that because of the earlier incidence of capital gains tax, the net yield is less than 6%, although this rate was used as the net yield basis for both purchasers. This is in contrast to the situation that applies when only income tax is involved.

EXAMPLE 8.3.3

A certain irredeemable stock bears interest at $5\frac{1}{2}\%$ per annum, payable quarterly on 31 March, 30 June, 30 September, and 31 December. On 31 August of a certain year, an investor bought a quantity of the stock at a price of £49.50; he sold it exactly 1 year later at a price of £57.71. Given that the investor was liable to capital gains tax at the rate of 30% and income tax at the rate of 40%, find the net yield per annum he obtained.

Solution

Consider the purchase of £100 nominal of the stock. This cost £49.50 and provided net annual income of £3.30 (i.e.,

0.6 × £5.50). The net sale proceeds, after payment of capital gains tax, were

$$57.71 - 0.3 \times (57.71 - 49.50) = 55.25$$

The net annual yield is that rate of interest for which

$$49.50 = 3.3v^{30/365}\ddot{a}_{\overline{1}|}^{(4)} + 55.25v$$

We obtain $i = 0.1895$, or 18.95% by interpolation.

8.4 OPTIONAL REDEMPTION DATES

In Section 7.7 we considered the consequences for a lender, subject only to income tax, of the borrower having a choice of redemption dates. In fact, even with the additional complication of capital gains tax, a lender should adopt the same strategy when valuing a loan for which the borrower has optional redemption dates. In the latter situation, however, a little more care is needed with the argument to see that this is, indeed, the case.

Suppose then that a person, liable to both income and capital gains tax, wishes to realize a net annual yield of at least i on a given loan. Suppose further that, subject to specified conditions, the borrower (i.e., the issuer) has a choice as to when he repays the loan. What price should the lender offer to achieve a net annual yield of at least i?

As before, we assume that interest is payable p times per annum at the rate g per annum on the capital outstanding (measured by cash, rather than nominal, amount). Note first that, if $i^{(p)} = g(1 - t_1)$, the lender should pay C. In this case, whenever the loan is repaid, the lender will achieve a net annual yield of i, and no capital gains tax will be due.

If $i^{(p)} < g(1 - t_1)$, it follows from Eq. 8.1.9 that the price to provide a net yield of i is

$$A* = K + (1 - t_1)I$$

$$= K + \frac{g(1 - t_1)}{i^{(p)}}(C - K)$$

$$= \frac{g(1 - t_1)C}{i^{(p)}} - \left[\frac{g(1 - t_1)}{i^{(p)}} - 1\right]K$$

Since, by hypothesis $i^{(p)} < g(1 - t_1)$, it follows that the least value of A^* will occur when K takes its greatest value. Therefore, the lender should value the loan on the basis that no capital gains tax will be due (from condition 8.1.3), and the borrower will choose that option for which the capital payments have the *greatest* possible value. In particular, if the entire loan must be repaid at one time, the lender should assume that the loan will be repaid as soon as possible. If this option is, in fact, chosen, the lender's net yield will be i; otherwise, it will exceed i.

If $i^{(p)} > g(1 - t_1)$, capital gains tax will be due (by condition 8.1.4) and the price to provide a net yield of i is (see Eq. 8.1.9) $f(K)$, where K is the value at rate i of the gross capital payments and

$$f(x) = \frac{(1 - t_2)x + [g(1 - t_1)/i^{(p)}](C - x)}{1 - t_2 x/C} \qquad (8.4.1)$$

It is readily verified from this last equation that

$$f'(x) = (1 - t_2)\left[1 - \frac{g(1 - t_1)}{i^{(p)}}\right] \Big/ \left(\frac{x}{C}\right)^2$$

Since, by hypothesis $i^{(p)} > g(1 - t_1)$, it follows that $f'(x) > 0$, so f is an increasing function. Hence, the smaller the value of K, the smaller the value of $f(K)$. Therefore, if $i^{(p)} > g(1 - t_1)$, the lender should value the loan on the basis that capital gains tax will be due and the borrower will choose that option for which the capital payments have the *least* possible value. In particular, if the entire loan must be repaid at one time, the lender should assume that the loan will be repaid as late as possible. This strategy will ensure that the actual net yield to the lender is at least i.

EXAMPLE 8.4.1

A loan of nominal amount £100,000 is to be issued bearing interest payable half-yearly in arrears at the rate of 8% per annum. The terms of the issue provide that the borrower shall repay the loan (at par) in ten consecutive annual installments each of nominal amount £10,000, the first repayment being made any time (at the borrower's option) between 10 and 25 years from the issue date.

An investor, liable to income tax at the rate of 40% and to capital gains tax at the rate of 30%, wishes to purchase the entire loan on the issue date at a price to guarantee him a net yield of at least 7% per annum.

(a) What price should he pay?

(b) Given that he paid the price as determined here and that the first capital repayment will actually be made after 14 years, find the net yield which the lender will in fact achieve on this transaction.

Solution

(a) Using the standard notation, we have $C = 100,000$, $p = 2$, $g = 0.08$, $t_1 = 0.4$, $t_2 = 0.3$, and $i = 0.07$. Since $g(1 - t_1) = 0.048$, which is clearly less than $0.07^{(2)}$, the lender should assume that capital gains tax will be due and the capital repayments have the least possible value. Since the repayments must be made in 10 consecutive years, the lender should assume that

Continued

EXAMPLE 8.4.1   *(cont'd)*

redemption occurs as late as possible, i.e., the first repayment will be after 25 years. On this basis

$$K = 10{,}000_{24}|a_{\overline{10}|} \quad \text{at } 7\%$$
$$= 10{,}000(a_{\overline{34}|} - a_{\overline{24}|})$$
$$= 13{,}846.75$$

The value of the net interest payments is therefore (by Makeham's formula)

$$\frac{0.08 \times (1 - 0.4)}{0.07^{(2)}}(100{,}000 - 13{,}846.75) = 60{,}092.87$$

Hence, the price to be paid is A'', where

$$A'' = 13{,}846.75 + 60{,}092.87 - \frac{0.3 \times (100{,}000 - A'')}{100{,}000}$$
$$\times 13{,}846.75,$$

from which we obtain

$$A'' = 72{,}810.14$$

Therefore, he should pay £72,810.14 for the loan. If the borrower delays repaying the loan until the latest permitted date, the lender's net annual yield will be 7%. If redemption occurs earlier than this, the net annual yield to the lender will exceed 7%.

(b) Now suppose that the price paid for the entire loan was £72,810.14 and that the first capital repayment will be made after 14 years. Each tranche of the loan is deemed to have cost £7,281.01, so that the capital gains tax payable with each redemption is 0.3 (10,000 − 7,281.01), i.e., £815.70. The net proceeds of each redemption payment are therefore £9,184.30. Note that the value on the issue date (at rate i) of the gross redemption monies is

$$10{,}000(a_{\overline{23}|} - a_{\overline{13}|}) \quad \text{at rate } i$$

Hence, using Makeham's formula, we obtain the net yield to the lender as that interest rate for which

$$72{,}810.14 = 9{,}184.30\left(a_{\overline{23}|} - a_{\overline{13}|}\right) + \frac{0.08 \times (1 - 0.4)}{i^{(2)}}$$
$$\times \left[100{,}000 - 10{,}000\left(a_{\overline{23}|} - a_{\overline{13}|}\right)\right]$$

The reader should verify that the yield is 7.44%.

8.5 OFFSETTING CAPITAL LOSSES AGAINST CAPITAL GAINS

Until now we have considered the effects of capital gains tax on the basis that it is *not* permitted to offset capital gains by capital losses. In some situations, however, it may be permitted to do so. This may mean that an investor, when calculating his liability for capital gains tax in any year, is allowed to deduct from his total capital gains for the year the total of his capital losses (if any). If the total capital losses exceed the total capital gains, no "credit" will generally be given for the overall net loss, but no capital gains tax will be payable.

A detailed treatment of this topic is beyond the scope of this book, but the following elementary example, 8.5.1, indicates the kind of situation that may arise.

Example 8.5.1 indicates how the investor may exploit the offsetting of gains by losses to obtain a greater net yield than he could obtain from either stock alone.

EXAMPLE 8.5.1

Two government stocks each have an outstanding term of 4 years. Redemption will be at par for both stocks. Interest is payable annually in arrears at the annual rate of 15% for the first stock and 8% for the second stock. Interest payments have just been made, and the prices of the stocks are £105.80 and £85.34, respectively.

(a) Verify that an investor, liable for income tax at the rate of 35% and capital gains tax at the rate of 50%, who purchases either of these stocks (but *not* both) will obtain a net yield on his transaction of 8% per annum.

(b) Assume now that the investor is allowed to offset capital gains by capital losses. Show that, if the proportion of his available funds invested in the 8% stock is such that the overall capital gain is zero, he will achieve a net yield on the combined transaction of 8.46% per annum.

Solution

(a) It is clear that there will be no capital gain on the 15% stock and a capital gain on the 8% stock. Both these stocks can be valued at $i = 8\%$ using the appropriate version of Eq. 8.1.9 in each case.

For the 15% stock, $g = 0.15$ and

$$A = 100v^4 + \left(1 - 0.35\right) \times \frac{g}{i}\left(100 - 100v^4\right) = 105.80\%$$

For the 8% stock, $g = 0.08$ and

$$A = \frac{\left(1 - 0.5\right) \times 100v^4 + \left(1 - 0.35\right)\frac{g}{i}\left(100 - 100v^4\right)}{1 - 0.5v^4}$$

$$= 85.34\%$$

as required.

(b) Assume that the investor uses a fraction λ of his funds to purchase the 8% stock (and a fraction $1 - \lambda$ to the 15% stock). We shall work on the basis of a total investment of £1,000. The *nominal* amounts of each stock purchased are then

$$\frac{1,000\lambda}{85.34} \times 100 \qquad \text{of the 8\% stock}$$

$$\frac{1,000(1 - \lambda)}{105.80} \times 100 \quad \text{of the 15\% stock}$$

The net income received each year is

$$0.65\left[\frac{1,000\lambda}{85.34}8 + \frac{1,000(1 - \lambda)}{105.80}15\right] = 92.16 - 31.22\lambda$$

The total *gross* redemption proceeds are

$$\frac{1,000\lambda}{85.34} \times 100 + \frac{1,000(1 - \lambda)}{105.80} \times 100 = 945.18 + 226.60\lambda$$

If $\lambda = (1,000 - 945.18)/226.60$ (i.e., 0.2419), the total gross redemption proceeds will be 1,000. In this case the capital gain on the 8% stock will be exactly offset by the capital loss on the 15% stock. In this situation no capital gains tax will be payable, and the net redemption proceeds will be £1,000 (i.e., the amount originally invested). Since, for this value λ, the combined net annual income from the two stocks is

$$92.16 - (31.22 \times 0.2419) = 84.61$$

the net yield on the combined transaction is 8.461%, as required.

SUMMARY

- A full treatment of the yields and prices of fixed-interest security for tax-paying investors requires consideration of both *income tax* and *capital gains tax*.
- It is not always evident at the outset that a capital gain will be made, and it is necessary to determine this before the calculation can proceed.
- A modified version of *Makeham's formula* can be used to determine the yield and price of fixed-interest securities when both income and capital gains taxes are due.

- Similar arguments to those used in Chapter 7 can be used to price securities with optional redemption dates for investors subject to both income and capital gains tax.

EXERCISES

8.1 A loan of nominal amount £100,000 is to be issued bearing coupons payable quarterly in arrears at a rate of 5% per annum. Capital is to be redeemed at 103% on a single coupon date between 15 and 20 years after the date of issue, inclusive. The date of redemption is at the option of the borrower.

An investor who is liable to income tax at 20% and capital gains tax of 25% wishes to purchase the entire loan at the date of issue. Calculate the price which the investor should pay to ensure a net effective yield of at least 4% per annum.

8.2 A loan of nominal amount £100,000 is to be issued bearing coupons payable quarterly in arrears at a rate of 7% per annum. Capital is to be redeemed at 108% on a coupon date between 15 and 20 years after the date of issue, inclusive. The date of redemption is at the option of the borrower.

An investor who is liable to income tax at 25% and capital gains tax at 35% wishes to purchase the entire loan at the date of issue. Calculate the price which the investor should pay to ensure a net effective yield of at least 5% per annum.

8.3 An investment manager is considering investing in the ordinary shares of a particular company. The current price of the shares is 12 pence per share. It is highly unlikely that the share will pay any dividends in the next 5 years. However, the investment manager expects the company to pay a dividend of 2 pence per share in exactly 6 years' time, 2.5 pence per share in exactly 7 years' time, with annual dividends increasing thereafter by 1% per annum in perpetuity.

In 5 years' time, the investment manager expects to sell the shares. The sale price is expected to be equal to the present value of the expected dividends from the share at that time at a rate of interest of 8% per annum effective.

(a) Calculate the effective gross rate of return per annum the investment manager will obtain if he buys the share and then sells it at the expected price in 5 years' time.

(b) Calculate the net effective rate of return per annum the investment manager will obtain if he buys the share today and then sells it at the expected price in 5 years' time if capital gains tax is payable at 25% on any capital gains.

(c) Calculate the net effective real rate of return per annum the investment manager will obtain if he buys the share and then sells it at the expected price in 5 years' time if capital gains tax is payable at 25% on any capital gains and inflation is 4% per annum effective. There is no indexation allowance.

8.4 A loan is to be issued bearing interest of 9% per annum payable half-yearly in arrears. The loan will be redeemed after 15 years at 110%.

An investor, liable to income tax at the rate of 45% and to capital gains tax at the rate of 30%, is considering the purchase of part of the loan on the issue date.

(a) What price (per cent nominal) should he pay in order to achieve a net effective annual yield of 8%?

(b) Given that the price actually payable by the investor will be £80%, find his net effective annual yield.

8.5 A loan is to be redeemed in 15 annual installments of equal nominal amount, the first installment being paid 5 years after the issue date. The redemption price is 120%, and the loan will bear interest of 8.4% per annum payable half-yearly *in advance*.

An investor, liable to income tax at the rate of 30% and to capital gains tax at the rate of 25%, purchased the entire loan on the issue date at a price to obtain a net effective annual yield of 7%. What price per cent nominal did the investor pay?

8.6 Investors A and B are both liable to capital gains tax at the rate of 40%, but neither is liable to income tax.

Investor A bought a bond of nominal amount £1,000 bearing interest of 6% per annum payable half-yearly in arrears. The bond was to be redeemed at par 10 years after the date of purchase, and the price paid by A was such that, if he had held the bond until it was redeemed, he would have obtained a net yield on his investment of 10% per annum.

Five years after purchasing the bond (and immediately after receiving the interest then due), A sold it to B, paying capital gains tax on the excess of his selling price over his original price. The bond was held by B until it was redeemed.

(a) If the purchase price paid by B were such that he obtained a net yield on his investment of 10% per annum, find the net annual yield obtained by A over his completed transaction.

(b) If the purchase price paid by B were such that A obtained a net yield over his completed transaction of 10% per annum, find the net annual yield obtained by B on his investment.

8.7 A loan of nominal amount £100,000 is redeemable at 115% by triennial (i.e., every 3 years) payments of capital. The cash amount of the tth repayment is

$$£(15,000 + 1,000t(t-1))$$

and the first repayment is due 10 years after the loan is issued. Interest on the loan is payable half-yearly in arrears at the rate of 6% per annum.

Find the value of the whole loan to a lender, liable to tax on income at 30% and to tax on capital gains at 40%, who wishes to obtain a net yield of 7% per annum on his investment.

8.8 A redeemable stock was issued on 1 January 2005 in bonds of £100 nominal. Interest was payable half-yearly in arrears at 4% per annum for the first 10 years and at 3% per annum for the next 10 years. Redemption was at par after 20 years.

On 1 July 2009 the stock was quoted at a price which would have given a tax-free investor a yield of 6% per annum. A bond was bought on that date by an investor who was subject at all times to income tax at 40% and to capital gains tax at 20%. This investor held the bond until redemption. What net annual yield did he obtain?

8.9 A loan of £1,000,000 nominal has just been issued in bonds of £100 nominal bearing interest at 4% per annum payable half-yearly in arrears. The loan is redeemable by drawings at the end of each year of 400 bonds per annum for the first seven years and 600 bonds per annum thereafter. Redemption is at par during the first 7 years and at 115% thereafter. The issue price is such as to give the purchaser of the entire loan, which is a life office subject to income tax at 35% and to capital gains tax at 30%, a net yield of 6% per annum.

(a) Calculate the price per bond paid by the life office.

(b) Determine the number of individual bonds which will give the life office a net yield of less than 6% per annum.

8.10 A loan of nominal amount £500,000 is issued to be repaid by drawings of equal nominal amount at the end of each year for 20 years. Interest is payable monthly in arrears at the rate of 12% per annum, and redemption is at par for the first ten drawings and at 110% thereafter. An investor is liable to income tax at the constant rate of 40%, and to capital gains tax at the rate of 40% until the first five drawings have been made, and at the rate of 30% thereafter. He purchases the entire loan on the issue date at a price to obtain a net yield of 8% per annum. What price does he pay?

8.11 A loan of nominal amount £100,000 is to be repaid at par by 20 installments each of nominal amount £5,000, the first installment being payable at the end of 10 years and subsequent installments at intervals of 2 years thereafter. Interest is payable annually in arrears at the rate of 6% per annum during the first 30 years and at 7% per annum thereafter. Find the issue price paid to earn 8% per annum net by a purchaser of the whole loan who is subject to income tax at 35% and capital gains tax at 30%.

8.12 An insurance company has just purchased, at a price of 94.5%, a loan of £880,000 nominal bearing interest at $5\frac{1}{2}$% per annum payable half-yearly in arrears. The loan is repayable at par by annual installments, the first being in 5 years' time. The first installment is £50,000, and the amount of each subsequent installment is £6,000 more than the preceding one.

Find the net annual yield that will be obtained by the insurance company, given that it pays income tax at 37.5% and capital gains tax at 30%.

8.13 A loan of nominal amount £300,000 in bonds of nominal amount £100 is to be repaid by 30 annual drawings, each of 100 bonds, the first drawing being 1 year after the date of issue. Interest will be payable quarterly in arrears at the rate of 8% per annum. Redemption will be at par for the first 15 drawings and at 120% thereafter.

An investor is liable to income tax at the constant rate of 40%, and to capital gains tax at the rate of 40% for 10 years and 25% thereafter. He purchases the entire loan on the issue date at a price to obtain a net yield of 7% per annum. What price does the investor pay for each bond?

CHAPTER 9

Term Structures and Immunization

In this chapter we discuss various further topics relating to fixed-interest securities and the valuation of assets and liabilities. Some of these topics are of interest to private investors, life assurance offices (in connection with valuations, for example), pension funds, and other financial institutions. We begin by considering the *term structure of interest rates* with a discussion based on the Institute and Faculty of Actuaries' *CT1* core reading. We then consider the concepts of *matching* and *immunization*.

9.1 SPOT AND FORWARD RATES

Consider a particular type of fixed-interest security actively traded in the market. For a given coupon rate (or range of coupon rates), we may fit an appropriate curve to a set of observed redemption yields, regarded as a function of t, the **term to redemption**. The resulting curve is known as a **redemption yield curve** and shows what is commonly referred to as the **term structure of interest rates**, i.e., how the yield depends on term to redemption. In practice, because of the effects of taxation and other economic factors, the shape of the redemption yield curve can vary somewhat. This will be discussed further in Section 9.2. However, it is necessary to first introduce the concepts of *spot* and *forward* rates.

The equation of value for a general fixed-interest security, of course, requires each individual cash flow (coupon payments and redemption payment) to be discounted back to the issue date. If, however, the discount rate or yield to be used for discounting is dependent on the *discounting period*, we cannot assume a fixed rate of annual interest as we have in previous chapters. Instead, we are required to introduce new notation, the first of which is the **n-year spot rate**, y_n. This quantity is best thought of as the annualized yield on a unit zero-coupon bond with term n years. (Recall that a zero-coupon bond pays only a redemption payment at maturity.)

If the price of such a zero-coupon bond is P_n, we have

$$P_n = \frac{1}{(1 + y_n)^n} \qquad (9.1.1)$$

An Introduction to the Mathematics of Finance. http://dx.doi.org/10.1016/B978-0-08-098240-3.00009-6

and so

$$y_n = P_n^{-1/n} - 1$$

It is assumed that, in general, $y_s \neq y_t$ for $s \neq t$, i.e., there is a term structure to the interest rates. The curve of spot rates against n is an example of a yield curve, mentioned previously.

Every fixed-interest security can be regarded as a combination of zero-coupon bonds. For example, in the absence of tax, the nominal amount N of a security with price A, characterized in terms of a time to redemption T and coupon rate D, has an equation of value which can be written as

$$A = DN\left(\frac{1}{1+y_1} + \frac{1}{(1+y_2)^2} + \cdots + \frac{1}{(1+y_T)^T}\right) + \frac{RN}{(1+y_T)^T} \qquad (9.1.2)$$

$$= DN(P_1 + P_2 + \cdots + P_T) + RNP_T$$

This expression demonstrates the use of spot rates. It should be clear how much of our previous work can be amended to include spot rates.

The key characteristic of a spot rate is that it applies from time zero to a future time; a natural generalization of this is then the *forward rate*, $f_{t,r}$. The forward

EXAMPLE 9.1.1

Three bonds, paying coupons in arrears of 6%, are redeemable at £105 per £100 nominal and reach their redemption dates in exactly 1, 2, and 3 years' time, respectively. The price of each of the bonds is £103 per £100 nominal. Calculate the gross redemption yield of the 3-year bond. Calculate all possible spot rates.

Solution

The gross redemption yield for the 3-year bond is computed from solving the equation of value

$$103 = 6a_{\overline{3}|} + 105v^3$$

We try $i = 6\%$, and evaluate the right side to be 101.1981, indicating that we need a lower value of i. At $i = 5\%$ the right side is 107.0424. Using linear interpolation we find the gross redemption yield to be 6.44%. (Alternatively, Makeham's formula could be solved for the yield.)

To evaluate the spot rates, we begin with the 1-year bond

$$103 = (6 + 105) \times (1 + y_1)^{-1}, \text{ leading to } y_1$$
$$= 7.767\% \text{ per annum}$$

For the 2-year bond,

$$103 = 6 \times (1 + y_1)^{-1} + (6 + 105)(1 + y_2)^{-2}$$
$$= 6 \times (1.07767)^{-1} + 111 \times (1 + y_2)^{-2}$$

leading to $y_2 = 6.736\%$ per annum.

For the 3-year bond,

$$103 = 6 \times (1 + y_1)^{-1} + 6 \times (1 + y_2)^{-2}$$
$$\quad + (6 + 105)(1 + y_3)^{-3}$$
$$= 6 \times (1.07767)^{-1} + 6 \times (1.06736)^{-2}$$
$$\quad + 111 \times (1 + y_3)^{-3}$$

leading to $y_3 = 6.394\%$ per annum.

rate is the annual interest rate agreed at time zero for an investment made at t >0 for a period of r years. The n-year forward rates at time $t = 0$ are, of course, the n-year spot rates, i.e., $f_{0,n} = y_n$.

EXAMPLE 9.1.2

An investor agrees at time 0 to deposit £1,000 at time 4 for 5 years. If $f_{4,5} = 6\%$ per annum, calculate the value of the investment at time $t = 9$.

Solution

The spot rate $f_{4,5}$ is that annual rate which applies from time 4 for 5 years. When we use the concepts from Chapter 2, the accumulated value of the investment at $t = 9$ is then

$$1,000 \times (1 + 0.06)^5 = 1,338.23$$

Forward rates, spots rates, and zero-coupon bond prices are connected through the *principle of consistency* introduced in Chapter 2.

The accumulation at time t of a unit investment made a time zero is $(1+y_t)^t$. If we agree at time zero to invest this accumulated amount at time t for r years, we would earn an annual rate of $f_{t,r}$, and accumulate to

$$(1 + y_t)^t (1 + f_{t,r})^r$$

However, we could also have invested the unit amount at the $(t + r)$-spot rate over $t + r$ years. Under the principle of consistency, these accumulations must be equal and we therefore have

$$(1 + y_t)^t (1 + f_{t,r})^r = (1 + y_{t+r})^{t+r}$$

from which we can determine that

$$(1 + f_{t,r})^r = \frac{(1 + y_{t+r})^{t+r}}{(1 + y_t)^t} = \frac{P_t}{P_{t+r}} \qquad (9.1.3)$$

With this in mind, it is clear that the full term structure can be determined given the spot rates, the forward rates, or the zero-coupon bond prices.

It is useful to introduce the *n-year par yield*. This is defined as the coupon rate per unit time of the stock with term n, which has price 1 (per unit redemption money). Simply put, the par yield is the fixed-interest coupon rate per annum such that the security is redeemed at par. In terms of equations of value, if yc_n is the n-year par yield, then

$$1 = yc_n \times \left(\frac{1}{1 + y_1} + \frac{1}{(1 + y_2)^2} + \cdots + \frac{1}{(1 + y_n)^n} \right) + \frac{1}{(1 + y_n)^n}$$

EXAMPLE 9.1.3

Two bonds paying annual coupons of 5% in arrears and redeemable at par have terms to maturity of exactly 1 year and 2 years, respectively. The gross redemption yield from the 1-year bond is 4.5% per annum effective; the gross redemption yield from the 2-year bond is 5.3% per annum effective. You are informed that the 3-year par yield is 5.6% per annum. Calculate all zero-coupon yields and all 1-year forward rates implied by the stated yields.

Solution

It is possible to obtain 1-year, 2-year, and 3-year spot rates (i.e., zero-coupon yields) from the information provided. We do this first and then infer the 1-year forward rates at times 0, 1, and 2.

The gross redemption yield of the 1-year bond is, of course, equal to the 1-year spot rate, and we have $y_1 = 4.5\%$ per annum.

For the 2-year bond, we can determine the price from the gross redemption yield and equate this to that obtained from using the spot rates

$$\frac{5}{1.053} + \frac{105}{1.053^2} = \frac{5}{1 + y_1} + \frac{105}{(1 + y_2)^2}$$

Using y_1 from above, we can determine that $y_2 = 5.3202\%$ per annum.

Our approach for the 3-year spot rate is slightly different, as we are required to use the par yield. The equation of value is then

$$100 = 5.6 \times \left(\frac{1}{1 + y_1} + \frac{1}{(1 + y_2)^2} + \frac{1}{(1 + y_3)^3} \right) + \frac{100}{(1 + y_3)^3}$$

Using the values for the 1-year and 2-year spot rates, calculated above, we can determine that $y_3 = 5.6324\%$ per annum.

The 1-year forward rate at time 0 is $f_{0,1} = f_0 = y_1 = 4.5\%$ per annum.

The 1-year forward rate at time 1, $f_{1,1} = f_1 = 6.1469\%$ per annum, is obtained from

$$(1 + y_1)(1 + f_1) = (1 + y_2)^2$$

Finally, the 1-year forward rate at time 2, $f_{2,1} = f_2 = 6.2596\%$ per annum, is obtained from

$$(1 + y_2)^2(1 + f_2) = (1 + y_3)^3$$

The interval over which each rate applies is indicated in Figure 9.1.1.

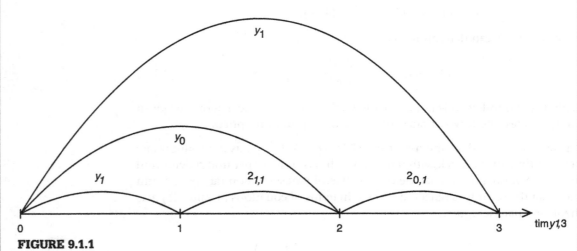

FIGURE 9.1.1
Spot and forward rates calculated in Example 9.1.3

Note that in Example 9.1.3 new notation was introduced for the one-period forward rate. In particular, the one-period forward rate at time t is often denoted $f_{t,1} = f_t$; this contraction of the notation is common practice in actuarial notation. Furthermore, we have $f_0 = y_1$.

When we employ the principle of consistency argument used previously, it is clear that we can relate the n-year spot rate to the 1-year forward rates in each prior year

$$(1 + y_t)^t = (1 + f_0)(1 + f_1)(1 + f_2) \cdots (1 + f_{t-1}) \qquad (9.1.4)$$

We have so far discussed spot and forward rates in the context of discretely paying fixed-interest securities, i.e., those that pay coupons at discrete time intervals. Indeed, the quantities y_t and $f_{t,r}$ are often referred to as *discrete time spot* and *discrete time forward rates*. However, as we have seen in previous chapters, it is possible to consider continuously paying fixed-interest securities, and we introduce the notion of *continuous time spot rates* and *continuous time forward rates*.

For consistency with the discrete time analogue in Eq. 9.1.1, we define the t-year spot *force of interest* or *continuously compounded spot rate*, Y_n, in terms of the price of a zero-coupon bond of term n as

$$P_n = e^{-Y_n n} \qquad (9.1.5)$$

and so

$$Y_n = -\frac{1}{n} \ln P_n \qquad (9.1.6)$$

It is clear that Y_n and its corresponding discrete annual rate y_n are related in the same way that a constant continuous force of interest δ and constant effective interest i are. It then follows that

$$Y_n = e^{Y_n} - 1$$

Similarly, a continuous time forward rate $F_{t,r}$ is the force of interest equivalent to the annual forward rate of interest $f_{t,r}$. We then have

$$f_{t,r} = e^{F_{t,r}} - 1$$

The relationship between the continuous time spot and forward rates may be derived by considering the accumulation of a unit investment at a continuous time spot rate, Y_t, for t years, followed by the continuous time forward rate, $F_{t,r}$, for r years. Equating this to the accumulation of a unit investment under the continuous spot rate Y_{t+r} for $t+r$ years leads to

$$e^{tY_t}e^{rF_{t,r}} = e^{(t+r)Y_{t+r}}$$

Considering the exponents and Eq. 9.1.6 leads us to

$$F_{t,r} = \frac{(t+r)Y_{t+r} - tY_t}{r}$$

$$= \frac{1}{r}\ln\left(\frac{P_t}{P_{t+r}}\right) \qquad\qquad (9.1.7)$$

With continuous time, it is possible to modify the definition of the forward rate to identify the **instantaneous forward rate**. In other words, this is the *instantaneous* rate of interest that applies at a time t in the future. Mathematically, it is obtained from the following limit:

$$F_t = \lim_{r \to 0} F_{t,r}$$

Using Eq. 9.1.7, we can express this in terms of the prices of zero-coupon bonds:

$$F_t = \lim_{r \to 0} \frac{1}{r}\ln\left(\frac{P_t}{P_{t+r}}\right)$$

$$= -\lim_{r \to 0} \frac{\ln P_{t+r} - \ln P_t}{r}$$

$$= -\frac{d}{dt}\ln P_t$$

$$= -\frac{1}{P_t}\frac{d}{dt}P_t$$

Integrating this expression and imposing the condition that $P_0 = 1$, i.e., the price of a unit zero-coupon bond at the moment of redemption is 1, we see that

$$P_t = e^{-\int_0^t F_s ds} \qquad\qquad (9.1.8)$$

When we compare Eq. 9.1.8 to Eq. 2.4.10, it is clear that the instantaneous forward rate can be thought of as the **forward force of interest**.

Spot and forward rates (both discrete and continuous) can be thought of as existing on the market; therefore, an investor can obtain access to these rates through investments. For example, the gross redemption yield obtained from a fixed-interest security is, in some sense, an average of the market rates at each term and reflects the relative weighing of the cash flows at each time.

9.2 THEORIES OF THE TERM STRUCTURE OF INTEREST RATES

A number of factors influence the distribution of spot rates with term. Many people believe that there is an underlying pure rate of interest available on fixed-interest securities, irrespective of term. For a government issue, for example, the pure rate would be related to the long-term growth in that particular economy. In times of economic stability, this rate has been typically around 2%–3% per annum. However, it is also widely accepted that this pure rate cannot exist for investments of all terms because there is a correlation between long-term investments and risk. The longer the term of the security, the greater the risk; investors would therefore demand a higher yield to compensate for this risk. This is the essence of *liquidity preference theory*, discussed next, and would suggest an increasing yield curve. However, increasing yield curves do not necessarily always occur in practice, and there are a number of explanations for this; the two most popular explanations are *expectations theory* and *market segmentation theory*.

Liquidity Preference Theory

Longer dated securities are more sensitive to interest rate movements than shorter dated bonds. This is true because, as we have seen elsewhere in this book, the effects of compound interest are much more significant over a long term. Furthermore, longer dated securities issued by non-government institutions are more sensitive to the fortunes of the institutions. Because of this increased risk with term, risk-averse investors will require compensation, in the form of high yields, for investing in longer dated securities. This leads to an increasing yield curve.

Expectations Theory

Under expectations theory, the relative attraction for short- and long-term investments will vary according to expectations of future movements in market interest rates. An expectation of a fall in prevailing rates will make short-term investments less attractive, as one is expecting to soon be able to borrow more cheaply. Conversely, long-term investments will become more attractive. The demand for short-term investments will then decrease and, consequently, their price will fall, leading to higher yields. The demand for long-term investments will rise, and their price will rise, leading to lower yields. An expectation of a rise in interest rates would have the opposite effect.

Under expectations theory, the demand at each term based on the market expectations will modify the tendency for an increasing yield curve that arises from liquidity preference theory.

Market Segmentation Theory

Under market segmentation theory, it is acknowledged that securities of different terms are attractive to different investors. For example, liabilities of many pension funds are long term, and consequently, they demand longer term assets. The liabilities of banks are very short term (investors may withdraw funds at short notice); hence, banks invest in very short-term assets. The demand for securities therefore differs for different terms. The supply of securities will also vary with term, as governments' and companies' financing requirements change. The market segmentation theory argues that a term structure will emerge from the action of supply and demand at each different term, and deviate from the tendency of the yield curve to increase under the liquidity preference theory.

In what follows we neglect the term structure of interest rates and, for ease of explanation, return to assuming that yields are independent of term. Of course, the material presented throughout this book can be generalized to include the effects of a term structure if required.

9.3 THE DISCOUNTED MEAN TERM OF A PROJECT

Consider a business project or investment that will produce a series of cash flows. The net present value of the project at a force of interest δ per annum is

$$NPV(\delta) = \sum c_t e^{-\delta t} \qquad (9.3.1)$$

where the summation extends over those values of t for which c_t, the net cash flow at time t, is non-zero. At this force of interest, if $NPV(\delta) \neq 0$, the *discounted mean term* of the project, $T(\delta)$, is defined to be

$$T(\delta) = \frac{\sum t c_t e^{-\delta t}}{\sum c_t e^{-\delta t}} \qquad (9.3.2)$$

Note that the discounted mean term is a weighted average of the future times at which there are cash flows. The 'weight' associated with any given time is the present value (at force of interest δ) of the net cash flow due at that time. The discounted mean term is sometimes known as the *Macauley duration*, or simply *duration*.

In the more general situation in which there is also a continuous cash flow, the net rate of payment being $\rho(t)$ per annum at time t years, the net present value of the project at force of interest δ per annum is

$$NPV(\delta) = \sum c_t e^{-\delta t} + \int_0^\infty \rho(t) e^{-\delta t} dt \qquad (9.3.3)$$

and the discounted mean term of the project is defined to be

$$T(\delta) = \frac{\sum tc_t e^{-\delta t} + \int_0^{\infty} t\rho(t)e^{-\delta t}\,dt}{\sum c_t e^{-\delta t} + \int_0^{\infty} \rho(t)e^{-\delta t}\,dt} \tag{9.3.4}$$

provided that $NPV(\delta) \neq 0$.

EXAMPLE 9.3.1

A mine owner estimates that the net cash flow from his mining operations will be as follows:

Time (Years)	Net Cash Flow (£)
1	500,000
2	400,000
3	300,000
4	200,000
5	100,000

Calculate the discounted mean term of the project at an effective interest rate of 6% per annum.

Solution

By Eq. 9.3.2, the discounted mean term is

$$\frac{5v + (2 \times 4v^2) + (3 \times 3v^3) + (4 \times 2v^4) + 5v^5}{5v + 4v^2 + 3v^3 + 2v^4 + v^5} \quad \text{at } 6\%$$

$$= \frac{4.7170 + 7.1200 + 7.5566 + 6.3367 + 3.7363}{4.7170 + 3.5600 + 2.5189 + 1.5842 + 0.7473}$$

$$= \frac{29.4666}{13.1274} = 2.245 \text{ years}$$

EXAMPLE 9.3.2

Ten years ago an investor purchased an annuity-certain payable continuously for 25 years. Throughout the tth year of payment ($0 \leq t \leq 25$), the rate of payment of the annuity is £t per annum. Find the duration of the *remaining* annuity payments on the basis of an effective annual interest rate of (a) 5% and (b) 10%.

Solution

Measure time in years from the present. Note that over the coming year, the rate of annuity payment is £11 per annum, in the following year £12 per annum, etc. In the final year of payment (i.e., the year commencing 14 years from now), the rate of payment is £25 per annum.

Using Eq. 9.3.4, we may express the duration (i.e., discounted mean term) of the future payments, at force of interest δ per annum, as

$$\frac{\int_0^1 t \times 11e^{-\delta t}\,dt + \int_1^2 t \times 12e^{-\delta t}\,dt + \int_2^3 t \times 13e^{-\delta t}\,dt + \cdots + \int_{14}^{15} t \times 25e^{-\delta t}\,dt}{\int_0^1 11e^{-\delta t}\,dt + \int_1^2 12e^{-\delta t}\,dt + \int_2^3 13e^{-\delta t}\,dt + \cdots + \int_{14}^{15} 25e^{-\delta t}\,dt}$$

The reader should verify that on simplification this expression becomes

$$\frac{\frac{11}{\delta^2} + \frac{1}{\delta}(Ia)_{\overline{14}|} + \frac{1}{\delta^2}a_{\overline{14}|} - 25\left(\frac{15v^{15}}{\delta} + \frac{v^{15}}{\delta^2}\right)}{10\bar{a}_{\overline{15}|} + (I\bar{a})_{\overline{15}|}}$$

which equals

$$\frac{\frac{11}{\delta} + (Ia)_{\overline{14}|} + \frac{1}{\delta}a_{\overline{14}|} - 25\left(15v^{15} + \frac{v^{15}}{\delta}\right)}{i[10\bar{a}_{\overline{15}|} + (I\bar{a})_{\overline{15}|}]}$$

Evaluating the last expression at rates of interest 5% and 10%, we obtain the answers (a) 7.656 years and (b) 6.809 years, respectively.

9.4 VOLATILITY

Consider an investment or business project with net cash flows as described in Section 9.3. Suppose that the net present value of the project is currently determined on the basis of an annual force of interest δ_0, but that this may change at very short notice to another value, δ_1 say, not very different from δ_0. If this were to happen, the *proportionate* change in the net present value of the project would be

$$\frac{\Delta NPV(\delta_0)}{NPV(\delta_0)} = \frac{NPV(\delta_1) - NPV(\delta_0)}{NPV(\delta_0)}$$

$$\simeq (\delta_1 - \delta_0)\frac{NPV'(\delta_0)}{NPV(\delta_0)} \tag{9.4.1}$$

Note that we use the notation $NPV'(\delta)$ to denote the derivative with respect to δ of $NPV(\delta)$. We also assume that $NPV(\delta_0) \neq 0$.

We now define the *volatility* (with respect to δ) of the project at force of interest δ_0 to be

$$\frac{-NPV'(\delta_0)}{NPV(\delta_0)} \tag{9.4.2}$$

With this definition, Eq. 9.4.1 may be expressed in words as

$$\left(\begin{array}{l}\text{proportionate change}\\ \text{in net present value}\end{array}\right) \simeq - (\text{change in force of interest}) \times (\text{volatility})$$

$$\tag{9.4.3}$$

Note that the definition of the volatility at force of interest δ_0 (Eq. 9.4.2) may be expressed as

$$\left[-\frac{1}{NPV(\delta)}\frac{d}{d\delta}NPV\right]_{\delta=\delta_0} = \left[-\frac{d}{d\delta}\ln NPV(\delta)\right]_{\delta=\delta_0} \tag{9.4.4}$$

By combining Eq. 9.4.3 with our definition 9.4.2, we see that, at force of interest δ, the volatility of the project equals

$$\frac{\sum tc_t e^{-\delta t} + \int_0^\infty t\rho(t)e^{-\delta t}dt}{\sum c_t e^{-\delta t} + \int_0^\infty \rho(t)e^{-\delta t}dt} \tag{9.4.5}$$

It is clear from Eqs 9.3.4 and 9.4.5 that, at any force of interest, we have the identity

$$\text{discounted mean term} \equiv \text{volatility} \tag{9.4.6}$$

We shall therefore use the symbol $T(\delta)$ to refer to the discounted mean term, (Macauley) duration, or the volatility of an investment or business project.

Returning to Eq. 9.4.1, we see that the proportionate change in the net present value on a small immediate change in the force of interest is approximately equal to $-(\delta_1 - \delta_0)T(\delta_0)$. Therefore, the proportionate profit or loss on a small change in interest rates depends more or less directly on

(a) the size of the change in the force of interest, which is similar to the size of the change in the rate of interest, and

(b) the volatility of the investment; if interest rates rise, there will, of course, be a loss; and if they fall, there will be a profit to the investor.

It follows that the volatility (i.e., duration) of a fixed-interest investment is of some interest to actual and prospective investors, and often appears in the financial press. In practical applications the volatility of a fixed-interest stock is calculated by Eq. 9.3.2, although in theoretical work, other formulae are also useful. In the next section we shall give expressions for the volatility of particular fixed-interest stocks and discuss the variation of volatility with the coupon rate and the term to redemption.

From the preceding description, it is clear that the volatility as defined by Eq. 9.4.2 gives the sensitivity of the net present value of a cash flow to changes in the force of interest, δ. However, it is also useful to consider how this is related to the sensitivity of the net present value to changes in the discretely compounding interest rate, i. By similar arguments to those used previously, it is clear that this sensitivity at interest rate i_0 (i.e., that value consistent with δ_0) can be expressed as

$$-\frac{1}{NPV(i_0)}\frac{dNPV(i_0)}{di} = \frac{d\delta}{di}\bigg|_{i_0} T(\delta_0)$$

where T is the duration as defined by Eq. 9.3.4. By noting that $\delta = \ln(1+i)$, we can write the preceding expression as

$$v(i_0) = \frac{T(i_0)}{1+i_0} \tag{9.4.7}$$

The quantity $v(i_0)$ is often referred to as the **effective duration** or the **volatility with respect to i**.

EXAMPLE 9.4.1

If the duration of a cash flow stream at $\delta_0 = 5\%$ per annum is $T = 10.629$ years, calculate the *effective duration*.

$$v(5.127\%) = \frac{10.629}{1.05127} = 10.111 \text{ years}$$

Solution

If $\delta_0 = 5\%$, $i_0 = e^{0.05} - 1 = 5.127\%$. By Eq. 9.4.7, the effective duration is then

9.5 THE VOLATILITY OF PARTICULAR FIXED-INTEREST SECURITIES

Consider a unit nominal amount of a fixed-interest stock, which bears interest at D per annum and is redeemable in n years' time at price R. Let $g = D/R$ (see Section 7.5) and let the annual force of interest implied by the current price of this stock be δ_0 (i.e., the current price of the stock equals the present value, at force of interest δ_0 per annum, of the future interest and capital payments). We suppose an investor has purchased the stock, so his future net cash flows are all positive.

If interest is payable annually in arrears and n is an integer, the discounted mean term of the stock at force of interest δ_0 is given by Eq. 9.3.2 as

$$T(\delta_0) = \frac{g(Ia)_{\overline{n}|} + ne^{-\delta n}}{ga_{\overline{n}|} + e^{-\delta n}} \qquad \text{at force of interest } \delta_0 \qquad (9.5.1)$$

In the particular case when g equals i_0, the rate of interest per annum corresponding to the annual force of interest δ_0, we obtain

$$T(\delta_0) = \frac{(\ddot{a}_{\overline{n}|} - nv^n) + nv^n}{(1 - v^n) + v^n} \qquad \text{at rate of interest } i_0$$

$$= \ddot{a}_{\overline{n}|} \qquad \text{at rate of interest } i_0 \qquad (9.5.2)$$

By letting n tend to infinity in Eq. 9.5.1, we obtain the volatility at force of interest δ_0 of a perpetuity as

$$T(\delta_0) = \frac{g(Ia)_{\overline{\infty}|}}{ga_{\overline{\infty}|}} \qquad \text{at force of interest } \delta_0$$

$$= \frac{1}{d} \qquad \text{at rate of interest } i_0 \qquad (9.5.3)$$

Note that the volatility does *not* depend on the coupon rate of the perpetuity.

Note also that for a *zero-coupon* bond redeemable in n years' time, it is an immediate consequence of Eq. 9.5.1 that

$$T(\delta_0) = n \qquad (9.5.4)$$

for *all* values of δ_0.

It is clear from Eq. 9.5.1 that the volatility at force of interest δ_0 of a fixed-interest stock depends on both the annual interest rate per unit indebtedness g and the term to redemption n of the stock. We wish to consider the variation of the volatility $T(\delta_0)$ with g and n. In order to simplify our calculations, we assume that interest is payable continuously. Equation 9.3.4 then implies that

$$T(\delta_0) = \frac{g(\overline{Ia})_{\overline{n}|} + nv^n}{g\overline{a}_{\overline{n}|} + v^n} \qquad \text{at force of interest } \delta_0 \qquad (9.5.5)$$

We first show that, for each fixed term n years, the volatility of the stock decreases as g increases. To prove this, we note that

$$\frac{\partial T(\delta_0)}{\partial g} = \frac{(\bar{I}\bar{a})_{\overline{n}|}\left(g\bar{a}_{\overline{n}|} + e^{-\delta_0 n}\right) - \bar{a}_{\overline{n}|}\left[g(\bar{I}\bar{a})_{\overline{n}|} + ne^{-\delta_0 n}\right]}{\left(g\bar{a}_{\overline{n}|} + e^{-\delta_0 n}\right)^2}$$

$$= \frac{e^{-\delta_0 n}\left[(\bar{I}\bar{a})_{\overline{n}|} - n\bar{a}_{\overline{n}|}\right]}{\left(g\bar{a}_{\overline{n}|} + e^{-\delta_0 n}\right)^2} < 0$$

since

$$(\bar{I}\bar{a})_{\overline{n}|} = \int_0^n te^{-\delta_0 t}\,dt < n\int_0^n e^{-\delta_0 t}\,dt = n\bar{a}_{\overline{n}|}$$

This result is illustrated in Figure 9.5.1.

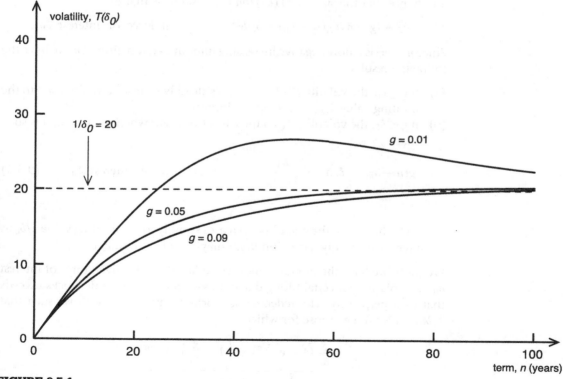

FIGURE 9.5.1
Variation of volatility with term

We now consider the variation of volatility with the term to redemption for a fixed value of g. Equation 9.5.5 implies that

$$\lim_{n \to \infty} T(\delta_0) = \frac{(\overline{Ia})_{\overline{\infty}|}}{\overline{a}_{\overline{\infty}|}} \qquad \text{at force of interest } \delta_0$$

$$= \frac{1}{\delta_0}$$

(9.5.6)

and also that

$$\lim_{n \to 0} T(\delta_0) = 0 \qquad\qquad (9.5.7)$$

Note that neither of these limiting values depends on the annual coupon rate. By straightforward (but somewhat tedious) calculation, it can be shown that

$$\frac{\partial T(\delta_0)}{\partial n} = \frac{\left[\delta_0 + g(g - \delta_0)n - (g - \delta_0)^2 \overline{a}_{\overline{m}|}\right]e^{-\delta_0 n}}{\delta_0(g\overline{a}_{\overline{m}|} + e^{-\delta_0 n})^2} \qquad \text{at force of interest } \delta_0$$

(9.5.8)

and hence that the sign of $\partial T(\delta_0)/\partial n$ is the same as that of

$$\delta_0 + (g - \delta_0)[g(n - \overline{a}_{\overline{m}|}) + \delta_0 \overline{a}_{\overline{m}|}] \qquad \text{at force of interest } \delta_0$$

Since $n - \overline{a}_{\overline{m}|}$ is a non-negative increasing unbounded function of n, we have the following results.

(a) If $g \geq \delta_0$, the volatility $T(\delta_0)$ increases steadily from zero when $n = 0$ to the limiting value $1/\delta_0$ as n tends to infinity.

(b) If $g < \delta_0$, the volatility $T(\delta_0)$ increases from zero when $n = 0$ until n is such that

$$g(n - \overline{a}_{\overline{m}|}) + \delta_0 \overline{a}_{\overline{m}|} = \frac{\delta_0}{\delta_0 - g} \qquad \text{at force of interest } \delta_0 \qquad (9.5.9)$$

after which term the volatility decreases steadily to the limiting value $1/\delta_0$ as n tends to infinity. Equation 9.5.9 may be solved numerically.

We therefore have the possibly unexpected fact that, at a given force of interest δ_0, the volatility of certain long-dated, low-coupon redeemable stocks exceeds that of a perpetuity. The redeemable stocks in question are those such that $T(\delta_0) > (1/\delta_0)$, i.e., those for which

$$\frac{\dfrac{g}{\delta_0^2}\left(1 - e^{-\delta_0 n}\right) + \left(1 - \dfrac{g}{\delta_0}\right)ne^{-\delta_0 n}}{\dfrac{g}{\delta_0}\left(1 - e^{-\delta_0 n}\right) + e^{-\delta_0 n}} > \frac{1}{\delta_0}$$

This inequality holds if and only if

$$\frac{g}{\delta_0^2}\left(1 - e^{-\delta_0 n}\right) + \left(1 - \frac{g}{\delta_0}\right) n e^{-\delta_0 n} > \frac{g}{\delta_0^2}\left(1 - e^{-\delta_0 n}\right) + \frac{e^{-\delta_0 n}}{\delta_0}$$

i.e., if and only if

$$\delta_0 > g + \frac{1}{n} \qquad\qquad (9.5.10)$$

The last inequality may be written in either of the forms

$$n > \frac{1}{\delta_0 - g}$$

$$g < \delta_0 - \frac{1}{n}$$

Therefore, if g and δ_0 are given with $0 < g < \delta_0$, a redeemable stock with interest rate g per unit indebtedness will have greater volatility (at force of interest δ_0) than a perpetuity if and only if the term to redemption exceeds $1/(\delta_0 - g)$. Alternatively, if δ_0 and n are given, a redeemable stock with term n will have greater volatility (at force of interest δ_0) than a perpetuity if and only if the interest rate per unit indebtedness is less than $\delta_0 - (1/n)$.

The variation of volatility with term (when $\delta_0 = 0.05$) is illustrated in Figure 9.5.1.

An explanation of the fact that certain low-coupon, long-dated securities are more volatile than a perpetuity is that, in the case of the redeemable stock, the

---

EXAMPLE 9.5.1

A speculator buys large quantities of a fixed-interest security when he expects interest rates to fall, with the intention of selling in a short time to realize a profit. At present, he is choosing between the following two secure fixed-interest stocks:

- *Security 1*, which bears interest at 5% per annum payable annually in arrears and is repayable at par in 5 years' time;
- *Security 2*, which bears interest at 11% per annum payable annually in arrears and is repayable at par in 6 years' time.

These stocks always have the same gross yield per annum, at present 10%. Which should he buy in order to obtain the larger capital gain on a small fall in interest rates? (Ignore taxation.)

Solution

We calculate the volatilities of each of the securities using Eq. 9.5.1.

The volatility of Security 1 is

$$\frac{0.05 \times (Ia)_{\overline{5}|} + 5v^5}{0.05 a_{\overline{5}|} + v^5} \qquad \text{at 10\% interest}$$

$$= 4.488$$

The volatility of Security 2 is

$$\frac{0.11 \times (Ia)_{\overline{6}|} + 6v^6}{0.11 a_{\overline{6}|} + v^6} \qquad \text{at 10\% interest}$$

$$= 4.725$$

Hence, the speculator should buy Security 2 because it is more sensitive to changes to interest rates.

main 'weight' attaches to the repayment of capital, which is very distant, whereas for the perpetuity the main 'weights' attach to the next few years' interest payments. Hence, the discounted mean term of the redeemable stock may be expected to be greater than that of the perpetuity.

EXAMPLE 9.5.2

Suppose that the force of interest per annum for valuing secure fixed-interest stocks is δ. When $\delta = 0.08$, the prices per unit nominal of two secure fixed-interest stocks, A and B, are equal. Furthermore, it is known that when $0.05 \leq \delta \leq 0.08$, the volatility of stock A is not less than that of stock B. Prove that, if the force of interest changes immediately from 0.08 per annum to 0.05 per annum, then the new price of stock A will not be less than the new price of stock B.

Solution

Let $V_A(\delta)$ and $V_B(\delta)$ denote the prices per unit nominal of stocks A and B, respectively, when the force of interest is δ per annum. We are given that

$$V_A(0.08) = V_B(0.08)$$

and so

$$\ln V_A(0.08) = \ln V_B(0.08)$$

Also, by Eq. 9.4.4,

$$-\frac{d}{d\delta}[\ln V_A(\delta) - \ln V_B(\delta)] \geq 0 \qquad 0.05 \leq \delta \leq 0.08$$

It follows that if we set

$$f(\delta) = \ln V_A(\delta) - \ln V_B(\delta)$$

then $f(0.08) = 0$ and $f'(\delta) \leq 0$ for $0.05 \leq \delta \leq 0.08$. Hence, by the mean value theorem of elementary calculus, $f(0.05) \geq 0$, we have

$$\ln V_A(0.05) \geq \ln V_B(0.05)$$

i.e.,

$$V_A(0.05) \geq V_B(0.05)$$

as required.

9.6 THE MATCHING OF ASSETS AND LIABILITIES

In a general business context, the *matching* of assets and liabilities of a company requires that the company's assets be chosen as far as possible in such a way as to make the assets and liabilities equally responsive to the influences which affect them both. In this wide sense, matching includes the matching of assets and liabilities in terms of currencies and the degree of inflation linking. In this chapter, however, we shall consider both assets and liabilities to be in money terms. Accordingly, we shall use the word 'matching' to refer only to a suitable choice of the terms of the assets in relation to the terms of the liabilities, so as to reduce the possibility of loss arising from changes in interest rates.

The *liabilities* of a business are the amounts it has contracted to pay in the future. Let S_t denote the liability at time t (i.e., the money which must be paid out at that time). Let P_t be the money to be *received* by the company at time t

from its business operation, excluding any investment proceeds due at that time. We define the *net liability-outgo* at time t to be

$$L_t = S_t - P_t \qquad (9.6.1)$$

The net liabilities of the business are the sums $\{L_t\}$. The net cash flow at time t is $c_t = -L_t$ for all t. We shall assume that all the liabilities and receipts are discrete, although the arguments given here may easily be generalized to include continuous cash flows.

EXAMPLE 9.6.1

Consider a life office that issued a 20-year capital redemption policy 10 years ago with a sum assured of £10,000 and an annual premium of £288.02. Measuring time in years from the present and assuming that the premium now due has been paid and that expenses may be ignored, calculate the net liabilities $\{L_t\}$ in respect of this policy.

Solution

We have

$$\begin{aligned} P_t &= 288.02 & \text{for } t = 1, 2, \ldots, 9 \\ S_t &= 10{,}000 & \text{when } t = 10 \end{aligned}$$

and so

$$L_t = \begin{cases} -288.02 & \text{for } t = 1, 2, \ldots, 9 \\ +10{,}000 & \text{for } t = 10 \end{cases}$$

The proceeds at time t of the company's investments, whether capital, interest, or both, are referred to as the *asset-proceeds* at time t years, and are denoted by A_t. The *assets* of the business are the collection of asset-proceeds (or, in ordinary usage, the securities providing them).

Suppose first that L_t is never negative and that the asset-proceeds are such that, for *all* values of t,

$$A_t = L_t \qquad (9.6.2)$$

i.e., the asset-proceeds are equal to the net liability-outgo at each time. It is clear that in this case the business will always have exactly sufficient cash to meet the net liabilities, no matter what the pattern of interest rates may be at present or in the future. The business is therefore said to be *absolutely matched*. If the net cash flow is sometimes positive and sometimes negative (as in Example 9.6.1), absolute matching is not possible, as some investment must take place in the future (at unknown interest rates) to provide for the liabilities in the final stages of the project.

Assume that at present the annual force of interest for valuing assets and liabilities is δ. The present value of the net liabilities is then

$$V_L(\delta) = \sum L_t v^t \qquad \text{at force of interest } \delta \qquad (9.6.3)$$

and the present value of the assets is

$$V_A(\delta) = \sum A_t v^t \qquad \text{at force of interest } \delta \qquad (9.6.4)$$

We shall assume for the remainder of this section that $V_L(\delta) > 0$ for all δ, or at any rate for all values of δ under consideration.

If the force of interest per annum will remain constant at δ_0, at least until all payments have been made and received, then it is clear that any set of secure assets satisfying the condition

$$V_A(\delta_0) = V_L(\delta_0) \tag{9.6.5}$$

will be exactly sufficient to meet the net liabilities as they emerge in the future.

Consider, for example, the policy given in Example 9.6.1 and assume that the force of interest per annum will remain constant at $\delta_0 = \ln 1.05$ (which corresponds to a rate of interest of $i = 5\%$ per annum). If the office possesses any fixed-interest assets of value

$$
\begin{aligned}
V_L(\delta_0) &= 10{,}000v^{10} - 288.02a_{\overline{5}|} \qquad\qquad \text{at } 5\% \\
&= 6{,}139.13 - 2{,}047.19 = 4{,}091.94
\end{aligned}
$$

then it will have exactly sufficient cash to meet the liability under the policy.

In practice, however, it cannot normally be assumed that the force of interest will remain constant until all transactions have been completed, and the company will wish to make a profit (or, at any rate, not make a loss) if interest rates change. A degree of protection against changes in interest rates may be obtained by using the concept of *immunization*, which we shall discuss in the next two sections.

It is possible, of course, that the investor or management of the business may deliberately maintain an unmatched position in order to profit from an anticipated rise, or fall, in interest rates: this policy is known as 'taking a view of the market'. Some businesses are, however, essentially custodians of money belonging to others—life assurance companies and pension funds fall into this category—and such institutions are, as a rule, constrained by law or considerations of solvency into a largely defensive position with respect to changes in interest rates. (The real position is, however, more complicated: 'with-profits' policies may exist, and the life office may have shareholders, so an element of risk in investment policy may be acceptable.) The theory of immunization is of interest to many financial institutions (although certain modifications may be necessary before it can be applied in practice).

9.7 REDINGTON'S THEORY OF IMMUNIZATION

In 1952, F. M. Redington[1] presented a *theory of immunization*, under which the investor (in Redington's original paper a life office) is protected against small

---

[1] Redington, F. M. 'Review of the principles of life office valuations', *Journal of the Institute of Actuaries*, 1952, 78(3), pp. 286–315.

changes in the rate of interest. We now give a brief description of this theory, referring the reader to Redington's original paper for full details if required.

Consider a financial institution, which for definiteness we take to be a life office, which has net liability-outgo L_t and asset-proceeds A_t at time t years. We assume that the current force of interest is δ_0, and that the present value at this force of interest of the net liabilities equals that of the asset-proceeds, i.e.,

$$V_A(\delta_0) = V_L(\delta_0) \tag{9.7.1}$$

It is further assumed that the asset-proceeds may be altered without dealing costs, and that the following additional conditions are satisfied:

$$V'_A(\delta_0) = V'_L(\delta_0) \tag{9.7.2}$$

and

$$V''_A(\delta_0) > V''_L(\delta_0) \tag{9.7.3}$$

The notations $V'(\delta)$ and $V''(\delta)$ are used to denote the first and second derivatives with respect to δ of $V_A(\delta)$. We define $V'_L(\delta)$ and $V''_L(\delta)$ similarly.

If these conditions are satisfied, it follows by elementary calculus that the function

$$f(\delta) = V_A(\delta) - V_L(\delta)$$

equals zero when $\delta = \delta_0$ and has a minimum turning point there. Therefore, there is a neighbourhood of δ_0 such that, if δ lies within but is not equal to δ_0, then

$$V_A(\delta) > V_L(\delta) \tag{9.7.4}$$

The position is illustrated by Figure 9.7.1.

An investor whose investments are such that conditions 9.7.1, 9.7.2, and 9.7.3 hold is then said to be *immunized against small changes in the rate of interest*, as it is clear that any immediate small change in the rate of interest will lead to a surplus, in the sense that the present value of the assets will exceed that of the net liabilities. We now discuss how conditions 9.7.1, 9.7.2, and 9.7.3 may be interpreted in practice.

Since

$$L_t = S_t - P_t$$

Equations 9.7.1 and 9.7.2 are equivalent to

$$\sum (P_t + A_t)v^t = \sum S_t v^t \qquad \text{at force of interest } \delta_0 \tag{9.7.5}$$

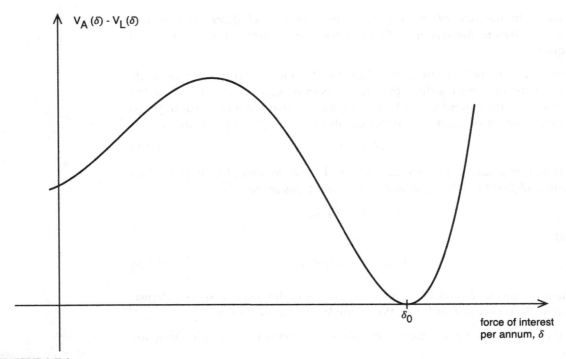

FIGURE 9.7.1
Immunization against small changes in rate of interest

and

$$\sum t(P_t + A_t)v^t = \sum tS_t v^t \qquad\qquad \text{at force of interest } \delta_0 \qquad (9.7.6)$$

These two equations imply that

$$\frac{\sum t(P_t + A_t)v^t}{\sum (P_t + A_t)v^t} = \frac{\sum tS_t v^t}{\sum S_t v^t} \qquad\qquad \text{at force of interest } \delta_0 \qquad (9.7.7)$$

Therefore, at force of interest δ_0, the discounted mean terms of the *total assets* (i.e., the receipts, P_t, plus the asset-proceeds, A_t) and of the *liabilities*, S_t, are equal. Let their common value be denoted by $T(\delta_0)$.

Note that condition 9.7.3 may be expressed as

$$\sum t^2 A_t v^t > \sum t^2 L_t v^t \qquad\qquad \text{at force of interest } \delta_0$$

i.e.,

$$\sum t^2 (A_t + P_t)v^t > \sum t^2 S_t v^t \qquad\qquad \text{at force of interest } \delta_0$$

Combined with Eqs 9.7.5 and 9.7.6, the last inequality is equivalent to the condition

$$\sum [t - T(\delta_0)]^2 (A_t + P_t) v^t > \sum [t - T(\delta_0)]^2 S_t v^t \quad \text{at force of interest } \delta_0$$

$$(9.7.8)$$

Hence, the 'spread' of the total assets about their discounted mean term must, if the conditions for immunization are satisfied, exceed that of the liabilities; that is to say, the spread of the receipts and asset-proceeds about the discounted mean term must exceed that of the liabilities. Since the net liabilities $\{L_t\}$ are predetermined, the asset-proceeds $\{A_t\}$ must be chosen in such a way as to satisfy conditions 9.7.5, 9.7.7, and 9.7.8. An equivalent formulation of these conditions is easily shown to be

$$\sum A_t v^t = \sum L_t v^t \qquad \text{at force of interest } \delta_0 \qquad (9.7.9)$$

$$\frac{\sum t A_t v^t}{\sum A_t v^t} = \frac{\sum t L_t v^t}{\sum L_t v^t} \qquad \text{at force of interest } \delta_0 \qquad (9.7.10)$$

$$\sum t^2 A_t v^t > \sum t^2 L_t v^t \qquad \text{at force of interest } \delta_0 \qquad (9.7.11)$$

The preceding arguments given here may easily be generalized to include continuous cash flows. In this case a more general statement of the conditions of Redington immunization is as follows:

$$V_A(\delta_0) = V_L(\delta_0) \qquad (9.7.12)$$

$$\left[-\frac{1}{V_A} \frac{dV_A}{d\delta} = -\frac{1}{V_L} \frac{dV_L}{d\delta} \right]_{\delta = \delta_0} \qquad (9.7.13)$$

$$\left[\frac{1}{V_A} \frac{d^2 V_A}{d\delta^2} > \frac{1}{V_L} \frac{d^2 V_L}{d\delta^2} \right]_{\delta = \delta_0} \qquad (9.7.14)$$

If we define the *convexity* of the cash flow by

$$\text{convexity} = \frac{1}{V} \frac{d^2 V}{d\delta^2}$$

it is clear that Redington's conditions for immunization can be written as

1. The present value of the assets at the starting rate of interest is equal to the present value of the liabilities;
2. The volatilities (or duration) of the asset and liability cash flows are equal; and
3. The convexity of the asset cash flow is greater than the convexity of the liability cash flow.

As an illustration of how the conditions for an immunized position may be satisfied, let us consider the capital redemption policy specified in Example 9.6.1. In order that Eqs 9.7.9 and 9.7.10 should hold, the value of the assets, at 5% interest, must be equal to the value of the *net* liabilities, i.e., £4,091.94 (see Example 9.6.1), and the discounted mean term of the assets, at 5% interest, must be equal to

$$
\frac{\sum t L_t v^t}{\sum L_t v^t} \qquad \text{at 5\% interest}
$$

$$
= \frac{10 \times 10{,}000 v^{10} - 288.02 (Ia)_{\overline{5}|}}{10{,}000 v^{10} - 288.02 a_{\overline{5}|}} \qquad \text{at 5\% interest} \qquad (9.7.15)
$$

$$
= 12.66 \text{ years.}
$$

Condition 9.7.11 must also be satisfied as a condition of immunization. Since Eqs 9.7.9 and 9.7.10 are satisfied, this last condition will be met if inequality 9.7.8 holds. This inequality states that the spread of the total assets (i.e., assets and receipts) about the common discounted mean term, $T(\delta_0)$, should be greater than the spread of the liabilities about $T(\delta_0)$. Since, in this case, all the liabilities are due at time 10 years, we have $T(\delta_0) = 10$, and the spread of the liabilities about this time is zero. It follows that *any* set of assets of present value £4,091.94 and discounted mean term 12.66 years will be such that the office will be immunized against any immediate small change in the rate of interest. If a large variety of secure fixed-interest stocks exist, an appropriate portfolio of assets may be chosen in one of many ways. For example, if the life office is able and willing to invest in a stock redeemable at par and bearing interest at 5% per annum payable continuously, the term n years should be such that

$$
12.66 = \text{the volatility of the stock}
$$

$$
= \frac{0.05 (I\bar{a})_{\overline{n}|} + n v^n}{0 \cdot 05 \bar{a}_{\overline{n}|} + v^n} \qquad \text{at an interest rate of 5\% effective}
$$

from which it follows that

$$
1.05^n - 0.003\ 088 n - 2.576\ 541 = 0
$$

Hence, $n \approx 19$ years.

Similar calculations may be performed for other coupon rates, and if a fixed term (of at least 12.66 years) is required, the appropriate coupon rate may be calculated.

EXAMPLE 9.7.1

A company has a liability of £400,000 due in 10 years' time. The company has exactly enough funds to cover the liability on the basis of an effective interest rate of 8% per annum. This is also the interest rate on which current market prices are calculated and the interest rate earned on cash. The company wishes to hold 10% of its funds in cash, and the balance in a zero-coupon bond redeemable at par in 12 years' time and a fixed-interest stock, which is redeemable at 110% in 16 years' time bearing interest at 8% per annum payable annually in arrears.

Calculate the nominal amounts of the zero-coupon bond and the fixed-interest stock that should be purchased to satisfy Redington's first two conditions for immunization. Calculate the amount that should be invested in each of the assets. Can Redington immunization be achieved?

Solution

The two conditions required are those stated in Eqs 9.7.1 and 9.7.2. We let X be the nominal amount of 12-year zero-coupon bond and Y the nominal amount of 8% bond.

If we work in units of £1,000, the present value of the liabilities is

$$V_L = 400v^{10} = 185.27740 \text{ at } i = 8\%$$

The present value of the assets is from the 10% of 185.2774 in cash, the zero-coupon bond, and the 8% bond; this is calculated as

$$V_A = 18.52774 + Xv^{12} + Y\left(0.08\, a_{\overline{16}|} + 1.1v^{16}\right)$$

Equating these present values, as required under Redington's first condition, leads to

$$166.74966 = 0.39711X + 1.02919Y$$

The second condition requires the first derivative of the present values with respect to δ. For the liabilities

$$V_L' = 4,000v^{10} = 1,852.7740$$

and for the assets,

$$V_A' = 12Xv^{12} + Y\left(0.08(Ia)_{\overline{16}|} + 1.1 \times 16v^{16}\right)$$

Equating these leads to

$$1,852.7740 = 4.76537X + 10.02650Y$$

Redington's conditions have led to a pair of simultaneous equations with two unknowns. These can be solved to find that $X = 254,583$ and $Y = 63,790$; i.e., the nominal amounts of the zero-coupon bond and the 8% bond are £254,583,000 and £63,790,000, respectively.

These nominal amounts are equivalent to investing $254,583,000v^{12} = £101,098,000$ and $185,277,000 - 18,528,000 - 101,098,000 = £65,651,000$ in the zero-coupon bond and 8% bond, respectively, at time zero.

For Redington immunization to occur, the third condition (given by Eq. 9.7.3) must also be satisfied. In this case it is clear that the spread of assets is greater than the spread of liabilities, as the liabilities are at a single point in time. All conditions for Redington immunization have therefore been satisfied.

As time goes by, the portfolio of investments must be varied—in theory continuously, but in practice at discrete intervals—to achieve an immunized position. If the annual force of interest used to value assets and net liabilities may change only at certain defined dates, for example times $t_1, t_2, \ldots$ years (as in Figure 9.7.2), then the portfolio of assets should be suitably chosen *just before* these times. Since no change in the force of interest can occur between these dates, no profit or loss may be made between the dates, and if the force of interest changes slightly at times $t_1, t_2, \ldots$, then a small profit will be made at each of these times.

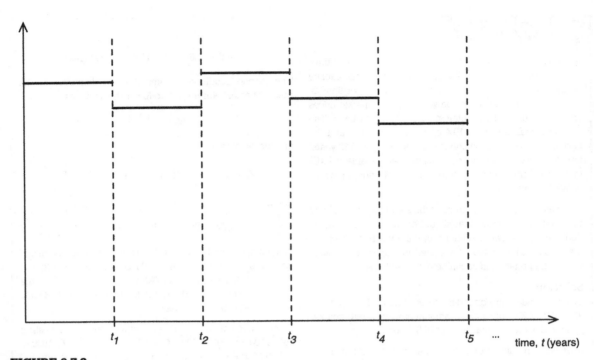

FIGURE 9.7.2
Force of interest per annum used at time t to value assets and liabilities

It is also true that, even if the preceding conditions hold, a loss may occur in the event of a *large* change in the rate of interest: Redington's theory covers only small changes. There is, however, another theory of immunization—which, to avoid confusion with the theory presented by Redington, we refer to as 'full' immunization—under which the investor makes a profit on *any* immediate changes in the rate of interest. We shall discuss this concept of 'full' immunization in the next section.

9.8 FULL IMMUNIZATION

Consider the investor to have a liability-outgo of S due at time t_1 years and to hold total assets (in the sense defined in Section 9.7) providing A at time $t_1 - a$ and B at time $t_1 + b$. This is illustrated in Figure 9.8.1. It is assumed that a and b are positive, and, for practical use, it is also necessary to assume that $a \leq t_1$. Suppose that exactly two of four values A, B, a, and b are known. It may be shown that in many cases it is possible to determine the other two values uniquely in such a way that the two equations

$$Ae^{\delta_0 a} + Be^{-\delta_0 b} = S \qquad (9.8.1)$$

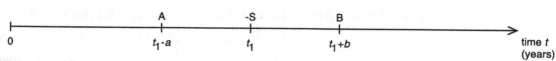

FIGURE 9.8.1

Cash flow diagram for full immunization

and

$$Aae^{\delta_0 a} = Bbe^{-\delta_0 b} \qquad (9.8.2)$$

are satisfied. There are exceptional cases when no solution exists, but if the known quantities are (a) a, b; (b) B, b; (c) A, a; or (d) A, b, then unique values of the other two quantities may certainly be found. See Question 9.15.

Let $V(\delta)$ denote the present value at time $t = 0$, at force of interest δ per annum, of the total assets less the present value of the liability, i.e., let

$$V(\delta) = e^{-\delta t_1}\left(Ae^{\delta a} + Be^{-\delta b} - S\right) \qquad (9.8.3)$$

It may easily be shown that conditions 9.8.1 and 9.8.2 are equivalent to $V(\delta_0) = 0$ and $V'(\delta_0) = 0$. Assume that these conditions are satisfied. We then have

$$
\begin{aligned}
V(\delta) &= e^{-\delta t_1}\left[Ae^{\delta_0 a}e^{(\delta-\delta_0)a} + Be^{-\delta_0 b}e^{-(\delta-\delta_0)b} - S\right] \\
&= e^{-\delta t_1}Ae^{\delta_0 a}\left[e^{(\delta-\delta_0)a} + \frac{a}{b}\,e^{-(\delta-\delta_0)b} - \left(1 + \frac{a}{b}\right)\right]
\end{aligned} \qquad (9.8.4)
$$

using Eq. 9.8.2 and the fact that, by Eqs 9.8.1 and 9.8.2,

$$Ae^{\delta_0 a}\left(1 + \frac{a}{b}\right) = Ae^{\delta_0 a}\left(1 + \frac{Be^{-\delta_0 b}}{Ae^{\delta_0 a}}\right) = S$$

Now consider the sign of the function

$$f(x) = e^{ax} + \frac{a}{b}\,e^{-bx} - \left(1 + \frac{a}{b}\right)$$

Note that $f(0) = 0$. Since

$$f'(x) = a\left(e^{ax} - e^{-bx}\right)$$

it follows that

$$f'(x)\begin{cases} > 0 & \text{for } x > 0 \\ = 0 & \text{for } x = 0 \\ < 0 & \text{for } x < 0 \end{cases}$$

and therefore that $f(x) > 0$ for all $x \neq 0$. Hence, by Eq. 9.8.4 (with $\delta - \delta_0 = x$),

$$V(\delta) > 0 \qquad \text{for } all \ \delta \neq \delta_0 \qquad (9.8.5)$$

In order that the theory of 'full' immunization may be applied in a more general context, we need to split up the liabilities in such a way that each item of liability-outgo is linked to two items of the total assets, the earlier of which may be a 'receipt' (an item of positive cash flow arising from the business itself) due *before* the item of liability-outgo. It then follows by the argument given previously that the asset associated with this item of liability-outgo (or the later asset, if there are two) may be varied in amount and date so that Eqs 9.8.1 and 9.8.2 hold. If this procedure is followed for *all* items of liability-outgo, the entire business will be immunized against *any* immediate change in the rate of interest.

EXAMPLE 9.8.1

Consider the 20-year capital redemption policy described in Example 9.6.1, and let the current rate of interest per annum be 5%. Show that the life office can, by suitable investment, achieve a fully immunized position (in which it will make a profit on any immediate change in the rate of interest).

Solution

As there is only one liability, and nine 'receipts' (i.e., premiums), we shall split the liability into nine equal sections, each of amount $10,000/9 = 1,111.11$. Consider the item of liability-outgo linked to the premium due at time r years.

We have the equations

$$288.02(1.05)^{10-r} + B(1.05)^{-b} = 1,111.11$$

and

$$288.02(10 - r)(1.05)^{10-r} = Bb(1.05)^{-b}$$

where B denotes the proceeds of a zero-coupon bond due at time $10 + b$ years from the present. By direct solution, we obtain

$$b = \frac{\left(\dfrac{288.02}{1,111.11}\right)(10 - r)(1.05)^{10-r}}{1 - \left(\dfrac{288.02}{1,111.11}\right)(1.05)^{10-r}}$$

and

$$B = \frac{1,111.11 - 288.02(1.05)^{10-r}}{(1.05)^{-b}}$$

$$= \quad \text{the proceeds of a zero-coupon}$$
$$\text{bond due at time } 10 + b \text{ years}$$

The investments may then consist of nine zero-coupon bonds of the following amounts maturing at the times shown:

r	Amount of Zero-coupon Bond, B (£)	Time to Maturity, $10 + b$ (years)
1	892.55	16.05
2	873.52	14.97
3	858.75	14.02
4	847.40	13.19
5	838.82	12.47
6	832.50	11.84
7	828.06	11.29
8	825.17	10.80
9	823.58	10.37

This portfolio of zero-coupon bonds is such that the office is fully immunized against any immediate change in the rate of interest. Note that we do not claim that this is the only such portfolio; many others can be constructed.

EXAMPLE 9.8.2

An investor has a single liability of £1,000 due in 15 years' time. The yield on zero-coupon stocks of any term is currently 4% per annum, and the investor possesses cash equal to the present value of his liability, i.e., $1,000v^{15}$ at 4% = £555.26. He wishes to invest in 10-year and 20-year zero-coupon stocks in such a way that he will make a profit on any immediate change in the force of interest. How much of each security should he buy, and how large a profit will he make if the rate of interest per annum immediately becomes 0.01, 0.02, 0.03, 0.05, 0.06, 0.07, or 0.08?

Solution

Let A and B be the redemption values of the 10-year and 20-year zero-coupon stock holdings, respectively. At time $t = 15$, Eqs 9.8.1 and 9.8.2 give

$$Ae^{5\delta_0} + Be^{-5\delta_0} = 1,000$$

and

$$5Ae^{5\delta_0} = 5Be^{-5\delta_0}$$

where

$$\delta_0 = \ln(1.04)$$

We therefore have the equations

$$A \times (1.04)^5 + Bv^5 = 1,000 \quad \text{at 4% interest}$$
$$A \times (1.04)^5 = Bv^5 \quad \text{at 4% interest}$$

These equations may easily be solved, giving $A = 410.96$ and $B = 608.32$. The quantities of zero-coupon stocks that he should buy are therefore those providing £410.96 at time 10 years and £608.32 at time 20 years. The profit on an immediate change in the rate of interest to i per annum is, by Eq. 9.8.3,

$$v^{15}\left[410.96(1+i)^5 + 608.32v^5 - 1,000\right] \quad \text{at rate } i$$

In the following table we give the present values of the liability $(1,000v^{15})$, the assets $(410.96v^{10} + 608.32v^{20})$, and the profit to the investor for each of the specified rates of interest:

Rate of Interest per Annum	Present Value of Liability	Present Value of Assets	Present Value of Profit
0.00	1,000.00	1,019.28	19.28
0.01	861.35	870.58	9.23
0.02	743.01	746.51	3.50
0.03	641.86	642.61	0.75
0.04	555.26	555.26	0.00
0.05	481.02	481.56	0.54
0.06	417.27	419.16	1.89
0.07	362.45	366.11	3.66
0.08	315.24	320.87	5.63

As in the case of Redington's theory of immunization, one must modify the assets as time goes by so as to achieve an immunized position. If the force of interest may change only at certain specified times, as in Figure 9.7.2, then suitable assets may be chosen just before these times. Various practical difficulties attend the application of this theory of immunization, notably the absence in the market of appropriate zero-coupon securities and possible variations in the force of interest with the term to redemption and the coupon rate.

SUMMARY

- The *n-year spot rate*, y_n, is the annual interest rate that applies from the present time for the next n years. This is in contrast to the *forward rate*, $f_{t,r}$, which applies over an r-year period starting at some future time t. Equivalent, but continuously compounding, rates are easily defined.
- The *n-year par yield* is the annual coupon rate required for an n-year fixed-interest security to be redeemed at par under the prevailing term structure.
- The **term structure of interest rates** indicates how the spot rates available on investments are distributed with term. The structure can be explained by the interplay of **liquidity preference theory**, **expectations theory**, and **market segmentation theory**.
- The *discounted mean term* (or [*Macauley*] *duration*) of a fixed-interest investment is the weighted average of the future times at which there are cash flows. The 'weight' associated with any given time is the present value of the net cash flow due at that time.
- The discounted mean term of a cash flow is mathematically equal to its *volatility*. This indicates the sensitivity of its present value to changes in the underlying force of interest.
- The *effective duration* indicates the sensitivity of the present value of a cash flow to changes in the underlying rate of interest.
- It is possible to *immunize* a surplus to *small* changes in interest rate using the theory developed by Redington.
- It is possible to immunize a surplus against *any* change in interest rate using *full immunization*.

EXERCISES

 **9.1** Consider the following bond issues:
Bond A: 5% 15-year bond
Bond B: 5% 30-year bond
Neither bond has an embedded option. Both bonds are trading in the market at the same yield. Which bond will fluctuate more in price when interest rates change? Why?

 **9.2** A portfolio manager wants to estimate the interest rate risk of a bond using duration. The current price of the bond is 82. A valuation model found that if interest rates decline by 30 basis points, the price will increase to 83.50, and if interest rates increase by 30 basis points, the price will decline to 80.75. *Estimate* the effective duration of this bond.

9.3 According to the expectations theory, what does a humped yield curve suggest about the expectations of future interest rates?

9.4 Assume the following Treasury spot rates:

Period	Years to Maturity	Spot Rate (%)
1	0.5	5.0
2	1.0	5.4
3	1.5	5.8
4	2.0	6.4
5	2.5	7.0
6	3.0	7.2
7	3.5	7.4
8	4.0	7.8

Compute
  (a) The 6-month forward rate 6 months from now,
  (b) The 6-month forward rate 1 year from now,
  (c) The 6-month forward rate 3 years from now,
  (d) The 2-year forward rate 1 year from now,
  (e) The 1-year forward rate 2 years from now.

9.5 The 1-year forward rate of interest at time $t = 1$ year is 5% per annum effective. The gross redemption yield of a 2-year fixed-interest stock issued at time $t = 0$, which pays coupons of 3% per annum annually in arrears and is redeemed at 102, is 5.5% per annum effective.
The issue price at time $t = 0$ of a 3-year fixed-interest stock bearing coupons of 10% per annum payable annually in arrears and redeemed at par is £108.9 per £100 nominal.
  (a) Calculate the 1-year spot rate per annum effective at time $t = 0$.
  (b) Calculate the 1-year forward rate per annum effective at time $t = 2$ years.
  (c) Calculate the 2-year par yield at time $t = 0$.

9.6 **(a)** Explain what is meant by the expectations theory for the shape of the yield curve.
  (b) Short-term, 1-year annual effective interest rates are currently 8%; they are expected to be 7% in 1 year's time, 6% in 2 years' time, and 5% in 3 years' time.
    (i) Calculate the gross redemption yields (spot rates of interest) from 1-year, 2-year, 3-year, and 4-year zero-coupon bonds assuming the expectations theory explanation of the yield curve holds.

(ii) The price of a coupon-paying bond is calculated by discounting individual payments from the bond at the zero-coupon bond yields in (i). Calculate the gross redemption yield of a bond that is redeemed at par in exactly 4 years and pays a coupon of 5% per annum annually in arrears.

9.7 (a) State the conditions that are necessary for an insurance company to be immunized from small, uniform changes in the rate of interest. An insurance company has liabilities to pay £100m annually in arrears for the next 40 years. In order to meet these liabilities, the insurance company can invest in zero-coupon bonds with terms to redemption of 5 years and 40 years.

(b) (i) Calculate the present value of the liabilities at a rate of interest of 4% per annum effective.

(ii) Calculate the duration of the liabilities at a rate of interest of 4% per annum effective.

(c) Calculate the nominal amount of each bond that the fund needs to hold so that the first two conditions for immunization are met at a rate of interest of 4% per annum effective.

(d) (i) Estimate, using your calculations in (b), the revised present value of the liabilities if there were a reduction in interest rates by 1.5% per annum effective.

(ii) Calculate the present value of the liabilities at a rate of interest of 2.5% per annum effective.

(iii) Comment on your results to (d) (i) and (ii).

9.8 (a) Calculate, at an effective annual interest rate of (i) 5% and (ii) 15%, the discounted mean term of the following set of payments:

£100 payable now;
£230 payable in 5 years' time;
£600 payable in 13 years' time.

(b) An investor is considering the purchase of an annuity, payable annually in arrears for 20 years. The first payment is £1,000. Using a rate of interest of 8% per annum, calculate the discounted mean term of the annuity when

(i) The payments remain level over the term;

(ii) The payments increase by £100 each year;

(iii) The payments increase at a rate of 8% per annum compound;

(iv) The payments increase at a rate of 10% per annum compound.

9.9 (a) At a given rate of interest, a series of payments has present value V_1 and discounted mean term t_1. At the same rate of interest, a second series of payments has present value V_2 and discounted mean term

t_2. Show that, if $V_1 + V_2 \neq 0$, at the given rate of interest, the discounted mean term of both series combined is

$$\frac{V_1 t_1 + V_2 t_2}{V_1 + V_2}$$

Generalize this result to the case of n different series of payments.

(b) An investor is entitled to receive an annuity of £10,000 per annum, payable annually in arrears for 10 years. The first payment will be made in 1 year's time. The investor is under an obligation to make two payments of £30,000, the first being due in 5 years' time and the second 10 years later.

On the basis of an interest rate of 10% per annum:

(i) Find the present value and the discounted mean term of (1) the payments to be received by the investor and (2) the payments to be made by the investor.

(ii) Using the result of (a), or otherwise, determine the discounted mean term of the investor's net cash flow.

9.10 A fixed-interest stock bears interest of 5% per annum payable continuously and is redeemable at par in n years' time, where n is not necessarily an integer.

(a) On the basis of a constant force of interest per annum of 0.07, determine the volatility of the stock if (*i*) $n = 20$ and (ii) $n = 60$.

(b) The volatility of the stock, on the basis of a specified constant force of interest per annum δ, may be considered to be a function of n. When $\delta = 0.07$, for what value of n is this volatility greatest, and what is its maximum value?

9.11 **(a)** A company is liable to make four payments at 5-yearly intervals, the first payment being due 5 years from now. The amount of the tth payment is £$(1,000 + 100t)$. The company values these liabilities at an effective rate of interest of 5% per annum. On this basis, find (i) the present value and (ii) the discounted mean term of these liabilities.

(b) An amount equal to the total value of the liabilities (on the basis of an effective annual interest rate of 5%) is immediately invested in two newly issued loans, one redeemable at the end of 10 years and the other at the end of 30 years. Each loan bears interest at 5% per annum payable annually in arrears. Both loans are issued and redeemable at par. Given that, on the basis of an effective annual interest rate of 5%, the discounted mean term of the asset-proceeds is the same as the discounted mean term of the liability-outgo, determine how much is invested in each of the loans.

9.12 In a certain country, investors may trade in zero-coupon stocks of any term. There are no dealing expenses, and all zero-coupon stocks are redeemable at par, exempt from taxation, and priced on the basis of the same constant force of interest.

An investor, who owes £1 million, due in 10 years' time, possesses exactly sufficient cash to meet this debt on the basis of the current constant market force of interest per annum δ_0. Show that, by purchasing suitable quantities of 5-year and 15-year zero-coupon bonds, the investor can be sure of making a profit on any immediate change in the constant force of interest used by the market to value zero-coupon stocks. Assuming further that $\delta_0 = 0.05$, and that the appropriate quantities of each bond are purchased, find the profit the investor will make now if this force of interest per annum changes immediately to (a) 0.07 and (b) 0.03.

9.13 An insurance company has a liability of £100,000 due in 8 years' time. The company, which has exactly sufficient money to cover the liability on the basis of a constant force of interest of 5% per annum (which is also the market basis for pricing all stocks), now wishes to invest this money in a mixture of the following securities:

a. Zero-coupon bonds redeemable at par in 20 years' time;

b. Very short-term deposits (which may be regarded as interest-bearing cash).

(i) The company requires that, on the basis of a constant force of interest of 5% per annum, the discounted mean term of the assets equal that of the liabilities. Find the amounts to be invested in each security.

(ii) Suppose that these investments are, in fact, made. Find the present value of the profit to the organization on the basis of a constant force of interest per annum of (i) 3% and (ii) 7%.

9.14 An insurance company has just issued a 15-year single premium capital redemption policy with sum assured £10,000. The amount of the single premium was calculated on the assumption that, for all values of $t \geq 0$, $\delta(t)$, the force of interest per annum in t years' time, would equal $\ln(1.08)$. Expenses were ignored.

The company immediately invested part of the single premium in cash and the balance in a 20-year zero-coupon stock (redeemable at par), the price of which was consistent with the constant value of $\delta(t)$ assumed above. The amount invested in cash was such that, on the basis of this constant value of $\delta(t)$, the discounted mean term of the assets equaled the discounted mean term of the liability.

(a) Calculate the single premium and the nominal amount of stock purchased.

(b) Immediately after the investment of the single premium, economic forces caused a change in market rates of interest. Calculate the present value of the profit or loss to the company on the assumption that future cash flows are discounted at the new market rates of interest for which

   (i) $\delta(t) = \ln(1.05)$ for all t;
   (ii) $\delta(t) = \ln(1.1)$ for all t;
   (iii) $\delta(t)$ is a linear function of t, with $\delta(0) = \ln(1.05)$ and $\delta(20) = \ln(1.1)$.

9.15 Suppose that δ_0 and S are known positive numbers. Suppose further that *two* of the four *positive* numbers a, A, b, and B are given. Consider the system of simultaneous equations

$$\left. \begin{array}{l} Ae^{\delta_0 a} + Be^{-\delta_0 b} = S \\ Aae^{\delta_0 a} = Bbe^{-\delta_0 b} \end{array} \right\}$$

for the two 'unknown' numbers from $\{a, A, b, B\}$.

(a) Show that if the given pair of values is either (i) a, b; (ii) B, b (with $Be^{-\delta_0 b} < S$); (iii) A, a (with $Ae^{\delta_0 a} < S$); or (iv) A, b (with $A < S$), then there exists a unique pair of *positive* values for the two 'unknown' numbers which satisfies the system of equations.

(b) Let $\delta_0 = 0.05$ and $S = 1$. Show that, if $a = 15$ and $B = 0.98$, then there are *two* distinct pairs of positive numbers (b, A) which satisfy the above equations.

CHAPTER 10

An Introduction to Derivative Pricing: Forwards and Futures

Derivatives are a class of financial instruments distinct from stocks (e.g., equities and fixed-interest securities) and commodities (oil, metals, food, etc.). There are many examples of derivatives, but the simplest examples can be considered as an agreement to trade an underlying asset (either stock or commodity) at a future date for a pre-agreed price. In a sense, a derivative sits above the underlying asset and *derives* its price from it. It is clear that derivative pricing is complicated, as a derivative's price depends not only on the terms of the particular contract, but also on the price evolution of the underlying asset during the term of the contract. Derivative pricing is a huge subject in the literature, and the vast majority of it is beyond the scope of this introductory text. However, it is possible to introduce the fundamental concepts common to many pricing techniques by considering the simplest derivatives: *forwards* and *futures*. We then proceed to consider *swaps* and *options*, which are more complicated examples of derivatives, in the next chapter.

In what follows we will assume that there are no transaction costs or taxes associated in buying, selling, or holding assets. Furthermore, we will simplify the description by assuming that the risk-free rate of interest is independent of the duration of investment. These are, of course, idealized assumptions, but they enable the development of methodologies that can easily be extended. In particular, the assumption regarding the term structure of the risk-free rates can be removed by replacing the risk-free rates with the appropriate risk-free spot or forward rates, as defined in Chapter 9. Rather than discussing this complication explicitly, we present the theory without reference to the term structure of interest rates but give some examples where it has been incorporated. Where appropriate, some of the descriptions given in this chapter are based on the *Faculty and Institute of Actuaries'* CT1 core reading.

CONTENTS

10.1 FUTURES CONTRACTS

A *futures contract*, often simply referred to as a *future*, is a standardized, exchange tradable contract between two parties to trade a specified asset on a set date in the future at a specified price. For example, the London Metal Exchange (LME) provides a market for the active trading of aluminum futures

An Introduction to the Mathematics of Finance. http://dx.doi.org/10.1016/B978-0-08-098240-3.00010-2

on standardized 25-tonne lots with maturity (i.e., trade) dates of 3, 15, or 27 months from the time of issue. Futures require two parties: the *buyer* and the *seller*. Under the futures contract, the buyer is committed to purchasing the underlying asset for the pre-arranged price, and the seller is responsible for delivering the asset in return for this price. The buyer is said to hold the *long forward position*, reflecting that his exposure to the asset is in the long term. The seller is said to hold the *short forward position*, reflecting that his exposure to the asset is restricted to the short term.

There are two main classes of futures: *commodity futures* and *financial futures*. The first futures contracts were traded on the Chicago Board of Trade (CBOT) in 1864 and were based on grain. These early *commodity futures* enabled consumers of grain to reduce their exposure to fluctuations in the market price and started a trend in futures contracts that spread to the trade of commodity futures and then financial futures on exchanges around the word. Examples of commodities on which futures are traded today include, but are not restricted to, oil, gold, silver, wheat, and coffee.

Rather than being based on an underlying physical commodity, *financial futures* are based on an underlying financial instrument. Financial futures exist in four main categories, which we introduce now.

Bond Futures

Bond futures are based on fixed-interest securities—for example, gilts and bonds—and require the physical delivery of a security at the maturity date. However, rather than always being based on a particular named asset, many bond futures allow the seller to deliver one of a range of bonds or gilts.

In the case that the bond future is specified in terms of a *particular* security, the seller (short position) is required to simply deliver the required amount of that stock. However, if the contract is specified in terms of a *notional stock*, the

---

EXAMPLE 10.1.1

A bond future is written on £100 of a notional 20-year fixed-interest security with 5% annual coupon. The exchange lists that the following securities are eligible for delivery under the contract:

- £115 of a 19-year bond with 3% annual coupon per £100 of the notional bond;
- £90 of a 25-year bond with 7% coupons per £100 of the notional bond.

If, at the expiry date of the future, the market prices of the 19-year stock and 25-year stock are £80% and £103%, respectively, determine which security the seller will deliver.

Solution
The seller will deliver the cheaper of the two alternatives. The market price of £115 of the 19-year stock is £115 × 0.80 = £92, and the market price of £90 of the 25-year stock is £90 × 1.03 = £92.7. The seller should therefore opt to deliver £90 of the 19-year stock.

securities eligible for delivery are listed by the exchange. The seller will then deliver the security from the list that is cheapest to deliver. The price paid by the receiving party is adjusted to allow for the fact that the coupon rate may not be equal to that of the notional stock that underlies the contract.

Short Interest Rate Futures

Short interest rate futures are based on a particular interest rate and settled for cash. The structure of the quotation is such that as interest rates fall, the price rises, and vice versa. The price of a short interest rate future is stated as 100 minus the 3-month interest rate. For example, with an interest rate of 6.25%, the future is priced as 93.75. The contract is based on the interest paid on a notional deposit for a specified period from the expiry of the future. However, no principal or interest changes hands. On expiry, the purchaser will have made a profit (or loss) related to the difference between the final settlement price and the original dealing price. The party delivering the contract will have made a corresponding loss (or profit).

Stock Index Futures

Stock index futures are based on a notional portfolio of equities as represented by a particular index, for example the FTSE100 in the UK or S&P 500 in the US. In principle, the seller of the contract should deliver a portfolio of shares in the same proportions as the particular index. However, this is impractical, and a stock index future is typically settled in cash. The cash amount is calculated from the difference between the futures price agreed at the start of the contract and the index value at the date of the contract.

EXAMPLE 10.1.2

On 16 May 2012, an investor enters into the short position of $1,000 of 3-month S&P 500 stock futures with settlement price $1,410. If, on 16 August 2012, the value of the S&P 500 index is actually $1,415.84, calculate the profit/loss made by the investor from the future.

Solution
Under the contract, the investor is committed to provide $1,000 × 1,415.84 in return for $1,000 × 1,410. The investor therefore makes a loss of $5,840.

Currency Futures

Currency futures require the delivery of a set amount of a given currency on the specified date in return for an amount of another currency. Common currencies on which currency futures are based include, but are not limited to, sterling, euros, yen, US dollar, and Swiss francs.

10.2 MARGINS AND CLEARINGHOUSES

In principle, both parties to a future (both commodity or financial) face two kinds of risk: *market risk* and *credit risk.* The *market risk* faced by an investor is the risk that market conditions of the underlying asset change in an adverse way and the terms of the future become unfavorable. The *credit risk* faced by an investor is the risk that the counterparty will default on his obligations under the contract. The credit risk is linked to the market risk, as the party adversely affected by the market movements is more likely to default.

EXAMPLE 10.2.1

Two investors enter into a futures contract to trade 1,000 barrels of oil in 3 months' time. If the agreed delivery price is $100 per barrel and the current market price is $98, discuss possible price movements and how they would affect the risk exposures of each investor.

Solution

If, after 2 months, the price of oil had risen to $102 per barrel, the market has moved in favor of the long position (buyer). The reason is that he is committed to buy each barrel for $100 in 1 month, which, at the moment, is cheap relative to the market. However, this is bad news for the short position (seller), as he is committed to selling the oil for $100. The

seller is therefore likely to make a loss at expiry in 1 month and may be tempted to default on the contract. This increases the buyer's exposure to credit risk.

Alternatively, if, after 2 months, the price of oil had fallen to $95 per barrel, the market has moved in favor of the short position (seller). The buyer is therefore likely to make a loss on the contract at expiry, and the seller is exposed to the increased credit risk of his defaulting.

As the market could still move in either direction within the remaining month, *both* parties are still exposed to a market risk in each example.

To mitigate investors' exposure to credit risk, a *clearinghouse* sits between both parties of a futures contract and acts as the counterparty to both. The clearinghouse takes on the credit risk of both parties and controls its own exposure through the use of *margins*. Margins are effectively deposits paid by both parties to the clearinghouse during the course of the contract. Futures markets are very liquid, and an investor need not worry about finding a counterparty to his intended position in a contract. In practice, the investor would lodge a request with the exchange to enter into a futures contract (as either a buyer or a seller), and the exchange would pair the request with an opposite position and execute the contract. The clearinghouse (often operated by the exchange) is then placed between the counterparties and, to all intents and purposes, an individual investor sees only his account with the clearinghouse. If, at any time, an investor wishes to remove his exposure to a futures contract, he can request to enter an equal but opposite position and *close out* of the contract, thereby removing all obligation to trade.

When the futures contract is first executed, both parties deposit an *initial margin* to individual accounts held at the clearinghouse. Payments are then made daily from the margin accounts to/from the clearinghouse to reflect changes in the market price of the underlying stock. An amount reflecting the market movement is *taken from* the margin account of the party that the market has moved against, and that amount is *added* to the margin account of the party that the market has moved in favor of. This process of reconciling the changes in the market price of the underlying asset to the margin deposit accounts is called *marking to market*. Eventually, it may be that the market has moved significantly against one investor such that his account falls below a stated level, called the *maintenance margin*. At this point the investor will receive a *margin call* from the clearinghouse which requests that either the investor exits the contract or deposits a further *variation margin* with the clearinghouse to raise his margin account back to the initial margin amount. If the investor decides to leave the contract he is able to withdraw what remains in the margin account, resulting in a loss, and the clearinghouse pairs the remaining position with a new counterparty from the market. The margin account and variation margin are the means by which the clearinghouse ensures that each party has sufficient funds to pay for future adverse market movements, thereby mitigating its own exposure to credit risk.

EXAMPLE 10.2.2

An investor decides to enter into the short position of a 6-month crude oil future at $110 per barrel (each future represents 1,000 barrels). If the clearinghouse demands an initial margin of 10% of the future's value and marks to market at the end of every day, determine the cash flows in the trading account if the investor liquidates his position after 3 days. You are given that the maintenance margin level is $6,000 and the market price of oil proceeds as follows for the first 3 months:

End of Day	Market Price per Barrel
1	$117
2	$112
3	$114

Solution

The investor has entered into the short position, i.e., is obliged to sell each barrel for $110 at the end of the 6 months. An *increase* in the price above $110 is therefore an unfavorable movement.

The initial margin is 10% of $110,000, i.e., $11,000. After 1 day the price has moved unfavorably to $117 per barrel. This represents a loss of $7 per barrel, i.e., $7,000, which is subtracted from the investor's trading account, leaving $4,000. This amount is below the maintenance margin of $7,000, so a margin call is issued requesting that an additional $6,000 be deposited. The account now stands at $11,000 (the initial margin level).

After a further day, the price has fallen $5 per barrel, which is a favorable movement of $5,000 relative to the previous position. The $5,000 is therefore credited to the investor's trading account, taking it to $16,000. No maintenance margin is required.

After a yet further day, the price has moved unfavorably to $114 a barrel, which represents a loss of $2,000. This amount is subtracted from the investor's trading account, which then stands at $14,000. At this point the investor liquidates his position and withdraws the $14,000. He has made a loss of $4,000 over the 3 days.

10.3 USES OF FUTURES

Investors in futures typically have one of two aims: *speculation* or *hedging*. *Speculation* is a process in which an investor aims to benefit from the changes in the market price of the underlying asset, but has no interest in taking delivery of the actual asset. For example, an investor can enter into the short position of a futures contract under which he is committed to *sell* an asset for a price that he believes will be greater than the market price on the maturity date. At maturity, the asset would be bought on the market and sold to the counterparty at a profit. Alternatively, the investor could enter into the long position of a futures contract under which he is committed to *buy* an asset at a price he believes will be lower than will exist on the market at the maturity date. At contract maturity, he would buy the asset under the contract and immediately sell it for a higher price on the market.

Of course, the market can move in the opposite direction to the speculator's expectations and the trade result in a loss; speculation is therefore inherently risky. *Hedging*, however, is a process in which an investor enters into a futures contract in order to *reduce* his exposure to risk.

EXAMPLE 10.3.1

A UK manufacturing company has a major order from a US customer that will result in $500,000 being received in 3 months' time. The company is concerned about its exposure to adverse changes in the GBP–US$ exchange rate over this period and is seeking to fix the sterling value of the order in advance. Discuss how futures contracts could be used to do this.

Solution
To remove the exchange rate risk (i.e., currency market risk), the manufacturer could enter into a futures contract to sell

US$ in exchange for sterling in 3 months' time. This fixes the exchange rate in advance, and the company is no longer exposed to adverse movements in the exchange rate. If the standardized unit of a currency future is $1,000, the company would enter into 500 3-month futures contracts that require the delivery of US$ at expiry for a pre-agreed amount of sterling.

The downside of using hedging strategies involving futures is that, in eliminating the exposure to potential unfavorable market movements, investors also remove their exposure to potential *favorable* market movements. For example, in Example 10.3.1, the terms of the contract are fixed at the outset, and the manufacturer is obligated to sell the US$ at the pre-arranged exchange rate at maturity. However, the foreign exchange market may have moved by this time such that it would actually be favorable to exchange the US$ on the market. As we shall see in Example 11.6.1, *options* enable an alternative method of hedging that removes this downside of futures.

10.4 FORWARDS

Prior to the introduction of grain futures in 1864, the Chicago Board of Trade had provided a market for *non-standardized* grain trades, called *forward contracts*, since 1848. Like a future, a *forward contract*, or *forward*, is a contract between two parties that enter into an agreement to trade a specified asset for a specified price at a later date. Forwards are negotiated agreements and, unlike futures, are *not standardized*. The trading of forwards is said to be *over the counter* and does not occur on an exchange. As a result, there is no clearinghouse to guarantee the trade and both the credit and market risks remain with the investors. To enter into a forward contract, one needs to find a counterparty that wants the opposite position. Furthermore, unlike futures, the lack of standardization and a liquid market in forwards makes closing out of the trade prior to delivery difficult. Forwards are therefore more likely to result in delivery of the underlying asset. Investors in forwards who seek to close out must look for an additional counterparty to reverse their position in the forward prior to expiry; however, this essentially increases their exposure to credit risk, as they are now reliant on two counterparties to fulfill their obligations.

10.5 ARBITRAGE

Before we study pricing within future and forward contracts, it is necessary to first discuss the concept of *arbitrage*.

An *arbitrage opportunity* is said to exist if either

(a) An investor can make a deal that would give him an immediate profit, with no risk of future loss, or

(b) An investor can make a deal that has zero initial cost, no risk of future loss, and a non-zero probability of a future profit.

In essence, either occurrence of arbitrage requires the opportunity for a *risk-free trading profit*.

It is generally assumed that arbitrage opportunities *do not exist* in modern financial markets. This assumption is called the *no-arbitrage assumption*, and is fundamental to modern financial mathematics. It is not claimed that arbitrage cannot exist, merely that when it does, the market will act quickly to remove it. Indeed, *arbitrageurs* are investors who purposefully seek arbitrage opportunities and profit from them. When an arbitrage opportunity does exist within a market, it would soon be spotted by arbitrageurs, and they will trade to exploit it. In principle, an arbitrage opportunity requires the existence of an underpriced asset and an equivalent (in some respects) overpriced asset; the arbitrage profit would be made from buying the cheaper asset while simultaneously selling the more expensive asset. The buying and selling of these respective assets on the market would lead to price movements through the action of supply and demand that would close the pricing discrepancy. In a modern market, the trades would happen at such speed and volume that the arbitrage opportunity would be very quickly removed. It is therefore sensible, realistic, and prudent to assume that arbitrage does not exist.

EXAMPLE 10.5.1

Consider a simple securities market, consisting of just two securities, A and B. At time $t = 0$, the prices of the securities are P_0^A and P_0^B, respectively. The term of both securities is 1 year, and, at this time ($t = 1$), there are two possible outcomes:

(1) The market has risen, in which case the securities pay $P_0^A(u)$ and $P_0^B(u)$; or

(2) the market has fallen, in which case the securities pay $P_0^A(d)$ and $P_0^B(d)$.

Investors can buy securities, in which case they *pay* the price at time $t = 0$ and *receive* the proceeds at time $t = 1$. Or they can sell the securities, in which case they *receive* the price at time 0 and *pay* the proceeds at time $t = 1$. Identify any arbitrage opportunities if the payments are as follows:

Security	P_0	$P_1(u)$	$P_1(d)$
A	£6	£7	£5
B	£11	£14	£10

Solution

At $t = 0$, an investor can buy one unit of security B for price £11 and sell two units of security A for price £12, leading to a net income of £1 at that time.

At $t = 1$,

- In the event of the market rising, the investor will receive £14 and pay out £14;
- In the event of the market falling, the investor will receive £10 and pay out £10.

We see that the net cash flows are zero at time $t = 1$, irrespective of the market movement.

This is therefore an arbitrage opportunity of type (a), as defined previously.

It is clear that investment in security A in Example 10.5.1 is unattractive compared with investment in B, in particular, $P_0^A > P_0^B/2$. As a result, there will be little demand for A and excessive demand for B; this will cause pressure to reduce the price of A and to increase the price of B. Ultimately, there will be a balance such that $P_0^A = P_0^B/2$, and the arbitrage opportunity will be eliminated.

In Example 10.5.2, investors will naturally choose to buy A and will want to sell B. This will put upwards pressure on the price of A and downwards pressure on the price of B. The arbitrage opportunity will be eliminated when the prices have moved such that $P_0^A > P_0^B$.

If we assume that no arbitrage opportunities exist in a market, it follows that any two securities (or combinations of securities) that give exactly the same cash flows must have the *same price*. This is the *law of one price*, and means that we can price a security by creating a portfolio (of known price) that exactly replicates the payments from the security. This technique is often called the *replicating portfolio* strategy.

EXAMPLE 10.5.2

Consider the two securities as described in Example 10.5.1, but now with payments as follows:

Security	P_0	$P_1(u)$	$P_1(d)$
A	£6	£7	£5
B	£6	£7	£4

Identify an arbitrage opportunity.

Solution

At $t=0$ an investor can by one unit of asset A for price £6 and sell one unit of B for price £6, leading to a net cash flow of zero at that time.

At $t=1$,

- In the event of the market rising, the investor will receive £7 and have to pay out £7, i.e., a zero net cash flow;
- In the event of the market falling, the investor will receive £5 and pay out £4, i.e., a net cash flow of +£1.

We therefore see that, for a zero initial outlay, there is a zero risk of future loss and a non-zero probability of a future profit. This is an arbitrage opportunity of type (b), as defined previously.

10.6 CALCULATING THE FORWARD PRICE

Despite forwards and futures having some different properties, they are very similar in essence and can be studied using similar techniques. The main difference arises from the fact that futures require daily marking to market payments to a clearinghouse, and forwards do not. A full treatment of the subtleties required for the accurate pricing of futures is beyond the scope of this book, and we refer only to forwards in what follows. For short-term contracts, however, the differences are negligible and can be ignored.

Recall that a forward contract is an agreement between two parties at time $t=0$ to trade a specified amount S of an asset at a specified future date. We denote the market price of the amount of the underling asset at general time u to be S_u. The market price of the underlying at any future time $(u>0)$ will not be predictable in general; however, the price will be known at $t=0$ and we can assume that the price will vary more or less continuously.

We define the *forward price*, K, to be the price agreed at time $t=0$ to be paid at time $t=T$, where T is the mutually agreed *expiry* or *maturity date* of the contract, i.e., the date when the trade will happen. It is clear that to be *acceptable to both positions* (the long and the short), the value of the forward contract must be zero at time $t=0$; however, the aim is to calculate the value of K that is also acceptable to both positions and consistent with a zero initial value. We will see that the forward price depends on the current price of the underlying asset, S_0; on the maturity time, T; and on the prevailing continuously compounding risk-free rate of interest, r (i.e., the *risk-free force of interest*).

It is clear that a forward contract will generally lead to a non-zero profit (or loss) at maturity. At time $t=T$, the buyer is required to purchase the security (of current market price S_T) for price K. The payoff from the contract to the *long position* (*buyer*) at maturity is therefore

$$S_T - K$$

Similarly, the *short position* (*seller*) will receive K from the forward but could have realized S_T from the market. The payoff from contract to the *short position* is therefore

$$K - S_T = -(S_T - K)$$

We see that if payoff to the long position is positive, the payoff to the short position is equal, but negative, and vice versa. Figure 10.6.1 demonstrates the possible payoffs for either party of the forward contract as a function of the underlying stock price at maturity time.

Our strategy for determining the forward price will consist of forming a replicating portfolio for the long position of the forward contract, assuming no arbitrage, and invoking the law of one price.

At this stage, we take care to distinguish between forwards on non-income-paying securities and forwards on income-paying securities. The crucial difference between these cases is that, prior to maturity of the contract, income generated by the underlying security is received by the short position, and *not*

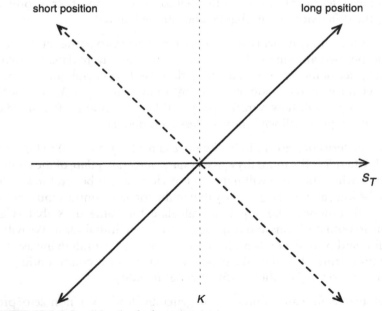

FIGURE 10.6.1 Payoff function for forward contract

by the long position. This has implications for each party's perceived value of the security at time $t = 0$.

Calculating the Forward Price for a Non-income-paying Security

We begin with the simplest case where the underlying security does not pay an income; examples include commodity forwards and forwards on zero-coupon securities. The market price S_u will be uncertain for all $u > 0$, but we now demonstrate how to use the assumption of no arbitrage and a replicating portfolio to find the forward price without requiring any assumptions about the explicit behaviour of S_u.

Consider the following two portfolios.

- *Portfolio A:* Enter into the long position of a forward contract on one unit of asset S, with forward price K, and maturity time T. Simultaneously, invest a cash amount of Ke^{-rT} at the risk-free rate.
- *Portfolio B:* Buy one unit of the asset at the current market price, S_0.

At time $t = 0$, the value of Portfolio A is Ke^{-rT} (as the price of the forward is zero). The value of Portfolio B is S_0.

At time $t = T$, the cash flows from Portfolio A are

(a) An amount K *received* from the risk-free investment accumulated at rate r for time T; and
(b) An amount K *spent* under the forward contract leading to the receipt of one unit of security S.

The net result from holding Portfolio A is receipt of S at time $t = T$. The result from holding Portfolio B is one unit of asset S at time $t = T$. We therefore see that the future proceeds from Portfolio A are identical to those from Portfolio B, i.e., A and B are mutual replicating portfolios. If we assume no arbitrage, the law of one price implies that the prices of the two portfolios must be identical at time $t = 0$. We therefore conclude that

$$Ke^{-rT} = S_0$$

and the forward price can be obtained by

$$K = S_0 e^{rT} \tag{10.6.1}$$

Equation 10.6.1 enables one to calculate the forward price that would be acceptable to both parties of a forward contract, based on the maturity time, the prevailing risk-free rate of return, and the current market price of the underlying (non-income-paying) security.

EXAMPLE 10.6.1

A 1-year forward contract on a commodity is entered into when the commodity price is £25. If the 1-year risk-free force of interest is 5% per annum, calculate the forward price for the commodity under the contract. You should assume no arbitrage.

Solution

The forward price is given by Eq. 10.6.1 with $S_0 = £25$, $T = 1$, and $r = 0.05$.

$$K = 25e^{0.05} = £26.28$$

EXAMPLE 10.6.2

Calculate the forward price within a 7-year forward contract on a non-income-paying security with current market price £92. You are given that the risk-free rate of interest is 3% per annum for investments of all terms.

Solution

Assuming no arbitrage, the forward price is given by Eq. 10.6.1 with $S_0 = 92$, $T = 7$, and $r = \ln(1.03)$.

$$K = 92 \times (1.03)^7 = £113.15$$

Note that you are given the risk-free *rate* of interest, not the continuously compounding value.

For forwards on an underlying security that pays income, it is important to first establish how the income is paid before generating the replicating portfolio. For example, if the income is a fixed cash amount regardless of the market price of the security at the payment date (as they would be for *fixed-interest securities*), we can assume that the income is invested at the risk-free rate. Alternatively, if the income is proportional to the market price of underlying security (e.g., as in the case of a *dividend yield*), we can assume that the income is reinvested in the security. The reason is that when the payment is proportional to the stock price we will know how many units of stock it will purchase, but we do not know how much cash is paid out. We can therefore only predict the amount of stock held at maturity if we assume the dividends are reinvested in the stock. However, with a cash-paying security, we will not know how much stock could be bought, but we do know how much cash would be generated from investment at the risk-free rate. We consider these two cases separately.

Calculating the Forward Price for a Security with Fixed Cash Income

We use the same notation as introduced previously, but assume now that at some time t_i $(i = 1, 2 \dots, n)$ with $0 \leq t_i \leq T$, the underlying security pays a *cash* amount c_i to the current holder.

Consider the following two portfolios.

- *Portfolio A:* Enter into the long position of a forward contract on one unit of asset S, with forward price K, and maturity time T. Simultaneously invest an amount $Ke^{-rT} + \sum_{i=1}^{n} c_i e^{-rt_i}$ at the risk-free rate.
- *Portfolio B:* Buy one unit of the asset at the current price S_0. Invest the income c_i at the risk-free rate at each time t_i.

At time $t = 0$, the value of Portfolio A is $Ke^{-rT} + \sum_{i=1}^{n} c_i e^{-rt_i}$. The value of Portfolio B is S_0.

At time $t = T$, the cash flows for Portfolio A are

(a) An amount $K + \sum_{i=1}^{n} c_i e^{r(T-t_i)}$ received from the risk-free investments accumulated at r; and

(b) An amount K spent under the forward contract leading to the receipt of one unit of asset.

The net result for Portfolio A is therefore receipt of S and a cash amount $\sum_{i=1}^{n} c_i e^{r(T-t_i)}$ at time T.

For Portfolio B, the net result at time $t = T$ is one unit of asset S and a cash amount $\sum_{i=1}^{n} c_i e^{r(T-t_i)}$. We see that the future cash flows from Portfolio A are identical to those from Portfolio B and, if we assume no arbitrage, the law of one price implies that

$$Ke^{-rT} + \sum_{i=1}^{n} c_i e^{-rt_i} = S_0$$

at $t = 0$, and so

$$K = \left(S_0 - \sum_{i=1}^{n} c_i e^{-rt_i}\right) e^{rT}$$

If we define I to be the present value at time $t = 0$ of the income received from the underlying security over the term of the forward contract, the forward price can be written as

$$K = (S_0 - I)e^{rT} \tag{10.6.2}$$

The quantity $S_0 - I$ can be thought of as the fair price of the underlying security at $t = 0$ to the long position. The long position will not receive the income of value I prior to taking delivery at the maturity time, and so the price of the asset at $t = 0$ must be reduced accordingly. The *adjusted* price $S_0 - I$ is, of course, a modification based on a *discounted cash flow* model of security pricing. Equation 10.6.2 reduces to 10.6.1 when the underlying does not pay income.

EXAMPLE 10.6.3

Calculate the forward price in Example 10.6.2 if the underlying security is now expected to pay a coupon payment of £3 at the end of each of the next 5 years (the first payment being exactly 1 year from now).

Solution

Equation 10.6.2 can be used to calculate the forward price. In particular, we have

$$I = 3 \times (1.03)^{-1} + \cdots + 3 \times (1.03)^{-5}$$
$$= 3a_{\overline{5}|} = 13.7391$$

which leads to

$$K = (92 - 13.7391) \times (1.03)^{7}$$
$$= 96.25$$

This forward price is significantly lower than that calculated in Example 10.6.2.

EXAMPLE 10.6.4

An investor wishes to purchase a 1-year forward contract on a risk-free 3-year bond that has a current market price of £90 per £100 nominal. The bond pays coupons at a rate of 6% per annum half-yearly. The next coupon payment is due in exactly 6 months, and the coupon payment following that is due just before the forward contract matures. If the 6-month risk-free spot force of interest is 5% per annum and the 12-month risk-free spot force of interest is 6% per annum, calculate the forward price of the bond.

Solution

Here we have the added complication that the risk-free rates are dependent on the term of investment. We consider £100 nominal of the bond and have $S_0 = 90$ and $T = 1$. Two payments are made during the term of the contract: the first of amount 3 at time $t = 0.5$, for which $r_{0.5} = 0.05$ per annum; and the second is also of amount 3 at time $t = 1$, for which $r_1 = 0.06$ per annum.

The forward price is therefore

$$K = \left(90 - 3e^{-0.5 \times 0.05} - 3e^{-1 \times 0.06}\right)e^{1 \times 0.06} = £89.46$$

i.e., £89.46 per £100 nominal.

Calculating the Forward Price for a Security with Known Dividend Yield

If the security underlying the forward contract is one that pays a continuously received dividend that can be immediately *reinvested in the security*, we calculate the forward price using a replicating portfolio as follows. The approach is relevant to forwards on equities (shares) with known dividend yield, for example.

Let D be the known dividend yield per annum, such that if we have a unit holding of the security at time $t = 0$, the accumulated holding at later time T (following reinvestment of the proceeds) would be e^{DT}. We consider the following two replicating portfolios.

- *Portfolio A:* Enter into a long position of a forward contract on asset S, with forward price K, and maturity time T. Simultaneously, invest an amount Ke^{-rT} at the risk-free rate.

- *Portfolio B:* Buy e^{-DT} units of the asset at the current price S_0. Reinvest the dividend income in the asset immediately as it is received.

At time $t=0$, the value of Portfolio A is Ke^{-rT}. The value of Portfolio B is $e^{-DT}S_0$.

At time $t=T$, the cash flows for Portfolio A are

(a) An amount K received from the risk-free investment accumulated at rate r; and

(b) An amount K spent under the forward contract leading to the receipt of one unit of asset S.

The net result is therefore receipt of S.

For Portfolio B, the result at time $t=T$ is also one unit of asset S. Assuming no arbitrage, the law of one price can be used to write that

$$Ke^{-rT}=S_0e^{-DT}$$

at $t=0$, and so

$$K=S_0e^{(r-D)T} \tag{10.6.3}$$

EXAMPLE 10.6.5

Calculate the forward price in Example 10.6.2 if the underlying security is now expected to pay a continuously compounding dividend yield of 2% per annum.

Solution

Assuming no arbitrage, the forward price is given by Eq. 10.6.3 with $S_0=92$, $T=7$, $r=\ln(1.03)$, and $D=0.02$.

$$K = 92(1.03)^7 e^{-0.02\times7} = £93.37$$

EXAMPLE 10.6.6

Calculate the forward price in Example 10.6.2 if the underlying security has paid and will continue to pay annually in arrears a dividend of 1.5% per annum of the market price of the security at the time of payment.

Solution

Here we have discrete dividends that depend on the market price of the security. The replicating portfolio must therefore be established under the assumption that the dividends are reinvested back into the holding of the security. The value of the holding relevant to the long position requires 7 years of this reinvestment to be "removed", and the forward price is given by

$$K = 92(1.015)^{-7}(1.03)^7 = £101.95$$

The three expressions for the forward price (10.6.1–10.6.3) can be summarized by defining a new quantity, $S_{m,0}$, which denotes the *modified market price* of the underlying within the replicating portfolio at time $t = 0$. The modified market price is taken to be the present value of the asset as *perceived by the long position*, i.e., the price that takes into account the income missed over the term of the contract. It is clear that

$$S_{m,0} = \begin{cases} S_0 \text{ for a non-income-paying asset} \\ S_0 - I \text{ for an asset that pays known cash amounts} \\ S_0 e^{-DT} \text{ for an asset that pays a known dividend yield} \end{cases} \qquad (10.6.4)$$

Using this equation, we can state the three expressions for the forward price as

$$K = S_{m,0} e^{rT} \qquad (10.6.5)$$

Expression 10.6.5 can be read as saying that, under the condition of no arbitrage, the forward price can be computed by assuming that the current value of underlying security (as perceived by the long position) grows at the risk-free rate. If this were not true, there would exist the opportunity for arbitrage. For example, if $K < S_{m,0}e^{rT}$, it would be possible to make profit from entering into the long position of the forward contract at $t = 0$ and simultaneously investing an amount $S_{m,0}$ at the risk-free rate.

10.7 CALCULATING THE VALUE OF A FORWARD CONTRACT PRIOR TO MATURITY

In Section 10.6 we discussed how to calculate the forward price that would be acceptable to both parties at the time the contract was struck. In doing this, we needed to insist that the value of the contract to both parties was zero at time $t = 0$. However, at an interim time prior to maturity, the market price for the underlying asset will have moved and the terms of the forward contract will naturally favor either the long or the short position. At that time, the forward will have a positive value to the favored position and an equal, but negative, value to the counterparty. We now discuss how to calculate this value at a general interim time, $0 < u < T$.

Consider a forward contract agreed at time $t = 0$. Let K_0 be the forward price under the contract for one unit of security S, and T be the maturity time. At the start of the contract, the value to both the long position (buyer) and short position (seller) is zero. At maturity, the value to the parties is given by their respective payoffs:

- $S_T - K_0$, for the long position; and
- $-(S_T - K_0)$, for the short position.

In order to determine the value of the long forward contract at time u, $V_l(u)$, consider two portfolios purchased at that time. (For illustration, we assume that the underlying stock does not generate income.)

- Portfolio A: Consisting of
  - (a) The existing long forward contract (entered into at time $t = 0$) with current value $V_l(u)$; and
  - (b) An investment of $K_0 e^{-r(T-u)}$ made at time u at the risk-free rate for $T - u$ years.
- Portfolio B: Consisting of
  - (a) A new long forward contract (on the same underlying) entered into at time u, with maturity time T, and forward price $K_u = S_u e^{r(T-u)}$; and
  - (b) An investment of $K_u e^{-r(T-u)}$ made at time u at the risk-free rate for $T - u$ years.

At time $t = u$, the value of Portfolio A is $V_l(u) + K_0 e^{-r(T-u)}$, and the value of Portfolio B is $K_u e^{-r(T-u)}$ (since the value of the new forward contract in Portfolio B is zero at this time).

At time $t = T$, the cash flows from Portfolio A are

- (a) K_0 received from the risk-free investment; and
- (b) An amount K_0 spent under the forward contract, leading to the receipt of one unit of asset S.

The net result of holding Portfolio A is therefore receipt of S at time $t = T$.

For Portfolio B, the cash flows at $t = T$ are

- (a) K_u received from the risk-free investment; and
- (b) An amount K_u spent under the new forward contract, leading to the receipt of one unit of asset S.

The net result of holding Portfolio B is therefore also receipt of S.

If we assume that arbitrage does not exist, the law of one price enables us to equate the value of the two portfolios at time $t = u$,

$$V_l(u) + K_0 e^{-r(T-u)} = K_u e^{-r(T-u)}$$

Further rearrangement enables this to be written as

$$V_l(u) = (K_u - K_0)e^{-r(T-u)} \qquad (10.7.1)$$

From the reasoning given previously, the value of the short forward contract is $V_s(u) = -V_l(u)$ and we have

$$V_s(u) = (K_0 - K_u)e^{-r(T-u)} \qquad (10.7.2)$$

Substituting $K_0 = S_0 e^{rT}$ and $K_u = S_u e^{r(T-u)}$ (from Eq. 10.6.1), the preceding can be rearranged to give the value of the original forward contract at some time prior to maturity in terms of the asset prices at time $t=0$ (when the contract was entered into) and that time u,

$$V_l(u) = S_u - S_0 e^{ru} \tag{10.7.3}$$

$$V_S(u) = S_0 e^{ru} - S_u \tag{10.7.4}$$

As previously discussed, Eq. 10.6.1 can be interpreted as showing that the forward price is equal to the current market price accumulated at the risk-free rate of interest. With this in mind, Eqs 10.7.2 and 10.7.3 are intuitive. In particular, they can be read as saying that the value of the forward contract at some time prior to maturity is the difference between the market price of the underlying at that time and the expected value of the underlying as assumed under the original calculation of the forward price. If, at time u, the market price of the underlying asset had increased above $S_0 e^{ru}$, the market has acted in favor of the long position, and Eqs 10.7.2 and 10.7.3 would show a positive value for the long position and negative for the short. Alternatively, if the market price of the underlying asset was to fall lower than $S_0 e^{ru}$, the market has acted in favor of the short position, and we see a positive value to the short position and negative for the long position.

EXAMPLE 10.7.1

A 1-year forward contract on a commodity was written when the market price of the commodity was £30. Six months later, the market price of the commodity has risen to £32. If the risk-free force of interest has and will be 5% per annum on investments of any term, calculate the value of the forward contract to both the long and short position at this time. Assume no arbitrage.

Solution

We apply Eq. 10.7.2 with $u=0.5$, $S_u=37$, $r=0.05$, and $S_0=30$, leading to

$$V_l(0.5)=32-30e^{0.5\times0.05}=£1.24$$

It can then be immediately stated that $V_{s(0.5)}=-1.24$.

Taking limits of Eqs 10.7.2 and 10.7.3, we see that

$$\lim_{u\to0} V_l(u) = S_0 - S_0 e^0 = 0$$
$$\lim_{u\to0} V_s(u) = S_0 e^0 - S_0 = 0 \tag{10.7.5}$$

and

$$\lim_{u\to T} V_l(u) = S_T - S_0 e^{rT} = S_T - K_0$$
$$\lim_{u\to T} V_s(u) = S_0 e^{rT} - S_0 = K_0 - S_T \tag{10.7.6}$$

Equation 10.7.4 shows that the value of a forward contract tends to zero as time approaches the time at which the contract was first written. This is entirely consistent with the reasoning of Section 10.6. Furthermore, Eq. 10.7.5 shows that the value tends to the payoffs for the appropriate position. This is as discussed previously.

The preceding discussions can be extended to consider the value of forwards on income-paying securities using general reasoning similar to that used in Section 10.6. From an intuitive point of view, it is clear that we can generalize Eqs 10.7.2 and 10.7.3 to

$$V_l(u) = S_{m,u} - S_{m,0}e^{ru}$$
$$V_S(u) = S_{m,0}e^{ru} - S_{m,u}$$

where $S_{m,u}$ is the price relevant to the long position at time $t = u$ (i.e., takes into account the income missed between times $t = u$ and $t = T$). The value of $S_{m,u}$ is obtained from an expression equivalent to Eq. 10.7.4, depending on the type of income paid by the underlying. In practice, the original market price and income of the underlying security may not be known, and it is therefore sensible to recast the preceding equations in terms of the original forward price.

$$V_l(u) = S_{m,u} - K_0 e^{-r(T-u)}$$
$$V_S(u) = K_0 e^{-r(T-u)} - S_{m,u}$$

$$(10.7.7)$$

EXAMPLE 10.7.2

An investor entered into a long forward contract on an underlying security 5 years ago with forward price £90. The contract is due to mature in 7 years' time. If the risk-free force of interest is 3% per annum throughout the 12-year period, calculate the value of the contract now if the security will pay dividends of £3 in 2 years' time and £4 in 4 years' time, and the current market price is £120. Assume no arbitrage.

Solution

We use the expression for $V_l(u)$ from Eq. 10.7.6 with $u = 5$, $T - u = 7$, $r = 0.03$, and $K_0 = 90$,

$$V_l(5) = S_{m,5} - 90e^{-r(T-u)}$$

The current modified price of the security is obtained as

$$S_{m,5} = 120 - 3e^{-2r} - 4e^{-4r} = 113.63$$

leading to

$$V_l(5) = 113.63 - 90e^{-7 \times 0.03} = £40.67$$

EXAMPLE 10.7.3

Calculate the value of the contract in Example 10.7.2 if the security has paid and will continue to a continuously compounding dividend yield of 2% per annum.

Solution

The modified price of the security holding is now

$$S_{m,5} = 120e^{-7 \times 0.02} = 104.32$$

and the value of the long position is calculated to be

$$V_l(5) = 104.32 - 90e^{-7 \times 0.03} = 31.37$$

EXAMPLE 10.7.4

A fixed-interest security is priced at £90 per £100 nominal, has a coupon rate of 6% per annum payable half-yearly, and has an outstanding term of 5 years. An investor holds a short position in a forward contract on £200 nominal of this security, with a delivery price of £98 per £100 nominal and maturity in exactly 1 year, immediately following the coupon payment then due. If the risk-free forces of interest for investments of terms 6 months and 1 year are 4.5% and 5.4%, respectively, calculate the value of this forward contract to the investor. Assume no arbitrage.

Solution

Consider £100 of the bond. The value of the short position at this time can be calculated from the expression for $V_s(t)$

in Eq. 10.7.6 with $K_0 = 98$, $T - u = 1$, $r_{0.5} = 0.045$, and $r_1 = 0.054$ as

$$V_s(u) = 98e^{-1 \times r_1} - S_{m,u}$$

The modified price of the security is

$$S_{m,r} = 90 - 3e^{-0.5 \times 0.045} - 3e^{-1 \times 0.054} = 84.52$$

leading to $V_s(u) = 8.62$. The total value of the forward contracts held is £17.24.

10.8 ELIMINATING THE RISK TO THE SHORT POSITION

An investor who agrees to *sell* an asset under a forward contract (i.e., holds the short position) need not hold the asset at the start of the contract; this is an example of *short selling*. However, by the end of the contract, he must own the asset ready to sell. If the investor waits until the end of the contract to buy the asset, the risk exists that the price will rise above the forward price, and he will have to pay more than he will receive. However, if he buys the asset at the start of the forward contract and holds it until the contract matures, there is a risk that the price will have fallen by the maturity date.

To hedge this *market risk*, the investor could enter into the short position and simultaneously borrow an amount Ke^{-rT} at the risk-free force of interest to purchase the asset. The value of this **hedge portfolio** at time $t = 0$ is $-Ke^{-rT} + S_0 = 0$. At maturity, the investor then owes K that is exactly covered by the price received from selling the asset under the forward contract. This way, if the investor holds the hedge portfolio, he is certain not to make a loss on the forward contract. However, as in other hedging strategies involving futures and forwards, there is also no chance of making a profit.

We have assumed here that there is no cash or dividend income associated with the underlying asset. However, the argument can be modified using the arguments presented in section in Section 10.7.

The hedge described here is called a *static hedge* since the hedge portfolio, which consists of the asset to be sold plus the sum borrowed at the risk-free rate, *does not change* over the term of the contract. For more complex financial instruments, the hedge portfolio is more complicated and requires (in principle) continuous rebalancing to maintain the hedged position. This is called a *dynamic hedge*.

SUMMARY

- *Futures* and *forwards* are the simplest examples of *derivative* contracts (contracts that *derive* their price from underlying securities). Both examples are agreements between two parties to trade a specified asset, for a specified price, at a specified future date. The seller holds the *short position*; the buyer holds the *long position.*
- Futures are standardized and exchange tradable contracts. Forwards are not standardized and are said to be traded *over the counter.*
- In either example, both parties are exposed to a *market risk* and a *credit risk*. A *clearinghouse* sits between each party of a *futures contract* and acts as counterparty to both, thereby removing the credit risk exposure. The clearinghouse mitigates its own credit risk exposure by requiring *margin payments*. No such facility exists for a *forward contract.*
- The *forward price* is chosen such that the value of the contract at the outset is zero to both parties. This is calculated as $K = S_0 e^{rT}$, where S_0 is the market price of the underlying *non-income-paying* asset at the time the contract is written, r is the risk-free force of interest, and T is the time to maturity. Similar expressions exist for income-paying assets.
- The *value* of the forward contract to the long and short positions at some time t prior to maturity is obtained from $V_l(t) = S_t - Ke^{-r(T-t)}$ and $V_s(t) = -V_l(t)$, respectively. Here S_t is the market price of the underlying *non-income-paying* asset at time t. Similar expressions exist for income-paying assets.
- The previous expressions are derived under the assumption of *no arbitrage*. *Replicating portfolios* are created and the *law of one price* used. Arbitrage is said to exist if there is an opportunity to make a risk-free trading profit.

EXERCISES

10.1 Assume Sun Microsystems expects to receive €20,000,000 in 90 days. A dealer provides a quote of $0.875 for a currency forward rate contract to expire in 90 days. Suppose that at the end of 90 days, the rate is $0.90. Assume that settlement is in cash. Calculate the cash flow at expiration if Sun Microsystems enters into a forward contract expiring in 90 days to buy dollars at $0.875.

 CFA Institute

 **10.2** Assume that a security is currently priced at $200. The risk-free rate is 5% per annum (discretely compounded).

    (a) A dealer offers you a contract in which the forward price of the security with delivery in 3 months is $205. Explain the transactions you would undertake to take advantage of the situation.

    (b) Suppose the dealer was to offer you a contract in which the forward price of the security with delivery in 3 months is $198. How would you take advantage of the situation?

 **10.3** Assume that you own a dividend-paying stock currently worth $150. You plan to sell the stock in 250 days. In order to hedge against a possible price decline, you wish to take a short position in a forward contract that expires in 250 days. The risk-free rate is 5.25% per annum (discretely compounding). Over the next 250 days, the stock will pay dividends according to the following schedule:

Days to Next Dividend	Dividends per Share ($)
30	1.25
120	1.25
210	1.25

    (a) Calculate the forward price of a contract established today and expiring in 250 days.

    (b) It is now 100 days since you entered the forward contract. The stock price is $115. Calculate the value of the forward contract at this point.

    (c) At expiration, the price of the stock is $150. Calculate the value of the forward contract at expiration.

 **10.4** The spot exchange rate for the British pound is $1.4390. The US interest rate is 6.3% per annum, and the British interest rate is 5.8% per annum (both with discrete compounding). A futures contract on the exchange rate for the British pound expires in 100 days.

    (a) Find the appropriate futures price for a US investor.

    (b) Suppose the actual price $1.4650. Is the future contract mispriced? If yes, how could an arbitrageur take advantage of the mispricing?

10.5 A bond is priced at £95 per £100 nominal, has a coupon rate of 5% per annum payable half-yearly, and has an outstanding term of 5 years. An investor holds a short position in a forward contract on £1 million nominal of this bond, with a delivery price of £98 per £100 nominal and maturity in exactly 1 year, immediately following the coupon payment then due. The continuously compounded risk-free rates of interest for terms of 6 months and 1 year are 4.6% per annum and 5.2% per annum, respectively.

Calculate the value of this forward contract to the investor assuming no arbitrage.

10.6 An investor is able to purchase or sell two specially designed risk-free securities, A and B. Short sales of both securities are possible. Security A has a market price of 20p. In the event that a particular stock market index goes up over the next year, it will pay 25p and, in the event that the stock market index goes down, it will pay 15p. Security B has a market price of 15p. In the event that the stock market index goes up over the next year, it will pay 20p and, in the event that the stock market index goes down, it will pay 12p.

(a) Explain what is meant by the assumption of "no arbitrage" used in the pricing of derivative contracts.

(b) Find the market price of B, such that there are no arbitrage opportunities and assuming the price of A remains fixed. Explain your reasoning.

10.7 An investor entered into a long forward contract for a security 5 years ago and the contract is due to mature in 7 years' time. Five years ago the price of the security was £95 and is now £145. The risk-free rate of interest can be assumed to be 3% per annum throughout the 12-year period. Assuming no arbitrage, calculate the value of the contract now if

(a) The security will pay dividends of £5 in 2 years' time and £6 in 4 years' time,

(b) The security has paid and will continue to pay annually in arrears a dividend of 2% per annum of the market price of the security at the time of payment.

10.8 A 1-year forward contract is issued on 1 April 2007 on a share with a price of 900p at that date. Dividends of 50p per share are expected on 30 September 2007 and 31 March 2008. The 6-month and 12-month spot, risk-free rates of interest are 5% and 6% per annum effective, respectively, on 1 April 2007. Calculate the forward price at issue, stating any assumptions.

10.9 An 11-month forward contract is issued on 1 March 2008 on a stock with a price of £10 per share at that date. Dividends of 50 pence per share are expected to be paid on 1 April and 1 October 2008. Calculate the forward price at issue, assuming a risk-free rate of interest of 5% per annum effective and no arbitrage.

10.10 An investor entered into a long forward contract for £100 nominal of a security 8 years ago, and the contract is due to mature in 4 years' time. Eight years ago the price per £100 nominal of the security was £94.50 and is now £143.00. The risk-free rate of interest can be assumed to be 5% per annum effective throughout the contract.

Calculate the value of the contract now if it were known from the outset that the security will pay coupons of £9 in 2 years and £10 in 3 years from now. You may assume no arbitrage.

10.11 An investor wishes to purchase a 1-year forward contract on a risk-free bond which has a current market price of £97 per £100 nominal. The bond will pay coupons at a rate of 7% per annum half-yearly. The next coupon payment is due in exactly 6 months, and the following coupon payment is due just before the forward contract matures. The 6-month risk-free spot interest rate is 5% per annum effective and the 12-month risk-free spot interest rate is 6% per annum effective.

Stating all necessary assumptions

(a) Calculate the forward price of the bond.

(b) Calculate the 6-month forward rate for an investment made in 6 months' time.

(c) Calculate the purchase price of a risk-free bond with exactly 1 year to maturity which is redeemed at par and which pays coupons of 4% per annum half-yearly in arrears.

(d) Calculate the gross redemption yield from the bond in (c).

(e) Comment on why your answer in (d) is close to the 1-year spot rate.

10.12 A security is priced at £60. Coupons are paid half-yearly. The next coupon is due in 2 months' time and will be £2.80. The risk-free force of interest is 6% per annum. Calculate the forward price an investor should agree to pay for the security in 3 months' time assuming no arbitrage.

10.13 A 9-month forward contract is issued on 1 March 2011 on a stock with a price of £9.56 per share at that date. Dividends of 20 pence per share are expected on both 1 April 2011 and 1 October 2011.

(a) Calculate the forward price, assuming a risk-free rate of interest of 3% per annum effective and no arbitrage.

(b) Explain why the expected price of the share in 9 months' time is not needed to calculate the forward price.

CHAPTER 11

Further Derivatives: Swaps and Options

In this chapter we discuss two further examples of standard derivative contracts: *swaps* and *options*. As we shall see, *swaps* are relatively simple contracts but are of significant practical importance. We introduce the main types of swap contracts and briefly discuss how they are used and priced. Our discussion of *options* will be significantly longer and is the main focus of this chapter. Although options arise from a seemingly simple modification to futures and forwards contracts, we will see that they are actually significantly more complicated derivatives to price. *Option pricing formulae* will be stated without proof, and we give a detailed discussion of standard *option trading strategies* and their importance in speculation and hedging.

11.1 SWAPS

A *swap* is a contract between two parties under which they agree to exchange (i.e., swap) a series of payments according to a prearranged formula. The simplest examples of swaps are *interest rate swaps* and *currency swaps*.

In the most common form of an **interest rate swap**, one party agrees to pay to the other a regular series of *fixed* amounts for a certain period, and, in exchange, a counterparty agrees to pay a series of *variable* amounts based on the level of a short-term interest rate. For example, the variable rate could be based on LIBOR, the *London Interbank Offered Rate*. Both sets of payments are in the same currency. Interest rate swaps are particularly useful to investors who have floating interest (e.g., LIBOR-linked) liabilities and fixed-interest assets, or vice versa. By entering into a swap, the investor is able to exchange fixed cash amounts for floating cash amounts, thereby aligning his assets and liabilities. This removes any exposure to unexpected changes in the floating rate that could act to adversely affect the value of the liabilities relative to the assets.

The market for swaps is very large and has grown exponentially since they were first formally used in 1981. The market for interest rate swaps alone (the most popular type) was estimated to be around US$600 trillion in 2012. Most swaps are traded *over the counter*, although a relatively small market for standardized swaps does exist on futures markets. The over-the-counter nature of swaps

235

An Introduction to the Mathematics of Finance. http://dx.doi.org/10.1016/B978-0-08-098240-3.00011-4
© 2013 Institute and Faculty of Actuaries (RC000243). Published by Elsevier Ltd. All rights reserved.
For End-of-chapter Questions: © 2013. CFA Institute. Reproduced and republished with permission from the CFA Institute. All rights reserved.

means that it is necessary for another investor with opposite requirements to exist who is willing to act as counterparty to the swap.

Another use of swaps is to exploit a *comparative advantage* that can exist between two companies that are each seeking to borrow money from third parties. For example, it could be that one company requires a loan with fixed-interest repayments (to align with its assets, for example), and another requires floating repayments (to align with its assets), but, for whatever reason, each has a "better" quotation for the other type of loan. In this case there is an advantage to be gained from both accepting the better quotations and entering into a swap to exchange the interest payments to the preferred type. This is best illustrated by Example 11.1.2.

In Examples 11.1.1 and 11.1.2, the fixed payments can be thought of as interest payments on a deposit at a fixed rate, while the variable payments are the interest payments on the same deposit at a floating rate. However, the deposit is purely *notional*, and no exchange of capital actually takes place.

EXAMPLE 11.1.1

A company takes out a 2-year loan of £5m on 1 January 2012 with biannual (i.e., twice yearly) interest payments based on a floating rate listed on that day. In order to reduce its exposure to the floating rate, the company enters into a swap under which it pays biannual interest equivalent to 5% per annum and receives biannual payments of the value of the floating rate on a notional £5m.

If, during the term of the swap, the floating rates due at each date are as follows, calculate the cash flows required from the company at each interest date.

Date	6-Month Floating Rate
1 January 2012	2.3%
1 August 2012	2.9%
1 January 2013	2.2%
1 August 2013	2.7%
1 January 2014	3.0%

Solution

Cash flows are required under the loan and under the swap. The cash flows under the loan are the original receipt of the capital at 1 January 2012, the regular interest payments every

6 months, and the return of the capital on 1 January 2014. Under the swap, the company pays a regular 2.5% of £5m and receives the floating rates. The resulting cash amounts are summarized as follows.

Date	Loan Cash Flow	Swap Cash Flow	Net Cash Flow
1 January 2012	+5m	N/A	+5m
1 August 2012	−0.145m	+0.145m −0.125m	−0.125m
1 January 2013	−0.110m	+0.110m −0.125m	−0.125m
1 August 2013	−0.135m	+0.135m −0.125m	−0.125m
1 January 2014	−0.150m −5m	+0.150m −0.125m	−5.125m

We see that by entering into the swap, the net cash flows for the company are equivalent to a £5m loan with interest paid biannually at 5% per annum, and the company has removed its exposure to the variable rate.

EXAMPLE 11.1.2

Two investors, X and Y, are seeking to borrow £100m over 3 years, and have been quoted the following interest rates for borrowing at fixed and floating rates.

	Fixed	Floating
X	5%	6-month LIBOR + 0.5%
Y	6.5%	6-month LIBOR + 1%

If X would like to borrow at the floating rate and Y at the fixed rate, demonstrate that there is a comparative advantage to be gained from X and Y entering into an interest rate swap. If, under the swap, X agrees to pay LIBOR to Y, calculate the fixed rate that Y should pay to X so that the comparative advantage is equally shared.

Solution

If both individually borrow as they would prefer, the total interest paid is

$$LIBOR + 0.5\% + 6.5\% = LIBOR + 7\%$$

However, if X borrows at the fixed rate and Y at the floating rate, the total interest paid would be

$$LIBOR + 1\% + 5\% = LIBOR + 6\%$$

The total interest paid is therefore lower by 1%. A swap should be established to share the 1% saving and enable the investors to borrow at their preferred floating or fixed rate. Under the swap

- X would borrow at the fixed rate of 5% and pay LIBOR to Y; and
- Y would borrow at LIBOR + 1% and pay X a fixed rate of $x\%$.

The value of x needs to be determined such that X gains a comparative advantage of 0.5%.

We require the net interest paid by X to be $(LIBOR + 0.5\%) - 0.5\% = LIBOR$; therefore,

$$LIBOR + 5\% - x = LIBOR$$

and so $x = 5\%$.

From Y's perspective, he is paying $LIBOR + 1\% + 5\% - LIBOR = 6\%$, which is 0.5% better than is available to him from the quotation. Both investors have then saved 0.5% under the swap.

As with forwards and futures, a swap is designed to have zero value at its outset, but market movements of the underlying rates (e.g., LIBOR) act to give it a positive or negative value at interim times. It is possible to value a swap at an interim time by evaluating the present value of the series of future cash flows at appropriate rates of interest; this is illustrated in Example 11.1.3.

Example 11.1.3 demonstrates that, in effect, the interest rate swap is valued as opposite positions on two *bonds*, one with fixed-interest payments and the other with floating. In the particular case of Example 11.1.3, the company could be considered to have *sold* a floating rate bond (on which it must pay variable interest) and bought a fixed rate bond (from which it receives fixed interest) of the same notional amount. Of course, buying and selling bonds involves the exchange of the capital amounts which are not required under interest rate swaps; however, these initial and terminal cash flows cancel and can be neglected. The pricing process therefore focuses on the relative cash flows at each interest date.

A *currency swap* is an agreement to exchange a fixed series of interest payments *and* a capital sum in one currency for a fixed series of interest payments and a capital sum in another. The exchange of capital is a key difference between

EXAMPLE 11.1.3

A company entered into a swap contract a number of years ago under which it pays a rate of 1-year LIBOR + 0.25% and receives 5% per annum fixed on a notional £10m. Interest is paid annually with the next due in exactly 1 year's time. If the remaining term of the swap is 2 years and the current continuously compounding LIBOR spot rates for 1 year and 2 years are 3.5% per annum and 4.0% per annum, respectively, calculate the current value of the swap to the company.

Solution

We define now as $t = 0$. At time $t = 1$, the company will receive a fixed-interest payment of 5% per annum and pay a floating rate based on the 1-year LIBOR rate + 0.25%, i.e., 3.75%. The net present value of the cash flow at time $t = 0$ is therefore

$$10 \times (0.05 - 0.0375)e^{-0.035} = 0.1207$$

In order to determine the net cash flow at time $t = 2$, we need to determine the 1-year forward (continuous compounding) rate of LIBOR at time $t = 1$; we denote this $R_{1,1}$. As the values stated are continuously compounding, the principle of consistency implies that

$$R_{1,1} = 2 \times 0.04 - 0.035 = 4.5\%$$

The net cash flow is therefore from a receipt of 5% per annum and payment of $R_{1,1} + 0.25\% = 4.75\%$. This leads to the net present value at time $t = 0$ of

$$10 \times (0.05 - 0.0475)e^{-2 \times 0.04} = 0.0231$$

The total value of the swap at $t = 0$ is therefore $0.1207 + 0.0231 = £0.1438$m. The value to the counterparty will be $- £0.1438$m.

interest rate and currency swaps. As with interest rate swaps, currency swaps are motivated by comparative advantages available between two parties. Currency swaps can be used to hedge exposure to foreign exchange markets, but it is important to realize that the hedge is not complete as the capital sums remain exposed.

EXAMPLE 11.1.4

A UK company decides to issue $15m of US$ denominated 5-year Eurobonds with a 5% per annum coupon rate paid bi-annually. A US company simultaneously decides to issue £10m of sterling denominated 5-year Eurobonds with a 5% per annum coupon rate paid biannually. Explain how the two companies could limit their respective exposures to the foreign exchange markets with respect to the interest payments. You should assume the current exchange rate is 1:1.5 (£:US$).

Solution

The two companies could agree to enter into a currency swap consisting of an initial capital exchange of $15m for

£10m and an agreement to exchange the regular coupon payments. In effect, each company now takes on the liability for paying the coupon payments of the other, with the advantage that these are in their domestic currency. After 5 years, the two parties exchange back the original capital amounts resulting in a holding of foreign currency that can be converted on the market at that time.

Although the exchange risk associated with the coupon payments is removed by the swap, a significant exchange risk remains on the principal amounts.

As with forwards, both counterparties in a swap (of any type) are exposed to a *market risk* and a *credit risk*. These two classifications of risk were discussed in Chapter 10. Note, however, that the credit risk might be considered as less significant under a swap. The reason is that any loss of future income resulting from the default of counterparty will be compensated for by the removal of the liability under the swap.

A variation on the swap agreements discussed previously is the *puttable swap*. In particular, a *puttable interest rate swap* is a swap agreement in which the fixed rate receiver has the right to terminate the swap on one or more dates prior to its scheduled maturity. This early termination provision is designed to protect the party from the adverse effects of large changes in the prevailing fixed rates.

11.2 OPTIONS

In contrast to futures and forwards where each party is committed to the trade (see Chapter 10), the **holder** of an **option** has the right *but not the obligation* to buy or sell a specified asset for a specified *strike* price at a future date. An option is said to be **written** by the party that initially offers the option for sale to the **holder**. It is the **writer** (i.e., the option's short position) who has the responsibility to fulfill the trade if the holder (i.e., the option's long position) opts to action it. A **call option** gives the right, but not the obligation, to the holder to *buy* the underlying asset from the writer for the strike price. A **put option** gives the right, but not the obligation, to the holder to *sell* the underlying asset to the writer for the strike price.

Unlike forwards and futures, the option has an initial premium (i.e., cost) that the holder must pay to the writer for the option; this is the upside for the writer. Since the option to exercise the trade remains with the holder, his exposure to the *market risk* is limited to the initial premium. The writer, however, is fully exposed to the market risk, and the potential downside of writing an option is unlimited. The writer therefore poses a *credit risk* to the holder, and clearing-houses are involved with standardized, exchange traded option contracts to remove this risk. Unlike futures, with an option, only the writer is required to deposit a margin, and the process of marking to market is similar to that discussed in Section 10.2. More exotic options can be traded over the counter, and the holder is exposed to the full credit risk.

Options can be further classified in terms of the timing of the trade relative to the *expiry date* after which the option is said to have expired. The main classifications are *European options* and *American options*. A **European option** is an option that can be exercised only on the expiry date, whereas an *American option* can be exercised on any date before its expiry at the option of the holder. Collectively, European and American options are often referred to as *vanilla options*, meaning that they are the most common variety. However, more exotic options are actively traded; for example, a *Bermudan option* enables the holder to action the trade at any one of a number of discretely spaced dates before expiry. Other examples include, but are not limited to, *Canary options*, *shout options*, and *compound options*.

11.3 OPTION PAYOFF AND PROFIT

We begin by defining the option *payoff* at expiry time, T, as the difference between the strike price, K, and the market value of the asset at expiry, S_T, as relevant to the option holder. Clearly, the payoff for a call option is different for a put option, and, as the option to exercise remains with the holder, the payoff is limited to non-negative values.

EXAMPLE 11.3.1

An investor buys a European option to buy 10 tonnes of coffee at a price of £130 per 50 kg in exactly 3 months' time. Discuss whether the investor would exercise the option if the market price of coffee at the expiry time is as stated and give the payoff in each case:

(a) £150 per 50 kg;
(b) £100 per 50 kg;
(c) £130.20 per 50 kg.

Solution

It is clear that the investor has purchased a 6-month *European call option* with strike price £130. He will exercise the option only if the payoff from the trade is positive, i.e., the market price at expiry is *higher* than the strike price.

(a) At £150 per 50 kg, the market price is higher than the strike price and the investor should exercise the option. When he does this, the option generates a payoff of £20 per 50 kg, which is £4,000.

(b) At £100 per 50 kg, the market price is lower than the strike price, and the investor would not exercise the option. The payoff is zero.

(c) At £130.20 per 50 kg, the market price is higher than the strike price, and the investor should exercise the option. When he does this, the option generates a payoff of £0.20 per 50 kg, which is £40.

EXAMPLE 11.3.2

An investor buys a European option to sell 50g of gold at a price £34 per gram in exactly 6 months' time. Discuss whether the investor would exercise the option if the market price of gold at the expiry time is as stated and give the payoff in each case:

(a) £40 per gram;
(b) £30 per gram;
(c) £33 per gram.

Solution

It is clear that the investor has purchased a 6-month *European put option*. He will exercise the option only if the market price at expiry is *lower* than the strike price.

(a) At £40 per gram, the market price is higher than the strike price, and the investor should sell the gold on the market. He will then not action the option, and the payoff is zero.

(b) At £30 per gram, the market price is lower than the strike price, and the investor should exercise the put option. When he does this, the option generates a payoff of £4 per gram, which is £200.

(c) At £33 per gram, the market price is lower than the strike price, and the investor should exercise the option. When he does this, the option generates a payoff of £1 per gram, which is £50.

As demonstrated in Examples 11.3.1 and 11.3.2, the payoff, $f(S_T)$, at expiry time $t = T$ for a European option is given by

$$f_{\text{call}}(S_T) = \max\{S_T - K, 0\}$$
$$f_{\text{put}}(S_T) = \max\{K - S_T, 0\} \qquad (11.3.1)$$

American options can be exercised at any time $t \leq T$ or not at all. We therefore define the time U as the time at which the American option is exercised by the holder or ∞ if it is not exercised, and write the payoff at that time to be

$$f_{\text{call}}(S_U) = \begin{cases} S_U - K \text{ if } U \leq T \\ 0 \qquad \text{if } U = \infty \end{cases}$$

$$f_{\text{put}}(S_U) = \begin{cases} K - S_U \text{ if } U \leq T \\ 0 \qquad \text{if } U = \infty \end{cases} \qquad (11.3.2)$$

The shapes of the payoff function at expiry are given in Figures 11.3.1 and 11.3.2 for call and put options, respectively.

The payoff expressions imply that it would be instructive to monitor the price of the underlying asset relative to the strike price once the option has been entered. Of course, this information is the basis for deciding whether to exercise

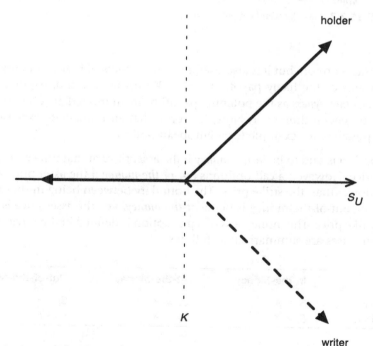

FIGURE 11.3.1 Payoff function for a call option

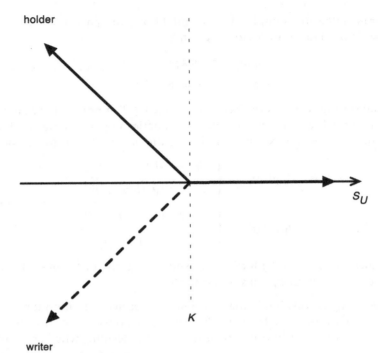

FIGURE 11.3.2 Payoff function for a put option

an American option, but it is also useful for the holders of European options as an indicator of the likely payoff at expiry. With this in mind, we define the concept of *moneyness* as the potential payoff from an immediate exercise of an option at general time $t \leq T$. Note, however, that an immediate exercise may not be possible, for example for a European option.

A call option is said to be *in-the-money* if the asset price at that time exceeds the strike price. However, a call option is *out-of-the-money* if the asset price at that time is lower than the strike price. The boundary between being in-the-money and being out-of-the-money is being *at-the-money*, i.e., the asset price is equal to the strike price. The moneyness of a put option is defined in the obvious way and both cases are summarized as follows:

	In-the-Money	At-the-Money	Out-of-the-Money
Call option	$S_t > K$	$S_t = K$	$S_t < K$
Put option	$S_t < K$	$S_t = K$	$S_t > K$

An option is said to be *deep-in-the-money* or *deep-out-of-the-money* if the potential payoff or loss, respectively, from immediate exercise is large.

However, *large* is a subjective term, and there is no formal definition of *deep* in this context.

Now that we have defined the payoff from an option from the perspective of the option holder (i.e., the long position), we can turn to the *profit* made by either the holder or writer. The *payoff* and *profit* are different because the holder of the option pays the writer an *option premium* at the outset.

The profit for the *holder* of an option is defined as the payoff minus the premium. Since the payoff function has a floor of zero, it is clear that the holder's loss is limited to the premium paid. The holder's profit has no upper limit. The profit for the *writer* of the option is negative that of the profit for the holder. The writer's profit is therefore capped at the value of the premium, and the loss is potentially unlimited.

It is often useful to plot the profit/loss resulting from entering into an option contract as a function of the underlying asset price at the time of exercise, U. Such plots are called *position diagrams* (or *profit diagrams*), and examples are given in Figure 11.3.3. Note that the figure does not distinguish between European and American options: for European options, $U = T$; for American options, $U \leq T$.

The position diagram in Figure 11.3.3 demonstrates that the holder of a call option will make a profit if the market price of the asset exceeds the value of the

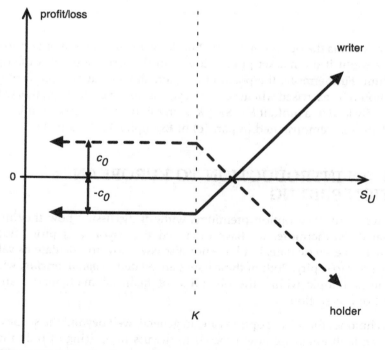

FIGURE 11.3.3 Position (profit) diagrams for the long and short positions on a call option

EXAMPLE 11.3.3

If, in Example 11.3.1, the holder has paid an option premium of £60, calculate the profit made by both parties in each case.

Solution

(a)   At £150 per 50 kg, the payoff was calculated to be £4,000. The holder therefore makes a profit of £3,940, and the writer a loss of £3,940.

(b)   At £100 per 50 kg, the payoff was calculated to be zero. The holder therefore makes a loss of £60, and the writer a profit of £60.

(c)   At £130.20 per 50 kg, the payoff was calculated to be £40. Although this is a non-zero payoff, the holder still makes a loss of £20, and the writer a profit of £20. Despite leading to a loss, the holder is still correct to exercise the option as the positive payoff acts to offset the premium already paid.

EXAMPLE 11.3.4

If, in Example 11.3.2, the holder paid an option premium of £100, calculate the profit made by both parties in each case.

Solution

(a)   At £40 per gram, the payoff was calculated to be zero. The holder therefore makes a loss of £100, and the writer a profit of £100.

(b)   At £30 per gram, the payoff was calculated to be £200. The holder therefore makes a profit of £100, and the writer a loss of £100.

(c)   At £33 per gram, the payoff was calculated to be £50. The holder, however, still makes a loss of £50, and the writer a profit of £50.

strike price plus the option premium. Similarly, the holder of a put option will make a profit if the market price is less than the asset value plus the option premium. Furthermore, the position diagram shows that an option of either type should be exercised whenever the payoff is positive; although this will not necessarily lead to a profit if $K < S_T \leq K +$ premium, it does act to minimize the loss. This was demonstrated in part (c) of Examples 11.3.3 and 11.3.4.

11.4 AN INTRODUCTION TO EUROPEAN OPTION PRICING

We have mentioned option premiums without discussing how they may be calculated. Furthermore, we have discussed the payoff and profit from an option at its exercise date, but have not discussed how to calculate its value at a time prior to expiry. Both of these issues are related to *option pricing,* which is a significant topic within the discipline of financial mathematics and the subject of this section.

The techniques for pricing options are, in general, well beyond the scope of this book, and, although we now proceed to discuss the pricing of only the very simplest option type—European options on non-income-paying assets—the final result will be presented without proper derivation. The complication in general option pricing is that the potential payoff at (or before) expiry depends on the decision to exercise the option or not. This, in turn, depends on the price of the underlying asset relative to the strike price. Unlike when pricing forwards, we cannot bypass the explicit requirement for a model of the underlying asset price. *Stochastic models* for the underlying asset price take the random behaviour of market prices into account and are beyond the scope of this book on *deterministic models*. (Although simple stochastic models are introduced in Chapter 12).

The total value of an option at a time prior to expiry can be considered as having two components: the *intrinsic value* and the *time value*. The **intrinsic value** of an option at time $t \leq T$ is the value arising from its moneyness at that time, as defined in Section 11.3. For example, the intrinsic value of a call option at time t is $\max\{S_t - K, 0\}$. The intrinsic value is positive if the option is in-the-money and zero if at-the-money or out-of-the-money.

The actual value of an option is in its potential for payoff and profit at exercise. For this reason, an option's value at a particular time should reflect not only its intrinsic value but also the likelihood that the intrinsic value will increase by the expiry date. This component is called the **time value**. The value of an option at time t is then the sum of intrinsic value and time value at that time. How, though, can we compute the value (i.e., price) of an option at general time t?

Note that our attention is henceforth restricted to European options on non-income-paying assets. In what follows, we denote the price of a European call option by c_t and the price of a European put option by p_t. The price of these options at time $t = 0$ gives the **option premium** (c_0 or p_0). As in Chapter 10, we denote the risk-free force of interest as r.

Before going straight to explicit expressions for c_t and p_t, we demonstrate how the ideas of *no arbitrage* and *replicating portfolios* (discussed in Chapter 10) can be applied to option pricing. Although they will not lead to the explicit expressions for option prices, they will enable us to deduce upper and lower bounds.

Theoretical Bounds on Option Prices
European Call Options

Consider Portfolio A at time $t \leq T$ that consists of a European call option (with remaining term $T - t$ and strike price K) of value c_t and a sum of money equal to $Ke^{-r(T - t)}$ invested at the risk-free rate.

At the expiry time T, the cash investment will have accumulated to amount K, which is equal to the strike price of the option. The payoff from the option will depend on the market price of the asset at that time. In the event that $S_T < K$, the call option will not be exercised, and the payoff from the cash investment K is greater than that from the option. In the event that $S_T > K$, the call option will be exercised and the cash investment used to fund the trade. Portfolio A therefore produces a payoff at time T that is at least as great as the underlying asset.

If we assume no arbitrage, the law of one price implies that the portfolio must have a value greater than or equal to S_t at time t—that is $c_t + Ke^{-r(T-t)} \geq S_t$ —and so

$$S_t - Ke^{-r(T-t)} \leq c_t \tag{11.4.1}$$

This gives a lower bound for the price of a European call option at time $t \leq T$.

In order to determine the upper limit on the price note that the payoff is $\max\{S_T - K, 0\}$ and so is always less than the price of the asset at the expiry time T. We therefore have

$$c_t \leq S_t \tag{11.4.2}$$

Combining Eqs 11.4.1 and 11.4.2, we see that

$$S_t - Ke^{-r(T-t)} \leq c_t \leq S_t \tag{11.4.3}$$

which gives the upper and lower bounds for the value or price of a European call option on a non-income-paying asset at time $t \leq T$.

European Put Options

Consider Portfolio B at time $t \leq T$ that consists of a European put option (with remaining term $T - t$ and strike price K) of value p_t and the underlying asset of value S_t.

The value of the portfolio at expiry time T depends on the market price of the asset at that time. In particular, if the asset price is $S_T \leq K$, the put option will be exercised leading to the receipt of cash amount K. If the asset price is $S_T > K$, the option will not be exercised, and the asset will remain in the portfolio. Portfolio B is therefore always worth at least K at time T, which has a present value of $Ke^{-r(T-t)}$ at time t. Assuming no arbitrage and invoking the law of one price enables us to conclude that $p_t + S_t \geq Ke^{-r(T-t)}$ and so

$$Ke^{-r(T-t)} - S_t \leq p_t \tag{11.4.4}$$

This gives a lower bound on the price of the put option at time $t \leq T$.

Equation 11.3.1 demonstrates that the maximum payoff a European put option can achieve at expiry is the strike price K. This is achieved only if the underlying asset price has fallen to zero. It is therefore clear that the value of the put option at time t cannot exceed the present value of the strike price at that time, and we have

$$p_t \le Ke^{-r(T-t)} \qquad (11.4.5)$$

Combining Eqs 11.4.4 and 11.4.5, we see that

$$Ke^{-r(T-t)} - S_t \le p_t \le Ke^{-r(T-t)} \qquad (11.4.6)$$

which gives the upper and lower bounds for the value or price of a European put option on a non-income-paying asset at time $t \le T$.

EXAMPLE 11.4.1

Three-month European options with strike price $16,000 are to be written on one tonne of nickel. If the current market price of nickel is $16,128 per tonne and the risk-free force of interest is 3% per annum, calculate the upper and lower theoretical bounds for European call and put option premiums.

Solution

European call: We apply Eq. 11.4.3 at time t with $K = 16,000$, $S_t = 16,128$, $r = 0.03$, and $T - t = \frac{3}{12}$, leading to

$$247.55 \le c_0 \le 16,128$$

European put: We apply Eq. 11.4.6 at time t with $K = 16,000$, $S_t = 16,128$, $r = 0.03$, and $T - t = \frac{3}{12}$, leading to

$$-247.55 \le p_0 \le 15,880.45$$

which is rewritten as $0 \le p_0 \le 15,880.45$ since the put cannot have negative value in practice.

As demonstrated in Example 11.4.1, the theoretical bounds obtained from expressions 11.4.3 and 11.4.6 can lead to very large ranges in the value of an option, including negative values in theory. A negative theoretical bound reflects the poor position that the holder of such an option would be in under the assumption that the asset price grows at the risk-free rate. In practical applications, the theoretical bounds are of limited use, and an alternative approach is required to compute actual prices, c_0 and p_0.

Put-Call Parity

Before we move to explicit expressions for option prices, it is important to consider the link between the prices of call and put options with identical expiry dates and strike prices, written on the same underlying non-income-paying asset. This is possible by combining the replicating portfolio arguments used to determine the previous theoretical price bounds.

We consider the two portfolios A and B, described previously. Recall that, at time t, Portfolio A consists of a European call option and a cash amount of $Ke^{-r(T-t)}$ invested at the risk-free rate r, and Portfolio B consists of a European put option and the underlying asset of value S_t. It is clear that the payoff of either portfolio at expiry is $\max\{K,S_T\}$. Furthermore, as European options cannot be exercised prior to expiry, invoking the law of one price (under the assumption of no arbitrage) leads to the situation that the two portfolios must have the same value at all $t < T$. This results in the *put-call parity* relationship for non-income-paying underlying assets:

$$c_t + Ke^{-r(T-t)} = p_t + S_t \qquad (11.4.7)$$

EXAMPLE 11.4.2

A 6-month European call option with strike price £12 was written on a commodity. If, after 2 months, the market price of the commodity is £10 and the price of the call option is 10p, calculate the price of an equivalent put option written on the same commodity at that date. The risk-free force of interest is 3% per annum.

Solution

We assume no arbitrage and use the put-call parity relationship 11.4.7 at time $t = \dfrac{2}{12}$, with $c_t = 0.10$, $K = 12$, $S_t = 10$, $T - t = \dfrac{4}{12}$, and $r = 0.03$. This leads to

$$p_{2/12} = 0.10 + 12e^{-0.03 \times 4/12} - 10 = 1.98$$

The price of the equivalent put option is £1.98.

EXAMPLE 11.4.3

A European call option and a European put option on the same underlying non-income-paying asset are to expire in exactly 1 year and have the same strike price. The current asset price is £2. If the risk-free force of interest is 2% per annum and both options currently have the same price, calculate the common strike price. Assume no arbitrage.

Solution

The put-call parity relationship connects the two prices. We have $c_t = p_t$, $S_t = 12$, $T - t = 1$, $r = 0.02$ and Eq. 11.4.7 therefore gives

$$\begin{aligned} K &= (p_t - c_t + s_t)e^{r(T-t)} \\ &= (0 + 12)e^{0.02} \\ &= 12.24 \end{aligned}$$

The common strike price must be £12.24.

The put-call parity relationship provides a useful check for arbitrage opportunities. Since it is based on the assumption of no arbitrage, if a pair of options can be found that violate Eq. 11.4.7, there must be an opportunity for risk-free profits. This is illustrated in Example 11.4.4.

As demonstrated in Example 11.4.4, the profit from an individual arbitrage trade could be small. However, the arbitrageur could expand this strategy to any

EXAMPLE 11.4.4

Two European options, one call and one put, are written on a particular commodity. Both options have a strike price of $20 and expire in 6 months' time. The market is currently pricing the underlying stock at $19, the call option at $1, and the put option at $1.50. If the risk-free force of interest is 3% per annum, determine if an arbitrage opportunity exists and, if so, how an arbitrageur could generate a risk-free profit.

Solution

The put-call parity relationship 11.4.7 gives the price of the European put option relative to the call option as

$$p_t = c_t + Ke^{-r(T-t)} - S_t$$
$$= 1 + 20e^{-0.03 \times 0.5} - 19$$
$$= 1.70$$

This value is *greater* than the current market price of the put option ($1.50) and the put can be considered as underpriced relative to the call. An arbitrage opportunity therefore exists.

Since the put is underpriced, the strategy to create a risk-free profit should involve *purchasing* the put at $1.50 and *selling* (i.e., writing) a call for $1. At the same time, the underlying asset should be purchased for $19. In order to fund this net expenditure, the arbitrageur can borrow an amount $1.50 + $19 − $1 = $19.50 at the risk-free rate, r.

If, at expiry, the asset price is greater than the strike price, $S_T > K$, the put option would not be exercised, leading to a zero payoff. The call option would, however, be exercised by the holder, leading to a negative payoff for the arbitrageur of $K - S_T = 20 - S_T$. This would be offset by the asset holding of value S_T, leading to an income of $20.

If, at expiry, the asset price were not greater than the strike price, $S_T \leq K$, the call option would not be exercised by the holder. This would lead to a zero payoff. The arbitrageur would, however, exercise the put option, leading to a positive payoff of $K - S_T = 20 - S_T$. With the asset holding, this would generate an income of $20.

We therefore see that the portfolio generates a payoff of $20, irrespective of the asset price at expiry. This should then be used to pay back the amount borrowed, leading to an arbitrage profit at expiry of $20 - 19.50e^{0.03 \times 0.5} = 0.21 irrespective of the future market movement.

number of trades, thereby magnifying the profit. For example, if 100,000 units of the portfolio in Example 11.4.4 were purchased, the arbitrage profit would be $21,000. This volume of trades from an individual arbitrageur or collectively from a large number of arbitrageurs performing smaller numbers of trades would have an effect on the market, as discussed in Chapter 10. The market prices of the options and underlying asset would then quickly shift to close the opportunity.

11.5 THE BLACK–SCHOLES MODEL

As demonstrated in Section 11.4, simple arguments involving the no arbitrage assumption and the law of one price can only get us so far in option pricing. In order to derive explicit expressions for the theoretical prices of European call and put options, we are forced to introduce pricing models for the underlying assets over the term of the option contract. Such models require the incorporation of random behaviour, i.e., are *stochastic* as opposed to *deterministic*, and a detailed treatment of such models is beyond the scope of this book. For this reason we merely state the option pricing expressions generated under the

Black–Scholes model and give some discussion of the underlying assumptions of the model. Interested readers are referred to textbooks on stochastic models in finance for a full treatment.

As with all applications of mathematics, the analysis of option pricing first requires one to make a number of assumptions. In particular, the Black–Scholes formulation requires the following assumptions to be made about the behaviour of underlying asset prices and the financial markets.

1. The underlying asset prices follow a *geometric Brownian motion* stochastic process.
2. The markets are *efficient*. By this, we mean that asset prices reflect all available information.
3. There are no arbitrage opportunities.
4. The risk-free rate of return r is constant and equal for all terms (i.e., has no term structure).
5. There are no transaction costs or taxes.
6. Asset trading is continuous.
7. Replicating strategies can always be generated.

Assumptions 2-7 are consistent with those previously made in the context of forward pricing. Assumption 1 is, however, an additional and significant assumption. It essentially means that the underlying asset price can be modeled as having a *log-normal distribution*. Such a model is discussed in the context of *stochastic interest rates* in Chapter 12.

In the particular case of European options on non-income-paying assets, the Black–Scholes model leads to the following expressions for option prices at time $t \leq T$:

$$c_t = S_t \Phi(d_1) - Ke^{-r(T-t)}\Phi(d_2)$$
$$p_t = Ke^{-r(T-t)}\Phi(-d_2) - S_t\Phi(-d_1)$$

(11.5.1)

Here, Φ is the cumulative distribution function of the *standard normal distribution*, and

$$d_1 = \frac{\ln\frac{S_t}{K} + \left(r + \frac{1}{2}\sigma^2\right)(T-t)}{\sigma\sqrt{T-t}} \quad \text{and} \quad d_2 = d_1 - \sigma\sqrt{T-t}$$

Expression 11.5.1 gives the theoretical price of a European call and put option at time $t \leq T$, in terms of a number of parameters, including the price of the underlying asset at that time, S_t; the strike price, K; the remaining term of the option, $t - T$; and the risk-free force of interest, r. These quantities are those required to derive theoretical bounds on the option prices in Section 11.4 (Eqs 11.4.3 and 11.4.6). However, the Black–Scholes pricing expressions also require the parameter σ, which denotes the *asset price volatility*. The quantity σ arises from the underlying asset-pricing model and is a parameter in the log-normal distribution function in Assumption 1.

EXAMPLE 11.5.1

An investor purchases a 1-year European call option on 1,000 non-dividend-paying shares, each with a current market price £1.50. If the strike price of each option is £1.70, the effective rate of risk-free interest is 5.127%, and the share price volatility is known to be 30%, calculate the total option premium paid.

Solution

We are required to evaluate c_0 from Eq. 11.5.1, with $S_0 = 1.50$, $K = 1.70$, $T - t = 1$, $r = \ln(1.05127) = 0.05$, and $\sigma = 0.30$. We begin by computing d_1 and d_2,

$$d_1 = \frac{\ln\frac{1.50}{1.70} + \left(0.05 + \frac{1}{2}0.30^2\right)}{0.30} = -0.10054$$

$$d_2 = d_1 - 0.30 = -0.40045$$

leading to

$$c_0 = 1.50 \times \Phi(-0.10054) - 1.70e^{-0.05} \times \Phi(-0.40054)$$
$$= 0.1330$$

The total call option premium is therefore around £133.

Note that values of $\Phi(-0.10054)$ and $\Phi(-0.40054)$ in Example 11.5.1 are obtained from statistical tables.

For example,

$$\Phi(-0.10054) = 1 - \Phi(0.10054) = 1 - 0.5400 = 0.4600$$

EXAMPLE 11.5.2

Compute the option premium for the corresponding put option in Example 11.5.1 using the Black–Scholes pricing model. Confirm that put-call parity is satisfied.

Solution

Using the expression for p_0 in Eq. 11.5.1 and the values of d_1 and d_2 from the solution to Example 11.5.1, we obtain

$$p_0 = 1.70e^{-0.05} \times \Phi(0.4005) - 1.50 \times \Phi(0.1005) = 0.2501$$

The total put option premium is therefore around £250.

Alternatively, one could use the put-call parity relationship 11.4.7 to calculate the put option premium.

$$p_0 = c_0 + Ke^{-r(T-t)} - S_0$$
$$= 0.1330 + 1.70 \times e^{-0.05} - 1.50$$
$$= 0.2501$$

This price agrees with the price from the Black–Scholes model, and we confirm that put-call parity is satisfied by the Black–Scholes pricing formulae.

Although the Black–Scholes pricing expressions are stated here without proof, it is important to note that they arise from the solution of a partial differentiation equation (PDE), the *Black–Scholes equation*:

$$\frac{\partial f}{\partial t} + \frac{1}{2}\sigma^2 S^2 \frac{\partial^2 f}{\partial S^2} + rS\frac{\partial f}{\partial S} - rf = 0$$

Here, $f(t)$ is the value of the particular option at time t (i.e., either c_t or p_t), and $S(t)$ is the value of the underlying asset at time t (i.e., S_t). The boundary condition for the solution of PDE at the expiry time is given by the payoff function for the particular type of option under consideration, i.e.,

$$f(T) = \max\{S - K, 0\} \text{ for a call}$$

$$f(T) = \max\{K - S, 0\} \text{ for a put}$$

Fischer Black and Myron Scholes first derived the Black–Scholes equation in the late 1960s. Their original derivation involved the formation of a continuous time hedging strategy with a carefully chosen replicating portfolio to permit the application of the law of one price. However, various people have since demonstrated that this is not the only way to derive the PDE. Alternative methods include the *Martingale Approach*, the *Binominal Model*, and the *Capital Asset Pricing Model*, for example, and all lead to the same expression. Interested readers are referred to textbooks on financial engineering for a detailed discussion of the various approaches to deriving the PDE and its solution leading to Eq. 11.5.1.

The underlying assumptions of the Black–Scholes model, as stated previously, and the restriction that the underlying asset does not pay income, clearly limit the practical application of the pricing formulae. However, despite the potential flaws in the model assumptions, analyses of market option prices do indicate that the Black–Scholes model is a very good approximation to the market, particularly for short-dated options. Furthermore, the restriction that the underlying does not pay income can be easily removed, leading to modified expressions

$$c_t = S_t e^{-q(T-t)} \Phi(d_1) - K e^{-r(T-t)} \Phi(d_2)$$
$$p_t = K e^{-r(T-t)} \Phi(-d_2) - S_t e^{-q(T-t)} \Phi(-d_1)$$

(11.5.2)

with

$$d_1 = \frac{\ln \frac{S_t}{K} + \left(r - q + \frac{1}{2}\sigma^2\right)(T-t)}{\sigma\sqrt{T-t}} \quad \text{and} \quad d_2 = d_1 - \sigma\sqrt{T-t}$$

Here, q is the **continuous rate of dividend payment**. Expressions 11.5.2 are often referred to as the **Garman-Kolhagen pricing formulae**. Note that expressions 11.5.2 reduce to the Black–Scholes pricing formulae when $q = 0$.

Further modifications to the Black–Scholes model can, of course, be made, but these are beyond the scope of this book. However, in practice, explicit expressions for the prices of more exotic options, and indeed other derivatives, cannot be found, and therefore **simulation techniques** are required. We

> **EXAMPLE 11.5.3**
>
> Determine the fair price for 2-year European call and put options written on the same dividend-paying share if both strike prices are £1.90 and the current market price of the share is £1.40. You are given that the risk-free force of interest is 3% per annum, the share price volatility is 20% per annum, and the continuously compounding dividend yield of the share is 5%. Assume no arbitrage.
>
> **Solution**
> We need to evaluate expressions 11.5.2 with $S_0 = 1.40$, $K = 1.90$, $T - t = 2$, $\sigma = 0.2$, $r = 0.03$, and $q = 0.05$. This leads to $c_0 = 0.023$ and $p_0 = 0.545$.

introduce simulation techniques in the context of stochastic interest rate modes in Chapter 12.

11.6 TRADING STRATEGIES INVOLVING EUROPEAN OPTIONS

The main uses of options are *speculation, arbitrage,* and *hedging*. These terms were discussed in Chapter 10, so we do not define them again here. In what follows, we consider various standard trading strategies involving one or more European options. As we will see, some trades can be considered as a means for speculation, some as hedging strategies, and others as a way to exploit arbitrage opportunities in the option markets. We begin with some straightforward applications of options and move to more complicated trades involving two or more options.

Simple Option Trades

Options have significant *financial leverage*, which makes them useful in *speculation*. *Financial leverage* refers to the fact that significant upside profits can be made from only a small initial outlay. In the particular case of equity (share) investment, for example, the option premium is significantly lower than the value of the equities on which the option is written. The leverage acts to magnify the effects of market movements.

As mentioned in Chapter 10, the downside of using futures or forwards in a *hedging* strategy to remove exposure to adverse market movements is that investors are then unable to benefit from favorable movements in the markets. The use of options removes this downside and can, in a sense, be thought of as providing *insurance* against adverse market movements while still allowing investors to benefit from favorable movements. The downside to using options in this context is that an option premium must be paid at the outset.

A further straightforward example of using options to protect the value of an investment is the **protective put** strategy, illustrated in Example 11.6.4.

EXAMPLE 11.6.1

An investor with £100 to invest believes that the share price of a particular company will *increase* from its current value of £5 to a value no less than £7 over the next 3 months. If the price of a 3-month call option on these shares with strike price £5 is 5p, explain how the investor could exploit the potential price rise by

(a) Purchasing shares,
(b) Purchasing call options on the shares.

Explain the downside risk in each case.

Solution

(a) The investor could purchase 20 shares for £100. If, after 3 months, the shares have appreciated to £7 or more, the investor would make a profit of at least £2 per share, i.e.,

a profit of £40 or more. The downside risk is that the share price might fall (however the investor would still own the shares and benefit from any future price rises).

(b) The investor could purchase 2,000 call options for £100. If, after 3 months, the shares have appreciated to £7 or more, the investor could exercise the call options and generate a payoff of £2 on each option, i.e., £4,000, leading to £3,900 profit. If the share price fails to appreciate by even 1p, the investor will not exercise the options and will lose the entire value of the investment. There is a further risk that the share price may not appreciate sufficiently for the option payoff to cover the option premium, leading to a loss.

EXAMPLE 11.6.2

An investor with £100 to invest believes that the share price of a particular company will *fall* over the next month. Explain how the investor could use a put option to profit from this potential fall in share price. Give an example of how this might work.

Solution

The investor could use the £100 to purchase put options on the shares with a strike price equal to, for example, their current market price. If the share price does, indeed, fall after 1 month, the investor would exercise the put options and profit by an amount equal to the fall in the share price for each option, minus the £100 expenditure.

For example, if the share price is currently £10 and the put option premium £1, the investor could purchase 100 1-month put options at a strike price of £10. If the market price falls to, say, £8, the investor could purchase the shares and exercise the put options, leading to a total *payoff* of £200. The *profit* would then be £100. The downside risk is that if the share price does not fall, the option premium will be lost, or, if the price falls by an amount less than £1, the payoff will not be sufficient to generate a profit. In both cases the strategy will result in a loss of, at most, £100 (i.e., the option premium).

EXAMPLE 11.6.3

Explain how the manufacturer in Example 10.3.1 could use options to hedge exposure to the exchange rate risk but retain the possibility of benefiting from favorable market movements.

Solution

Rather than entering into 3-month future contracts to trade $500,000 for sterling, the manufacturer could, for example,

buy 500 3-month put options, each based on $1,000. Upon expiry, if the terms of the option are favorable relative to the market conditions, the options can be exercised. If the terms of the option are unfavorable, the currency could be traded for GBP on the market. The downside is the loss of the option premium.

EXAMPLE 11.6.4

An investor owns 1,000 non-dividend-paying shares in a manufacturing company. On 1 July 2012, the shares had a value of £1.50 each, and had increased to £1.70 by 1 August 2012. If the investor is intending to sell his shares on 1 September 2012, explain how he could protect the value of the shares throughout August while still benefitting from any further price rises. If the price volatility of the shares is 30%, and the risk-free force of interest is 3% per annum, calculate the cost of this strategy.

Solution

The investor should purchase 1-month European *put* options on 1,000 shares with a strike price of £1.70 each. If the share price falls over August, he is able to exercise the put options and sell them at £1.70 each. If, however, the price had continued to appreciate, he could sell them on the market on 1 September for a price in excess of £1.70 each.

The cost of this strategy will be the initial option premium, $1,000p_0$, calculated from Eq. 11.5.1. We have $K = 1.70$, $S_0 = 1.70$, $T - t = \frac{1}{12}$, $r = 0.03$, and $\sigma = 0.3$; therefore, $1,000p_0 = £56.54$.

Synthetic Forwards

Simultaneously purchasing a call option and selling a put option, both on the same underlying, with the same strike price and expiry date, constructs a *synthetic forward*. It is clear that this holding mimics a forward contract, as the investor will take delivery of the underlying for the strike price at expiry, irrespective of the prevailing market price at that time. The advantage of using a synthetic forward over a regular forward is that, unlike a regular forward, a synthetic forward can be written with any strike price. Synthetic forwards can therefore be written with strike prices lower than would exist on the equivalent regular forward contracts. The downside is that there is a cost arising from the call option premium, offset by receipt of the put option premium (recall that there is no cost associated with entering into a regular forward contract). The put-call parity relationship 11.4.7 implies that this cost at $t = 0$ will be

$$c_0 - p_0 = S_0 - Ke^{-rT}$$

The strike price will always be lower that the regular *forward price*, $S_0 e^{rT}$ (Eq. 10.6.1), and so the cost is positive. (If the strike price was higher than the regular forward price, the investor would be better off using a regular forward contract.)

Collars

A *collar* is another example of a hedging strategy involving two options and a holding of the underlying asset. In particular, collars limit the returns from the underlying asset to a specified range, which is in contrast to a synthetic forward that locks onto a single, specified return. A collar can be constructed by an investor who owns the underlying asset by purchasing a put option on the

EXAMPLE 11.6.5

An investor wishes to purchase 50kg of coffee in 3 months' time at $231 per 50kg. If the current market price of coffee is $230 per 50kg and the risk-free force interest is 2.5% per annum, describe how this could be done. Determine any upfront cost of establishing this trade (you should assume the price volatility of coffee is 30% per annum).

Solution

The price of 50kg of coffee under a *regular forward contract* is $231e^{0.025 \times \frac{3}{12}} = 231.44$. The investor therefore wishes to purchase coffee at a price *below* the forward price and requires a *synthetic forward contract*. The investor should purchase 3-month call options on 50kg of coffee with a strike price of £231. Simultaneously, he should sell a 3-month put option on 50kg of coffee with the same strike price. The put-call parity relationship implies that this will cost $0.44.

At the expiry date, if the market price of coffee is above $231, the investor will exercise the call option to purchase the coffee at £231. If the market price of coffee is below $231, the holder of the put option will exercise it and the investor will purchase the coffee at £231.

underlying with strike price K_f and writing a call option on the same underlying with strike price K_c, both with identical expiry dates and on the same underlying. The difference between the two strike prices gives the *width* of the collar, $K_c - K_f$, with K_f the *floor* (i.e., lower bound) and K_c the *cap* (i.e., upper bound).

Collars are useful in **bear** markets (where prices are falling) because they limit the extent of any future loss on a stock holding without having to immediately sell the stock. It should also be noted that the initial cost of purchasing the put option is offset from the sale of the call option with a higher strike price; collars therefore often have low (or even *negative*) costs. This is an advantage over using single-sided *protective puts*, for example. A further advantage of collars is that investors can reduce their exposure to volatile markets and obtain a known return in the event of a large price shift in either direction.

EXAMPLE 11.6.6

An investor owns non-dividend-paying stock with a current market price of £300. Explain how a collar could be formed so that the gain/loss on this holding is constrained between ±£20 after 6 months.

Solution

A collar should be created with cap £320 and floor £280. This is formed from the purchase of a 6-month put option on the stock holding with strike price £280, and the writing of an equivalent 6-month call with strike price £320.

At expiry, one of three trades will happen.

1. If the stock price is above £320, the investor will not exercise the put option, but the holder of the call option will exercise that. The investor will then sell the stock for £320 under the call option, i.e., the price has been capped at £320.

2. If the stock price is below £280, the holder of the call option will not exercise it, but the investor will exercise the put option. The investor will then sell the stock for £280 under the put option, i.e., the price has been floored at £280.

3. If the stock price is between £280 and £320, neither the holder of the call nor the investor will exercise his respective option. The value of the stock can be realized from the market.

EXAMPLE 11.6.7

If the price volatility of the underlying stock in Example 11.6.6 is 40% and the risk-free force of interest is 4% per annum, calculate the price of entering into the collar. Assume no arbitrage.

Solution

The call option is priced from the Black–Scholes pricing formulae 11.5.1 with $K = 320$, $S_0 = 300$, $T - t = 0.5$, $r = 0.04$, and $\sigma = 0.40$, leading to $c_0 = 28.05$. Similarly, the put option with strike price 280 has price $p_0 = 21.15$. The cost of the collar is then $-£6.90$.

Note that here the cost of the put has been entirely offset from the income from writing the call. This is an example of a **negative cost collar**. Collars can also be *zero cost* or *positive cost*, depending on the net premium.

Spreads

A further for profiting from a bear market is called a *bear spread*, or alternately a *put spread*. A bear spread is constructed by purchasing a put option at strike price K_2 and simultaneously selling another put option at a *lower* strike price $K_1 (< K_2,)$ but at the same expiry date and on the same underlying asset. The resulting position is similar to a put option but with limited upside potential, as shown in Figure 11.6.1.

The cost of purchasing one put option within a bear spread is offset from the selling of the other; this is an advantage over, say, a protective put strategy. Bear

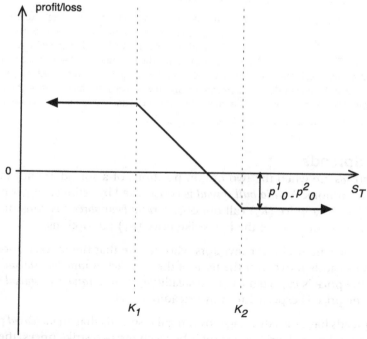

FIGURE 11.6.1 Position (profit) diagram for a bear (put) spread

spreads are of particular use to investors who either expect a moderate fall in market prices of an asset, or require a reduced exposure to a volatile bear market.

The equivalent formulation to exploit a *bull market* (where prices are rising) is the *bull spread* (or *call spread*). As one might expect, these are constructed from purchasing a call option at strike price K_1, and simultaneuously selling another call option at a *higher* strike price K_2 $(>K_1)$, but the same expiry date. The resulting position is like a call option but with limited upside potential, as shown in Figure 11.6.2.

EXAMPLE 11.6.8

An investor believes that a non-income-paying stock, currently trading at £20, will be subject to a moderate rally within 6 months, but will not exceed £26. To exploit this, he enters into a *bull spread* constructed from a long 6-month call with strike price £22 and a short 6-month call with strike price £26, each on 100 shares. If the long position costs £5 and the short position generates £3, discuss the resulting position if the market price of the stock after 6 months is

(a) £25,
(b) £27,
(c) £18.

Solution

The net cost of this spread is £2 at the outset.

(a) If the market price of the stock at expiry is £25, the investor will exercise the long call at strike price £22, generating a payoff of £3 per share. The short call will

not be exercised by the holder, and the overall profit for the investor will be $100 \times £3 - £2 = £298$.

(b) If the market price of the stock at expiry is £27, the investor will exercise the long call at strike price £22, generating a payoff of £5 per share. The short call will also be exercised at £26, leading to a negative payoff for the investor of £1 per share. The overall profit for the investor will be $100 \times (£5 - £1) - £2 = £398$.

(c) If the market price of the stock at expiry is £18, neither option will be exercised, leading to a total loss of £2 for the investor.

Note that the maximum profit for the investor is £398, and will occur if the market price of the stock is at any value *above* £26. This arises from the £4 price differential between the two call positions (which would both be exercised) and the initial £2 cost. The maximum loss for the investor is the initial £2 net premium and will arise if the market price of the stock is at any value *below* £22 and neither option is exercised.

Ratio Spreads

Ratio spreads are when the two option positions of a spread are not taken in equal proportions. A *ratio bull spread* is constructed by selling a larger number of the high strike price (K_2) call options. A *ratio bear spread* is constructed by selling a larger number of the low strike price (K_1) put options.

Ratio spreads are useful for investors who believe that the market price of an asset will steadily move over the term of the contract. A *ratio bull spread* will be used if the price is expected to move steadily higher; a *ratio bear spread* will be used if the price is expected to move steadily lower.

Ratio spreads have the advantages over regular spreads that an *increased* profit is generated if the stock price at expiry is between the two strike prices, the initial

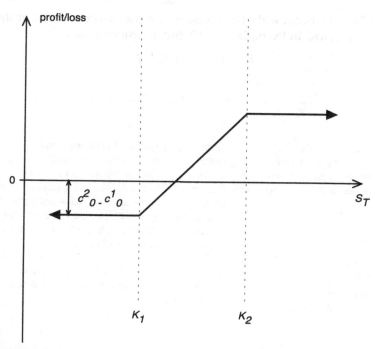

profit/loss

$c^2{}_0 - c^1{}_0$

S_T

K_1

K_2

FIGURE 11.6.2 Position (profit) diagrams for a bull spread

cost is significantly lower (perhaps even negative), and they result in a profit even if the underlying stock price has remained reasonably still on expiry.

Box Spreads

A *box spread* is a position formed from simultaneously entering into a *long bull spread* and a *long bear spread* (i.e., purchasing both spreads) on the same underlying with the same expiry date and the same pair of strike prices, K_1 and K_2. The box spread is a complicated position of *four* options and is best understood through Example 11.6.10.

Example 11.6.10 implies that a *guaranteed* payoff of £40 can be generated for an outlay of £39.01 if that particular box spread is used. This implies that a risk-free profit has been generated, but at what rate? The law of one price states that the return must be equal to the prevailing risk-free rate if no arbitrage opportunities exist within the particular trading strategy, but has this been the case? This is addressed in Example 11.6.11.

Examples 11.6.10 and 11.6.11 demonstrate an important consequence of using box spreads in an arbitrage-free environment: they return at a rate equal to the risk-free rate. This can be demonstrated theoretically using the put-call parity relationship. We begin with the expression for the theoretical cost of the box spread, P, as stated in Example 11.6.10, but at general time t

$$P = c_t^1 - c_t^2 + p_t^2 - p_t^1$$

EXAMPLE 11.6.9

Draw the position diagram for the resulting *ratio call spread* if, in Example 11.6.8, one low strike price call is purchased but two high strike price calls are written. Also, calculate the profit made for each of the closing prices given in the example.

Solution

The position diagram is shown in Figure 11.6.3

In this case, the net cost is £5 − 2 × £3 = −£1, i.e., the initial trade leads to a positive cash flow.

(a) If the market price of the stock at expiry is £25, the investor will exercise the long call at strike price £22,

generating a payoff of £3 per share. The two short calls will not be exercised by the holder, and the overall profit for the investor will be 100 × £3 + £1 = £301.

(b) If the market price of the stock at expiry is £27, the investor will exercise the long call at strike price £22, generating a payoff of £5 per share. The two short calls will also be exercised at £26, leading to a negative payoff for the investor of £2 per share. The overall profit for the investor will be 100 × (£5 − £2) + £1 = £301.

(c) If the market price of the stock at expiry is £18, no options will be exercised, leading to a total profit of £1 for the investor.

FIGURE 11.6.3 Position (profit) diagram for ratio call spread in Example 11.6.9 (not to scale)

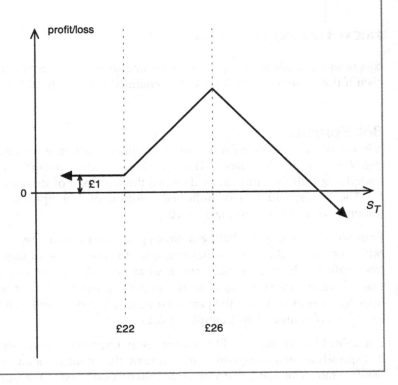

EXAMPLE 11.6.10

Consider a 6-month box spread with strike prices £80 and £120, constructed on an underlying non-income-paying stock with current price £100. If the risk-free force of interest is 5% per annum and the price volatility of the underlying is 20%, calculate the payoff at expiry and the upfront cost of forming the box spread.

Solution

The box spread is formed from purchasing simultaneous bear and bull spreads with strike price pairs K_1 and K_2 ($> K_1$). The bear spread is formed from a long put option at strike price K_2 and a short put at strike price K_1; the bull spread is formed from a long call option at strike price K_1 and a short call at strike price K_2. Each of the four options is written on the underlying stock and expires in 6 months' time. In this case, $K_1 = 80$ and $K_2 = 120$.

To calculate the payoff, we consider the following ranges for the stock price at the expiry date.

- $S_T < K_1 (= 80)$: Only the put options will be exercised. The net payoff to the investor will be £40.
- $(80 =) K_1 \leq S_T \leq K_2 (= 120)$: The call option with strike K_1 and the put option with strike K_2 will be exercised. The net payoff to the investor will be £40.

- $S_T > K_2 (= 120)$: Only the call options will be exercised. The net payoff to the investor is £40.

We therefore see that the payoff from the box spread is £40, irrespective of the price of the underlying at the expiry date.

The cost of forming the box spread is

$$c_0^1 - c_0^2 + p_0^2 - p_0^1$$

where the superscripts denote the strike price (K_1 or K_2) on which the call or put is written. Since we are assuming arbitrage does not exist, the option prices can be valued using the Black–Scholes pricing formulae 11.5.2, with $S_t = 100, T - t = 0.5, r = 0.05$, and $\sigma = 0.20$, to give

- $c_0^1 = £22.17,$
- $c_0^2 = £1.02,$
- $p_0^1 = £0.20,$
- $p_0^2 = £18.06.$

The cost of the box spread is therefore £18.06 − £0.20 + £22.17 − £1.02 = £39.01.

EXAMPLE 11.6.11

Calculate the return generated on the box spread strategy of Example 11.6.10. Comment on your answer.

Solution

We solve the following for δ:

$$39.01e^{\delta \times 0.5} = 40$$
$$\delta = 5\% = r$$

We therefore see that the return generated from the box spread is equal to the risk-free market rate, and no arbitrage opportunities exist within the collection of four options.

Of course, this is true as the arbitrage-free Black–Scholes pricing formulae were used to calculate the option prices in Example 11.6.10.

Grouping together terms with the same strike price and using the put-call parity relationship 11.4.7 leads to

$$P = c_t^1 - p_t^1 - (c_t^2 - p_t^2)$$
$$= S_t - K_1 e^{-r(T-t)} - S_t + K_2 e^{-r(T-t)}$$

and so

$$P = (K_2 - K_1)e^{-r(T-t)} \qquad (11.6.1)$$

We note that the payoff from a box spread is always equal the price differential of the two strike prices, and the result found in Example 11.6.11 is clear from Eq. 11.6.1.

EXAMPLE 11.6.12

Confirm Eq. 11.6.1 using the box spread described in Example 11.6.10.

Solution
We have $K_1 = 80$, $K_2 = 120$, and $r = 0.05$, and so Eq. 11.6.1 implies

$$P = (120 - 80)e^{-0.05 \times 0.5} = £39.01$$

which is identical to that obtained in Example 11.6.10.

The implications of the preceding description is that a return *greater* than the risk-free rate can be obtained from a box spread when arbitrage opportunities *do* exist within the options market.

EXAMPLE 11.6.13

Demonstrate how a profit can be made if the market prices of 6-month options on the same non-income-paying stock are

c_0^1	c_0^2	p_0^1	p_0^2
21.10	1.02	0.20	18.06

The superscripts indicate strike prices of $K_1 = 80$ and $K_2 = 120$. You are given that the risk-free force of interest is 5% per annum and the underlying stock has current price £100 and volatility 20%.

Solution
The market specifics are identical to those considered in Example 11.6.10, and we note that c_0^1 is currently trading at a price *lower* than the theoretical price of £22.17. This discrepancy can be exploited using the box spread discussed

in Example 11.6.10 where this option is required in the long position. The return on the box spread is £40 after 6 months, and the upfront cost, based on the preceding market data, is

$$c_0^1 - c_0^2 + p_0^2 - p_0^1 = 37.94$$

The return on this trade is then given by

$$37.94e^{\delta \times 0.5} = 40$$

and so

$$\delta = 10.6\%$$

This is greater than the risk-free return and arises from the existence of the arbitrage opportunity.

We have seen that box spreads can be used to exploit pricing discrepancies in the options market to form arbitrage profits. As discussed previously, the assumption that arbitrage does not exist within a market is crucial to the subject of financial mathematics; however, in reality, it can—and does—exist. Arbitrageurs are able to make large profits from seeking out these opportunities, and options are often useful in their strategies. The crucial point is that the action of the market will quickly act to remove the arbitrage opportunity. Unless one is explicitly looking for opportunities and is prepared to act quickly when such an opportunity is found, it is sensible, realistic, and prudent to assume that arbitrage does not exist.

Straddles

Straddles consist of two options, a put and a call, with the same expiry date and written on the same underlying asset. A particular feature of regular straddles is that they are *struck at-the-money*, i.e., are constructed on options with strike prices equal to the current market price of the underlying asset. There are two variations of regular straddles: *long straddles* and *short straddles*. A long straddle consists of long positions in both the call and put (i.e., they are purchased), and a short straddle consists of short positions in both the call and put (i.e., they are written and sold). The position diagrams are shown in Figure 11.6.4, and from

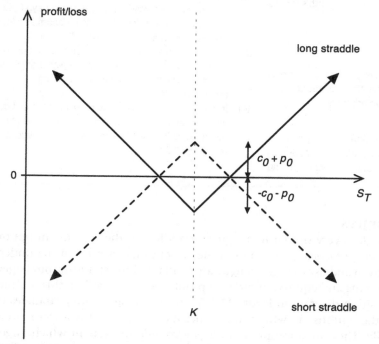

FIGURE 11.6.4 Position (profit) diagrams for long and short straddles

these, it is easy to understand how they are useful. The particular feature is that they are non-directional, meaning that the payoff depends on the *magnitude* of the price change, not whether it was up or down.

A *long straddle* gives a positive payoff on any change in price, but requires an upfront payment for the two options. Long straddles are therefore useful in volatile markets when market prices are expected to change significantly in either direction. The loss is limited to the price of the two options and the potential profit is unlimited.

As the short straddle involves writing two options and leads to a positive upfront cash flow but a negative payoff on any price change. Short straddles are therefore useful in a market where prices are *not* expected to change. The profit is limited to the option premiums and the potential loss is unlimited.

EXAMPLE 11.6.14

An investor is seeking to take advantage of a period of low volatility in the market price of a particular non-income-paying stock. Calculate the profit from taking a 1-month short straddle position if the price of the stock in 1 month's time is

(a) £101,
(b) £95,
(c) £110.

You are given that the current market price of the asset is £100, the price volatility is 10%, and the risk-free force of interest is 5% per annum. Assume no arbitrage.

Solution

The short straddle position involves writing a call option and a put option with strike price £100. Assuming no arbitrage, we value the options using the Black–Scholes pricing

formulae as $c = £1.37$ and $p = £0.95$. The upfront income generated is therefore £2.32.

(a) If the stock price at expiry is £101, only the call option will be exercised, leading to a negative payoff of £1 for the investor. This is less than the premium income, and the net *profit* from the short straddle is £1.32.

(b) If the stock price at expiry is £95, only the put option will be exercised, leading to a negative payoff of £5 for the investor. This is greater than the premium income, and the net *loss* on the short straddle is £2.68.

(c) If the stock price at expiry is £110, only the call option will be exercised, leading to a negative payoff of £10 for the investor. The net *loss* on the short straddle is therefore £7.68.

Strangles

A *strangle* is very similar to a regular straddle, but the constituent options are struck *out-of-the-money*, i.e., not at the current market price of the underlying. The action and use of *long strangles* are similar to long straddles, but larger price movements are required in order to profit from the position; this is clear from the position diagram in Figure 11.6.5. The advantage of long strangles is that, since the options are written out-of-the-money, the cost is lower than a long straddle. They are appropriate in *highly* volatile markets in which large price movements in either direction can be expected.

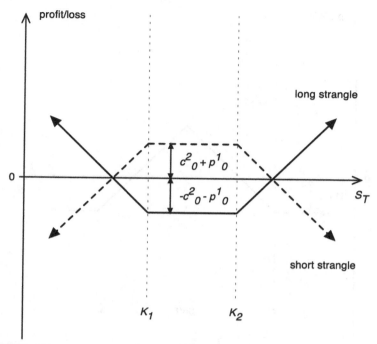

FIGURE 11.6.5 Position (profit) diagrams for long and short strangles

The action and use of *short strangles* are similar to short straddles, but the investor's profit from writing the position is protected for small changes in the underlying market price. The disadvantage is that short strangles are cheaper to write and the upfront income is lower.

EXAMPLE 11.6.15

The investor in Example 11.6.14 chooses instead to construct a 1-month *short strangle* with a spread of £5 in either direction around the current market price of £100. Using the same market data as in Example 11.6.14, calculate the profit from this position if the price of the stock in 1 month's time is

(a) £101,
(b) £95,
(c) £110.

Solution

The short strangle involves writing a call with strike price £105 and a put with strike price £95. Assuming no arbitrage,

the Black–Scholes prices of these options is $c = £0.08$ and $p = £0.03$. The upfront income generated is therefore £0.11.

(a) If the stock price at expiry was £101 neither of the options will be exercised. The net *profit* on the short strangle is therefore the full premium, £0.11.

(b) If the stock price at expiry was £95, again, neither option would be exercised. The net *profit* on the short strangle is therefore £0.11.

(c) If the stock price at expiry is £110, only the call option will be exercised, leading to a negative payoff of £5 for the investor. This is greater than the option premium and the net *loss* on the short strangle is £4.89.

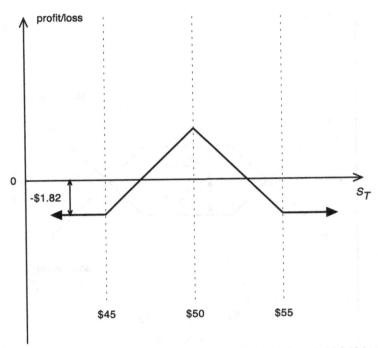

FIGURE 11.6.6 Position (profit) diagram for the long butterfly spread in Example 11.6.16 (not to scale)

Butterfly Spreads

The downside to using a short straddle or short strangle strategy is the *unlimited* loss that can result from movements either side of the strike price, even in a market with low volatility. To limit the exposure to such losses, one can form a *butterfly spread*. In particular, a *long butterfly spread* is formed from *four* call options: two short calls written at the current market price and a two long calls with strike prices either side of current market price. A position diagram of a long butterfly spread is shown in Figure 11.6.6. It is clear that the strategy limits the loss to the net value of the option premium.

EXAMPLE 11.6.16

If the current price of a non-income-paying asset is $50, devise a trading strategy that gives a positive payoff after 3 months if the price has moved by less than $5 in either direction and zero otherwise. Plot the position diagram of this strategy.

Solution

A *long butterfly spread* should be formed from two short 3-month calls with strike price $50, a long 3-month call with strike price $45, and another long 3-month call with strike price $55. The position diagram is then given in Figure 11.6.6.

EXAMPLE 11.6.17

If, in Example 11.6.16, the risk-free interest force of interest is 5% per annum and the price volatility of the underlying is 20%, calculate the profit if the market price of the asset after 3 months is

(a) $52,
(b) $46,
(c) $57.

Assume no arbitrage.

Solution

When we use the Black–Scholes pricing formulae, the constituent call options have the following prices:

c_0^{45}	c_0^{50}	c_0^{55}
5.84	2.31	0.60

The price of writing the butterfly spread is therefore $-2 \times 2.31 + 5.84 + 0.60 = \$1.82$.

(a) If the stock price at expiry was £52, the investor will exercise the $45 call, with payoff $7, and payout on the two short $50 calls, with payoff $-\$4$. The net payoff is $3, and the *profit* is $1.18.

(b) If the stock price at expiry was £46, the investor will exercise the $45 call, with payoff $1. The other options will not be exercised. The *loss* is $0.82.

(c) If the stock price at expiry is £57, all options will be exercised. The net payoff will be $12 + 2 - 2 \times 7 = 0$, and the *loss* is from the net premium price $1.82.

SUMMARY

- *Swaps* are derivative contracts whereby two parties agree to *swap* a series of payments according to a prearranged formula. They are mostly traded *over the counter,* and investors are exposed to both *market* and *credit* risks.
- *Interest rate swaps* involve the exchange of fixed-interest payments and variable interest payments on a *notional* principal amount. *Currency swaps* involve the exchange of regular cash flows and principal amounts of two currencies.
- Swaps are useful to investors wishing to exploit a *comparative advantage* or seeking to align their assets and liabilities.
- Interest rate swaps are constructed to have zero value to either party at the outset. Their value at a later time can be calculated by considering a long and short position in appropriate bonds.
- The *holder* of an *option* has *the right but not the obligation* to trade a specified asset, at a specified strike price, within a certain period of time. The option *writer* receives an *option premium* at the outset and is obligated to complete the trade at the option of the holder.
- The holder of a *call option* has the option to *purchase* the asset from the writer for the strike price K. The holder of a *put option* has the option to *sell* the asset to the writer for the strike price.
- A *European option* can be exercised only at the *expiry date* of the option. An *American option* can be exercised at any date prior to expiry. More exotic options exist.

■ The *payoff* to the holder of a European option at exercise is calculated as

$$f_{\text{call}}(S_T) = \max\{S_T - K, 0\}$$

$$f_{\text{put}}(S_T) = \max\{K - S_T, 0\}$$

where S_T is the market price of the underlying asset at the expiry time T (from the date it was written). The *profit* to either party is formed from the payoff and the option premium.

■ Option pricing is complicated and requires a pricing model for the underlying asset. The **Black–Scholes formulation** assumes the asset price has a log-normal distribution with price volatility σ. Under the no-arbitrage assumption, the formulation leads to the **pricing formulae** at general time $t < T$

$$c_t = S_t e^{-q(T-t)} \Phi(d_1) - K e^{-r(T-t)} \Phi(d_2)$$

$$p_t = K e^{-r(T-t)} \Phi(-d_2) - S_t e^{-q(T-t)} \Phi(-d_1)$$

with

$$d_1 = \frac{\ln \dfrac{S_t}{K} + \left(r - q + \dfrac{1}{2}\sigma^2\right)(T-t)}{\sigma\sqrt{T-t}} \quad \text{and} \quad d_2 = d_1 - \sigma\sqrt{T-t}$$

where q is the **continuously compounding dividend yield** generated by the underlying asset.

■ The **put-call parity** expression connects option prices for otherwise equivalent put and call options on non-income-paying assets:

$$c_t + K e^{-r(T-t)} = p_t + S_t$$

■ Various **option-trading strategies** exist for the purposes of hedging, speculation, and arbitrage. They involve single or multiple options.

EXERCISES

 CFA Institute

11.1 Consider European-style call and put options on a bond. The options expire in 60 days. The bond is currently at $1.05 per $1 nominal and makes no cash payments during the life of the option. The risk-free rate is 5.5% per annum (discretely compounding). Assume that the contract is on $1 nominal. Calculate the lowest and highest possible prices for the calls and puts with exercise prices of
(a) $0.95,
(b) $1.10.

11.2 You are provided with the following information on European put and
call options on a stock:
Call price, $c_0 = \$6.64$
Put price, $p_0 = \$2.75$
Exercise price, $K = \$30.00$
Days to option expiration $= 219$
Current stock price, $S_0 = \$33.19$
Put-call parity shows the equivalence of a call/bond portfolio
(also known as a *fiduciary call*) and a put/underlying portfolio
(i.e., a *protective put*). Illustrate put-call parity assuming stock prices at
expiration of $20 and of $40. Assume that the risk-free rate, r, is 4%
per annum.

11.3 A British company enters into a currency swap in which it pays a fixed
rate of 6% per annum in US$, and the counterparty pays a fixed rate of
5% per annum in sterling. The notional principals are £75 million and
$105 million. Payments are made semi-annually, and on the basis of 30
days per month and 360 days per year.
(a) Calculate the initial exchange of payments that takes place at the
beginning of the swap.
(b) Calculate the semi-annual payments.
(c) Calculate the final exchange of payments that takes place at the end
of the swap.

11.4 Suppose you believe that the price of a particular underling,
currently selling at $99, will decrease considerably in the next 6
months. You decide to purchase a put option expiring in 6 months
on this underlying. The put option has an exercise price of $95 and
sells for $5.
(a) Determine the profit for you under the following outcomes for the
price of the underlying 6 months from now:
  (i) $100;
  (ii) $95;
  (iii) $93;
  (iv) $90;
  (v) $85.
(b) Determine the breakeven price of the underlying at expiration.
Check that your answer is consistent with the solution to part (a).
(c) What is the maximum profit that you can have? At what expiration
price of the underlying would this profit be realized?

11.5 You simultaneously purchase an underlying priced at $77 and write
a call option on it with an exercise price of $80 and selling at $6. This
position is commonly called a *covered call*.
(a) Determine the value at expiration and the profit for your strategy
under the following outcomes:

        (i) The price of the underlying at expiration is $70;

      (ii) The price of the underlying at expiration is $75;

    (iii) The price of the underlying at expiration is $80;

    (iv) The price of the underlying at expiration is $85.

(b) Determine the following:

        (i) The maximum profit;

      (ii) The maximum loss;

    (iii) The expiration price of the underlying at which you would realize the maximum profit;

    (iv) The expiration price of the underlying at which you would incur the maximum loss.

(c) Determine the breakeven price at expiration.

 **11.6** You are bullish about an underlying that is currently trading at a price of $80. You choose to go long one call option on the underlying with an exercise price of $75 and selling at $10, and go short one call option on the underlying with an exercise price of $85 and selling at $2. Both the calls expire in 3 months.

(a) What is the term commonly used for the position that you have taken?

(b) Determine the value at expiration and the profit for your strategy under the following outcomes:

        (i) The price of the underlying at expiration is $89;

      (ii) The price of the underlying at expiration is $78;

    (iii) The price of the underlying at expiration is $70.

(c) Determine the following:

        (i) The maximum profit;

      (ii) The maximum loss.

(d) Determine the breakeven underlying price at expiration.

 **11.7** A stock is currently trading at a price of $114. You construct a butterfly spread using puts of three different strike prices on this stock, with the puts expiring at the same time. You go long one put with an exercise price of $110 and selling at $3.50, go short two puts with an exercise price of $115 and selling at $6, and go long one put with an exercise price of $120 and selling at $9.

(a) Determine the value at expiration and the profit for your strategy under the following outcomes:

        (i) The price of the stock at expiration of the puts is $106;

      (ii) The price of the stock at expiration of the puts is $110;

    (iii) The price of the stock at expiration of the puts is $115;

    (iv) The price of the stock at expiration of the puts is $120;

     (v) The price of the stock at expiration of the puts is $123.

(b) Determine the following:

        (i) The maximum profit;

    (ii) The maximum loss.

  (c) Determine the breakeven underlying price at expiration of the put options.

11.8 A stock is currently trading at a price of $80. You decide to place a collar on this stock. You purchase a put option on the stock, with an exercise price of $75 and a premium of $3.50. You simultaneously sell a call option on the stock with the same maturity and the same premium as the put option. This call option has an exercise price of $90.

 **CFA Institute**

  (a) Determine the value at expiration and the profit for your strategy under the following outcomes:

    (i) The price of the stock at expiration of the options is $92;

    (ii) The price of the stock at expiration of the options is $90;

    (iii) The price of the stock at expiration of the options is $82;

    (iv) The price of the stock at expiration of the options is $75;

    (v) The price of the stock at expiration of the options is $70.

  (b) Determine the following:

    (i) The maximum profit;

    (ii) The maximum loss;

    (iii) The stock price at which you would realize the maximum profit;

    (iv) The stock price at which you would incur the maximum loss.

  (c) Determine the breakeven underlying price at expiration of the put options.

11.9 You believe that the market will be volatile in the near future, but you do not feel particularly strongly about the direction of the movement. With this expectation, you decide to buy both a call and a put with the same exercise price and the same expiration on the same underlying stock trading at $28. You buy one call option and one put option on this stock, both with an exercise price of $25. The premium on the call is $4, and the premium on the put is $1.

 **CFA Institute**

  (a) What is the term commonly used for the position that you have taken?

  (b) Determine the value at expiration and the profit for your strategy under the following outcomes:

    (i) The price of the stock at expiration is $35;

    (ii) The price of the stock at expiration is $29;

    (iii) The price of the stock at expiration is $25;

    (iv) The price of the stock at expiration is $20;

    (v) The price of the stock at expiration is $25.

  (c) Determine the following:

    (i) The maximum profit;

    (ii) The maximum loss.

(d) Determine the breakeven underlying price at expiration of the options.

11.10 Spenser & Michael (S&M) is a UK-based food retailer, which is well known throughout Europe and the Far East, but largely unknown in the US. S&M has tried to borrow US$500m at a fixed rate of interest in US dollars, but the interest rates S&M can secure are prohibitively expensive. S&M has been quoted a 5-year fixed rate of 6% per annum for a sterling denominated loan.

BIM is a US-based food retailer and would like to borrow the sterling equivalent of US$500m over 5 years at a fixed rate of interest in sterling. Like S&M, BIM has been quoted prohibitively expensive rates for a sterling loan.

BIM has been quoted a 5-year fixed rate of 5.25% per annum for a US$ denominated loan. At the time of the transaction, the yield on 5-year government bonds is 5.25% in the UK and 4.75% in the US.

You are the head of the currency swap desk of a global investment bank.

(a) Describe, using the preceding information, the factors that will influence the design of a 5-year currency swap.

(b) Design a 5-year currency swap for S&M and BIM that will net the global investment bank 0.45% per annum over the life of the swap while ensuring that S&M and BIM have no exchange rate risk on their exchange of interest payments.

(c) Describe the risks that the global investment bank takes on in structuring this swap for BIM and S&M, and how the global investment bank can hedge these risks.

11.11 (a) Sketch, on the same diagram, the payoff at maturity of the following options on a quoted share assuming that they have the same strike price, same maturity date, and that the ratio of premiums is 2:1 with the call option being the more expensive option:

   (i) A put option;
   (ii) A call option.

(b) Using the same diagram as in part (a), sketch the payoff profile of an equally weighted portfolio consisting of options (a)(i) and (ii).

(c) Suggest an investment scenario in which the payoff profile in part (a)(ii) might be of interest to an investor.

11.12 A friend has decided that she would like to try to increase her wealth by investing in derivatives. After reading some articles, she realizes that she does not fully understand some of the terminology and has approached you for help.

(a) Explain the main uses of derivatives.

(b) For a derivative contract traded on an exchange:

   (i) Explain the term *margin*;

   (ii) Outline the different types of margin payments that are
payable;

   (iii) Explain why a clearinghouse requires these payments.

(c)   **(i)** Explain the difference between a put and a call.

   (ii) Explain the terms "European" and "American" in the context
of options.

The current share price of XYZ is 60p. Your friend has been offered the
following options in XYZ.

Strike Price	3 Month	6 Month
Call — 75p	5p	10p
Put — 85p	10p	5p

(d) Given this information, draw the pay-off charts associated with
each of these options clearly identifying the price when the option
is "in the money".

11.13 An employer contracts with his staff to give each of them 1,000 shares
in 1 year's time provided the share price increased from its current level
of £1 to at least £1.50 at the end of the year.
You may assume the following parameters:

Institute and Faculty of Actuaries

- Risk-free interest rate: 4% per annum continuously compounded;
- Stock price volatility: 30% per annum;
- Dividend yield: nil.

(a) Calculate the value of the contract with each employee by considering
the terms of the Black–Scholes formula.

The employer now wishes to limit the gain to each employee to £2,000.

(b) Calculate the value of this revised contract.

(c) An employee has said that he believes the original uncapped contract is
worth £300. He has determined this by saying that he believes there is
a 30% chance of the share price being at least £1.50; therefore, 30% ×
£1 = 30p.

  (i) Compare the approach taken by the employee and the approach
used in (a).

  (ii) Comment on the implications of the differences in (c)(i) if there
were a market in such contracts.

11.14 A risk-averse individual coming up to retirement age has around 1% of
his retirement funds invested in the shares of a highly respected bank.
The bank has historically been involved in low-risk activities, producing
steady returns. Since new management was put in place 3 years ago, the
bank has been involved in a number of high-profile risky investments

Institute and Faculty of Actuaries

that have gone wrong. As a result, there has been a sharp decline in the share price.

(a) Outline how the change in management has affected the risk and return profile of the individual's investment portfolio.

The individual had the opportunity to sell the stock 6 months ago but decided to hold onto the stock. Since then, the share price has fallen further.

(b) Outline the various reasons why the stock might not have been sold.

(c) Discuss whether the investment is suitable for the individual's circumstances.

The investor believes the share price has reached its lowest point and expects it to rise in the near future. The investor wants to try to make back some of his losses.

(d) Describe a technique, using the current share price, that the investor could use to make a profit on his holding without selling any shares.

The investor decides to use the technique in (d).

(e) Describe the effect on the investor's exposure to the bank if the bank's share price rose by 30%.

(f) Describe the effect on the investor's exposure to the bank if the bank's share price fell by 30%.

Institute
and Faculty
of Actuaries

11.15 The trustees of a pension fund decide to purchase a 3-year swap contract under which the pension fund will receive a fixed rate payment stream. The pension fund is required to pay a floating rate payment stream in return. The pension fund receives the following information about the swap and the likely payments:

- Term 3 years;
- Notional value of swap £50m;
- Payments are made in arrears semi-annually;
- The swap year calculations assume there are 360 days in a year.

Period	Number of Days in Period	Annual Forward Interest Rate
1	183	4.00%
2	181	4.25%
3	182	4.5%
4	182	4.75%
5	181	5.0%
6	183	5.25%

(a) Define the term "puttable swap".

(b) Explain why a pension fund may wish to purchase a puttable swap.

(c) State the type of swap that the trustees have purchased.

(d) Using the preceding information, calculate
  (i) Present value of the floating rate payments;
  (ii) The fixed rate of the swap.
The pension fund trustees proceed with the proposed contract for the payments described in this exercise, and the fixed rate of the swap is set at 4.75% per annum.

(e) Calculate the profit or loss to the pension fund at the end of the swap contract.

(f) Explain what difference there would have been to the profit/loss on the swap if interest rates had risen during the duration of the swap contract.

11.16 (a) (i) Define the terms "American call" and "European call".
     (ii) Explain which one is likely to attract a higher premium.

Institute and Faculty of Actuaries

  (b) Draw a diagram for each of the following strategies and explain why an investor may wish to undertake such strategies.
    (i) Long one call at a strike price of $(X - a)$;
        Short two calls at a strike price of X;
        Long one call at a strike price of $(X + a)$;
        All three have the same expiry date.
    (ii) Buying one call and one put with the same expiry and strike price.
    (iii) Buying call options of a certain strike price and selling the same number of call options at a lower strike price (in the money) with the same expiry date.

As part of an investor's portfolio there are 100 call options that have been written with an exercise price of £1.50 and an expiry date of November. The option premium received was £0.50 per option.

  (c) State the payoff for the investor.
  (d) Draw the payoff chart for the entire holding.
  (e) Calculate the profit or loss to the investor if the price of the share at expiry is
    (i) £0.75;
    (ii) £1.50;
    (iii) £2.15.
    State any assumptions made.

The derivatives exchange where the call options are traded requires an initial margin of 20% of the premium received. In addition a variation margin has to be paid equal to 100% of the option price movement. The value of the premium at the end of September was £0.55.

  (f) Calculate the total margin the investor has had to post to the exchange at the end of September.

11.17 Consider a stock paying a dividend at a rate δ and denote its price at any time t by S_t. The dividend earned between t and T, $T \geq t$, is $S_t(e^{\delta(T-t)} - 1)$.

Institute and Faculty of Actuaries

Let C_t and P_t be the price at time t of a European call option and European put option, respectively, written on the stock S, with strike price K and maturity $T \geq t$. The instantaneous risk-free rate is denoted by r.

Prove put-call parity in this context by adapting the proof of standard put-call parity that applies to put and call options on a non-dividend-paying stock.

11.18 A European call option on a stock has an exercise date 1 year away and a strike price of 320p. The underlying stock has a current price of 350p. The option is priced at 52.73p. The continuously compounded risk-free interest rate is 4% per annum.

(a) Estimate the stock price volatility to within 0.5% per annum, assuming the Black–Scholes model applies. (This is often referred to as the *implied volatility*.)

A new derivative security has just been written on the underlying stock. This will pay a random amount D in 1 year's time, where D is 100 times the terminal value of the call option capped at 1p (i.e., 100 times the lesser of the terminal value and 1p).

(b) (i) State the payoff for this derivative security in terms of two European call options.

(ii) Calculate the fair price for this derivative security.

11.19 A European call option and a European put option on the same stock with the same strike price have an exercise date 1 year away, and both are priced at 12p. The current stock price is 300p. The continuously compounded risk free rate of interest is 2% per annum.

(a) Calculate the common strike price, quoting any results that you use.

Assume the Black–Scholes model applies.

(b) Calculate the implied volatility of the stock.

(c) Construct the corresponding hedging portfolio in shares and cash for 5,000 of the call options.

CHAPTER 12

An Introduction to Stochastic Interest Rate Models

Throughout this book we have frequently remarked or implied that financial contracts are often of a long-term nature. Accordingly, at the outset of many contracts, there may be considerable uncertainty about the economic and investment conditions that will prevail over the duration of the contract. Therefore, for example, if it is desired to determine premium rates or security prices on the basis of one fixed rate of interest, it is nearly always necessary to adopt a conservative basis for the rate to be used in any calculations. An alternative approach to recognizing the uncertainty that exists in reality is provided by the use of a *stochastic interest rate model*. In such models no single interest rate is used, and variations in the rate of interest are allowed for by the application of probability theory. This is the subject of this final chapter.

An in-depth discussion of stochastic interest rate models is beyond the scope of this book. Although in this chapter we are able to present only a very brief introduction to the subject, it is important to recognize that, when we combine elementary financial concepts with probability theory, stochastic interest rate models offer a powerful tool for the analysis of financial problems. Moreover, this tool is fundamentally different from the deterministic approach considered throughout this book until this final chapter. We proceed to develop a stochastic model for interest rates before introducing the Brownian motion process, which features heavily in stochastic models within mathematical finance in the broad sense.

12.1 INTRODUCTORY EXAMPLES

Possibly one of the simplest stochastic interest rate models is that in which each year the rate of interest obtained is independent of the rates of interest in all previous years and takes one of a finite set of values, each value having a constant probability of being the actual rate for the year. Alternatively, the rate of interest may take any value within a specified range, the actual value for the year being determined by some given probability density function.

Even simple models, such as those just described, can be of practical use. For example, they may serve to illustrate the sometimes considerable financial

CONTENTS

An Introduction to the Mathematics of Finance. http://dx.doi.org/10.1016/B978-0-08-098240-3.00012-6

consequences to an office of the departure of its actual investment experience from that assumed in its premium bases.

Modern computers make it relatively simple to study considerably more sophisticated models. For example, it is possible (and probably more realistic) to remove the independence assumption and to build in some measure of dependence between the rates of interest in successive years (see Section 12.5).

For many stochastic interest rate models, the attainment of theoretically exact results is difficult. However, this is not a significant barrier to further progress. The use of *simulation techniques* often provides revealing insights of practical importance into the financial risks involved in many kinds of contracts.

At this stage we consider briefly an elementary example, which, although necessarily artificial, provides a simple introduction to the probabilistic ideas implicit in the use of stochastic interest rate models.

Suppose that a company wishes to issue a block of *single-premium capital redemption policies*, each policy having a term of 15 years. Note that under these policies the policyholder (or investor) pays a single individual premium at time $t = 0$ and expects to receive a specified return at time $t = 15$.

The company will invest the premiums received in a fund that grows under the action of compound interest at a constant rate throughout the term of the policy. This constant rate of interest is not known *now* but will be determined immediately after the policies have been issued.

Suppose that the effective annual rate of interest for the fund will be 2%, 4%, or 6%, and that each of these values is equally likely. In the probabilistic sense, the expected value of the annual interest rate for the fund is

$$E[i] = \left(\frac{1}{3} \times 0.02\right) + \left(\frac{1}{3} \times 0.04\right) + \left(\frac{1}{3} \times 0.06\right) = 0.04 \qquad (12.1.1)$$

or 4%.

Consider now a policy with unit sum assured, i.e., a return of 1 after 15 years. Let P be the single premium charged. If i denotes the annual rate of interest for the fund in which the premiums are invested, the accumulated profit (i.e., the company's profit at the time the policy matures) will be

$$P(1+i)^{15} - 1 \qquad (12.1.2)$$

This will equal $[P(1.02)^{15} - 1]$, $[P(1.04)^{15} - 1]$, or $[P(1.06)^{15} - 1]$, each of these values being equally likely. The expected value of the accumulated profit is therefore

$$E[\text{accumulated profit}] = \tfrac{1}{3}\Big[P(1.02)^{15} - 1\Big] + \tfrac{1}{3}\Big[P(1.04)^{15} - 1\Big]$$

$$+ \tfrac{1}{3}\Big[P(1.06)^{15} - 1\Big] \qquad (12.1.3)$$

$$= 1.847\ 79P - 1$$

Since the expected value of the annual yield is 4%, the company might choose to determine P on the basis of this rate of interest. In this case

$$P = v^{15} \quad \text{at } 4\% \quad = 0.555\ 26$$

and so, from Eq. 12.1.3,

$$E[\text{accumulated profit}] = 0.026\ 01$$

It should be noted that in this case the use of the average rate of interest in the premium basis does *not* give zero for the expected value of the accumulated profit. The reader should verify that the company's accumulated profit will be 0.330 72, 0, or −0.252 69, each of these values being equally likely. Although it is possible for the company to make a loss, it is equally possible for the company to make a greater profit.

An alternative viewpoint is provided by consideration of the net present value of the policy immediately after it is effected, i.e., the value at the outset of the policy of the company's contractual obligation less the value of the single premium paid. If i is the annual yield on the fund, the net present value is

$$(1 + i)^{-15} - P \qquad (12.1.4)$$

This will equal $(1.02^{-15} - P)$, $(1.04^{-15} - P)$, or $(1.06^{-15} - P)$, each of these values being equally likely. The expected value of the net present value is therefore

$$E[\text{net present value}] = \tfrac{1}{3}(1.02^{-15} - P) + \tfrac{1}{3}(1.04^{-15} - P) + \tfrac{1}{3}(1.06^{-15} - P)$$

$$= 0.571\ 88 - P$$

$$(12.1.5)$$

If, as before, P has been determined on the basis of the average rate of interest of 4% as 0.555 26, it follows that

$$E[\text{net present value}] = 0.016\ 62$$

Again, it should be noted that this is *not* zero.

Note also that the value of P for which the expected value of the accumulated profit is zero is *not* the value of P for which the expected net present value is zero.

12.2 INDEPENDENT ANNUAL RATES OF RETURN

In our introductory example in Section 12.1, the effective annual rate of interest *throughout the duration* of the policy was 2%, 4%, or 6%, each of these values being equally likely. A more flexible model is provided by assuming that over each *single year* the annual yield on invested funds will be one of a specified set of values or lie within some specified range of values, the yield in any particular year being independent of the yields in all previous years and being determined by a given probability distribution. For example, we might assume that each year the yield obtainable will be 2%, 4%, or 6%, each of these values being equally likely. Alternatively, we might assume that each year all yields between 2% and 6% are equally likely, in which case the density function for i is uniform on the interval $[0.02, 0.06]$.

We now formulate this scenario mathematically. Measure time in years and consider the time interval $[0, n]$ subdivided into successive periods $[0, 1]$, $[1,2]$, $\ldots$, $[n-1, n]$. For $t = 1, 2, \ldots, n$, let i_t be the yield obtainable over the tth year, i.e., the period $[t-1, t]$. Assume that money is invested only at the beginning of each year. Let F_t denote the accumulated amount at time t of all money invested before time t and let P_t be the amount of money invested at time t. Then, for $t = 1, 2, 3, \ldots, n,$

$$F_t = (1 + i_t)(F_{t-1} + P_{t-1}) \qquad (12.2.1)$$

It follows from this equation that a *single investment* of 1 at time 0 will accumulate at time n to

$$S_n = (1 + i_1)(1 + i_2) \cdots (1 + i_n) \qquad (12.2.2)$$

Similarly for a *series of annual investments*, each of amount 1, at times 0, 1, 2, ..., $n-1$ will accumulate at time n to

$$
\begin{aligned}
A_n = \; &(1 + i_1)(1 + i_2)(1 + i_3) \quad \cdots \quad (1 + i_n) \\
&+ (1 + i_2)(1 + i_3) \quad \cdots \quad (1 + i_n) \\
&+ \quad \vdots \qquad \vdots \qquad \vdots \\
&+ \quad (1 + i_{n-1})(1 + i_n) \\
&+ (1 + i_n)
\end{aligned} \qquad (12.2.3)
$$

Note that A_n and S_n are random variables, each with its own probability distribution function.

In general, a theoretical analysis of the distribution functions for A_n and S_n is somewhat difficult. It is often more useful to use simulation techniques in the study of practical problems (see Section 12.4). However, it is perhaps worth noting that the moments of the random variables A_n and S_n can be found relatively simply in terms of the moments of the distribution for the yield each year. This may be seen as follows.

EXAMPLE 12.2.1

If the yield each year is 2% per annum, 4% per annum, or 6% per annum, and each value is equally likely, demonstrate the possible values of the accumulation of a unit investment after 2 years.

Solution

It is clear that there are nine possible values that the accumulation S_2 could take. These are

Yield in Year 1	Yield in Year 2	S_2
2%	2%	$(1.02) \times (1.02) = 1.0404$
2%	4%	$(1.02) \times (1.04) = 1.0608$
2%	6%	$(1.02) \times (1.06) = 1.0812$
4%	2%	$(1.04) \times (1.02) = 1.0608$
4%	4%	$(1.04) \times (1.04) = 1.0816$
4%	6%	$(1.04) \times (1.06) = 1.1024$
6%	2%	$(1.06) \times (1.02) = 1.0812$
6%	4%	$(1.06) \times (1.04) = 1.1024$
6%	6%	$(1.06) \times (1.06) = 1.1236$

Note that the value of n-year unit accumulation, S_n, will be between 1.02^n and 1.06^n. Each of these extreme values will occur with probability $(1/3)^n$.

Moments of S_n

From Eq. 12.2.2 we obtain

$$(S_n)^k = \prod_{t=1}^{n} (1 + i_t)^k$$

and hence

$$E\left[S_n^k\right] = E\left[\prod_{t=1}^{n} (1 + i_t)^k\right]$$
$$= \prod_{t=1}^{n} E\left[(1 + i_t)^k\right] \tag{12.2.4}$$

since (by hypothesis) $i_1, i_2, \ldots, i_n$ are independent. Using this last expression, and given the moments of the annual yield distribution, we may easily find the moments of S_n.

For example, suppose that the yield each year has mean j and variance s^2. Then, letting $k = 1$ in Eq. 12.2.4, we have

$$E[S_n] = \prod_{t=1}^{n} E[1 + i_t]$$

$$= \prod_{t=1}^{n} (1 + E[i_t]) \tag{12.2.5}$$

$$= (1 + j)^n$$

since, for each value of t, $E[i_t] = j$.

With $k = 2$ in Eq. 12.2.4 we obtain

$$E[S_n^2] = \prod_{t=1}^{n} E[1 + 2i_t + i_t^2]$$

$$= \prod_{t=1}^{n} (1 + 2E[i_t] + E[i_t^2]) \tag{12.2.6}$$

$$= (1 + 2j + j^2 + s^2)^n$$

since, for each value of t,

$$E[i_1^2] = (E[i_t])^2 + \text{var}[i_t] = j^2 + s^2$$

The variance of S_n is

$$\text{var}[S_n] = E[S_n^2] - (E[S_n])^2$$

$$= (1 + 2j + j^2 + s^2)^n - (1 + j)^{2n} \tag{12.2.7}$$

from Eqs 12.2.5 and 12.2.6.

These arguments are readily extended to the derivation of the higher moments of S_n in terms of the higher moments of the distribution of the annual rate of interest.

EXAMPLE 12.2.2

For the yield distribution given in Example 12.2.1, compute values for $E[S_n]$ and var $[S_n]$ for $n = 2$ and 4.

Solution

We have $j = E[i] = \dfrac{0.02 + 0.04 + 0.06}{3} = 0.04$. Furthermore, $E[i^2] = \dfrac{0.02^2 + 0.04^2 + 0.06^2}{3} = 0.001867$, leading to $s^2 = 0.001867 - 0.04^2 = 0.000267$. Using Eqs 12.2.5 and 12.2.7, we then compute that

n	$E[S_n]$	var $[S_n]$
2	1.08160	0.00058
4	1.16986	0.00135

The reader is invited to confirm the values for $n = 2$ directly from the possible values for S_2 calculated in Example 12.2.1.

Moments of A_n

It follows from Eq. 12.2.3 (or from Eq. 12.2.1) that, for $n \geq 2$,

$$A_n = (1+i_n)(1+A_{n-1}) \tag{12.2.8}$$

The usefulness of this equation lies in the fact that, since A_{n-1} depends only on the values $i_1, i_2, \ldots, i_{n-1}$, the random variables i_n and A_{n-1} are independent. (By assumption the yields each year are independent of one another.) Accordingly, Eq. 12.2.8 permits the development of a recurrence relation from which may be found the moments of A_n. We illustrate this approach by obtaining the mean and variance of A_n.

Let

$$\mu_n = E[A_n]$$

and let

$$m_n = E[A_n^2]$$

Since

$$A_1 = 1 + i_1$$

it follows that

$$\mu_1 = 1 + j \tag{12.2.9}$$

and so

$$\begin{aligned} m_1 &= E[A_n]^2 + \text{var}[A_n] \\ &= 1 + 2j + j^2 + s^2 \end{aligned} \tag{12.2.10}$$

where, as before, j and s^2 are the mean and variance of the yield each year.

Taking expectations of Eq. 12.2.8 and using that i_n and A_{n-1} are independent, we obtain that

$$\mu_n = (1+j)(1+\mu_{n-1}) \quad n \geq 2 \tag{12.2.11}$$

This equation, combined with initial value μ_1, implies that, for all values of n,

$$\mu_n = \ddot{s}_{\overline{n}|} \qquad \text{at rate } j \tag{12.2.12}$$

The expected value of A_n is then simply $\ddot{s}_{\overline{n}|}$, calculated at the mean rate of interest. Recall from Chapter 3 that $\ddot{s}_{\overline{n}|}$ is the accumulated value (at $t = n$) of a unit annuity paid in advance. Since

$$A_n^2 = \left(1 + 2i_n + i_n^2\right)\left(1 + 2A_{n-1} + A_{n-1}^2\right)$$

by taking expectations we obtain, for $n \geq 2$,

$$m_n = \left(1 + 2j + j^2 + s^2\right)\left(1 + 2\mu_{n-1} + m_{n-1}\right) \tag{12.2.13}$$

EXAMPLE 12.2.3

A company plans to invest £50,000 now and at the start of the next 9 years at the interest rate used in Example 12.2.1. Calculate the expected value and standard deviation (s.d.) of the accumulated fund after 10 years.

n	μ_n	m_n	n	μ_n	m_n
1	1.040	1.082	6	6.898	47.617
2	2.122	4.503	7	8.214	67.524
3	3.246	10.544	8	9.583	91.907
4	4.416	19.513	9	11.006	121.248
5	5.633	31.748	10	12.486	156.070

Solution

We work in units of £1,000. The investments are made at times $t = 0, 1, 2, \ldots, 9$, and we are required to calculate $50\mu_{10} = 50 \times E[A_{10}]$ and $50 \times$ s.d.$[A_{10}]$. We have $j = 0.04$ and, by Eq. 12.2.2,

$$50\mu_{10} = 50\ddot{s}_{10} = 50 \times \frac{(1.04)^{10} - 1}{0.04/1.04} = 624.318$$

Equation 12.2.13 requires a recurrence approach to compute m_{10}. The multiplying factor $1 + 2j + j^2 + s^2$ has the value 1.08187, and we compute successive values of m_n as follows (noting that $\mu_0 = 0 = m_0$).

This leads to

$$50 \times \text{s.d.}[A_{10}] = (m_{10} - \mu_{10}^2)^{\frac{1}{2}}$$
$$= 20.051$$

The company would therefore expect to have an accumulated fund after 10 years with an expected value of £624,318 and a standard deviation of £20,051.

EXAMPLE 12.2.4

A company considers that, on average, it will earn interest on its funds at the rate of 4% per annum. However, the investment policy is such that in any one year the yield on the company's funds is equally likely to take any value between 2% and 6%. For both single and annual premium capital redemption policies with terms of 5, 10, 15, 20, and 25 years, and premium £1, find the mean accumulation and the standard deviation of the accumulation at the maturity date. You may ignore all expenses.

Solution

The annual rate of interest is uniformly distributed on the interval [0.02, 0.06]. The corresponding probability density function is constant and equal to 25 [i.e., 1/(0.06 − 0.02)]. The mean annual rate of interest is clearly $j = 0.04$, and the variance of the annual rate of interest is

$$s^2 = \int_{0.02}^{0.06} 25(x - 0.04)^2 = \frac{4}{3}10^{-4}$$

We are required to find $E[A_n]$, (var $[A_n])^{1/2}$, $E[S_n]$, and (var $[S_n])^{1/2}$ for $n = 5, 10, 15, 20,$ and 25.

Substituting the preceding values of j and s^2 in Eqs 12.2.5 and 12.2.7 we immediately obtain the results for the single premium policies. For the annual premium policies, we must use the recurrence relation 12.2.13 (with $\mu_{n-1} = \ddot{s}_{\overline{n-1}}$ at 4%) together with Eq. 12.2.14.

The results are summarized as follows. It should be noted that, for both annual and single premium policies, the standard deviation of the accumulation increases rapidly with the term.

	Single Premium £1		Annual Premium £1	
Term (years)	Mean Accumulation (£)	Standard Deviation (£)	Mean Accumulation (£)	Standard Deviation (£)
5	1.216 65	0.030 21	5.632 98	0.094 43
10	1.480 24	0.051 98	12.486 35	0.283 53
15	1.800 94	0.077 48	20.824 53	0.578 99
20	2.191 12	0.108 86	30.969 20	1.004 76
25	2.665 84	0.148 10	43.311 74	1.593 92

As the value of μ_{n-1} is known (by Eq. 12.2.12), Eq. 12.2.13 provides a recurrence relation for the calculation of successive $m_2, m_3, m_4, \ldots$. (Note that these may also be solved by the methods of difference equations.) The variance of A_n may be obtained as

$$\begin{aligned} \text{var}[A_n] &= E[A_n^2] - (E[A_n])^2 \\ &= m_n - \mu_n^2 \end{aligned} \qquad (12.2.14)$$

In principle, the preceding arguments are fairly readily extended to provide recurrence relations for the higher moments of A_n. However, a knowledge of the first few moments does not define the distribution, and it is generally more practicable to use simulation methods to gain further understanding of actual problems. We illustrate such methods in Section 12.4.

12.3 THE LOG-NORMAL DISTRIBUTION

We have already remarked that in general a theoretical analysis of the distribution functions for A_n and S_n is somewhat difficult, even in the relatively simple situation when the yields each year are independent and identically distributed. There is, however, one special case for which an exact analysis of the distribution function for S_n is particularly simple.

Suppose that the random variable $\ln(1 + i_t)$ is normally distributed with mean μ and variance σ^2. In this case, the variable $(1 + i_t)$ is said to have a *log-normal* distribution with parameters μ and σ^2. In this case, Eq. 12.2.2 is equivalent to

$$\ln S_n = \sum_{t=1}^{n} \ln(1 + i_t)$$

The sum of a set of independent normal random variables is itself a normal random variable. Hence, when the random variables $(1 + i_t)$ $(t \geq 1)$ are independent and each has a log-normal distribution with parameters μ and σ^2, the random variable S_n has a log-normal distribution with parameters $n\mu$ and $n\sigma^2$.

Since the distribution function of a log-normal variable is readily written down in terms of its two parameters, in the particular case when the distribution function for the yield each year is log-normal, we have a simple expression for the distribution function of S_n.

12.4 SIMULATION TECHNIQUES

All computers and calculators are able to generate "random numbers". In a strict sense, such numbers, when generated by computer, are *pseudo-random*, in that

EXAMPLE 12.3.1

The random variable $(1 + i)$ has a log-normal distribution with parameters μ and σ^2.

(a) Assuming standard results for the log-normal distribution, show that

$$E[i] = \exp\left(\mu + \frac{1}{2}\sigma^2\right) - 1$$

and

$$\mathrm{var}[i] = \exp\left(2\mu + \sigma^2\right)\left[\exp(\sigma^2) - 1\right]$$

(b) Let $E[i] = j$ and $\mathrm{var}[i] = s^2$. Show that

$$\sigma^2 = \ln\left[1 + \left(\frac{s}{1+j}\right)^2\right]$$

and that

$$\mu = \ln\frac{1+j}{\sqrt{1 + \left(\frac{s}{1+j}\right)^2}}$$

(c) **(i)** Given that $\mu = 0.08$ and $\sigma = 0.07$, find j and s.
   (ii) Given that $j = 0.08$ and $s = 0.07$, find μ and σ.

Solution

(a) We are given that $(1 + i)$ has a log-normal distribution with parameters μ and σ^2. From the standard properties of the distribution,

$$E[(1 + i)^n] = \exp\left(n\mu + \frac{1}{2}n^2\sigma^2\right)$$

Therefore,

$$1 + E[i] = E[1 + i] = \exp\left(\mu + \frac{1}{2}\sigma^2\right)$$

so that

$$E[i] = \exp\left(\mu + \frac{1}{2}\sigma^2\right) - 1$$

Also

$$
\begin{aligned}
\mathrm{var}[i] &= E\left[(1+i)^2\right] - (E[1+i])^2 \\
&= \exp(2\mu + 2\sigma^2) - \left[\exp\left(\mu + \frac{1}{2}\sigma^2\right)\right]^2 \\
&= \exp(2\mu + \sigma^2)\left[\exp(\sigma^2) - 1\right]
\end{aligned}
$$

(b) We have

$$1 + j = \exp\left(\mu + \frac{1}{2}\sigma^2\right)$$

and

$$s^2 = \exp(2\mu + \sigma^2)\left[\exp(\sigma^2) - 1\right]$$

Therefore,

$$\frac{s^2}{(1+j)^2} = \exp(\sigma^2) - 1$$

so that

$$\sigma^2 = \ln\left[1 + \left(\frac{s}{1+j}\right)^2\right]$$

Also

$$
\begin{aligned}
\ln(1 + j) &= \mu + \frac{1}{2}\sigma^2 \\
&= \mu + \frac{1}{2}\ln\left[1 + \left(\frac{s}{1+j}\right)^2\right]
\end{aligned}
$$

so that, as required,

$$\mu = \ln\frac{1+j}{\sqrt{\left[1 + \left(\frac{s}{1+j}\right)^2\right]}}$$

(c) **(i)** Using the expressions derived in this example, $j = 0.085\ 944$ and $s = 0.076\ 109$.
   (ii) Using the expressions derived in this example, $\mu = 0.074\ 865$ and $\sigma = 0.064\ 747$.

EXAMPLE 12.3.2

The yields on a company's fund in different years are independently and identically distributed. Each year the distribution of $(1 + i)$ is log-normal, and i has mean value 0.06 and variance 0.0003.

(a) Find the parameters μ and σ^2 of the log-normal distribution for $(1 + i)$.

(b) Let S_{15} be the random variable denoting the accumulation of a single premium of £1 for a period of 15 years. Show that S_{15} has a log-normal distribution with parameters 0.872 025 and 0.063 28^2. Hence, calculate values for $E[S_{15}]$ and var$[S_{15}]$.

(c) Confirm the values of $E[S_{15}]$ and var$[S_{15}]$ by Eqs 12.2.5 and 12.2.7.

(d) Find the probability that a single premium of £1,000 will accumulate over 15 years to (i) less than £2,100 and (ii) more than £2,700.

Solution

(a) In our earlier notation, $j = 0.06$ and $s^2 = 0.0003$, and so $\mu = 0.058\ 135$ and $\sigma^2 = 0.000\ 267$.

(b)
$$S_{15} = (1 + i_1)(1 + i_2)\cdots(1 + i_{15})$$

where i_t denotes the yield in year t. Hence,

$$\ln S_{15} = \ln(1 + i_1) + \ln(1 + i_2) + \cdots + \ln(1 + i_{15})$$

The terms on the right side of this equation are independent normally distributed random variables, each with mean μ and variance σ^2, as found in (a). This implies that $\ln S_{15}$ is normally distributed with mean $\mu' = 15\mu = 0.872\ 025$ and variance $\sigma'^2 = 15\sigma^2 = 0.004\ 005 = (0.063\ 28)^2$.

Hence (see Example 12.3.1),

$$E[S_{15}] = \exp\left(\mu' + \frac{1}{2}\sigma'^2\right) = 2.396\ 544$$
$$E\left[(S_{15})^2\right] = \exp(2\mu' + 2\sigma'^2) = 5.766\ 469$$

and so

$$\text{var}[S_{15}] = 5.766\ 469 - (2.396\ 544)^2 = 0.023$$

(c) By Eq. 12.2.5, $E[S_{15}] = 1.06^{15} = 2.3965$. By Eq. 12.2.7, var $[S_{15}] = (1.06^2 + 0.0003)^{15} - 1.06^{30} = 0.023$.

(d) (i)
$$S_{15} < 2.1 \Leftrightarrow \ln S_{15} < \ln 2.1 = 0.741\ 937.$$

Since $\ln S_{15}$ has a normal distribution with mean 0.872 025 and standard deviation 0.063 28 (as found previously), the required probability is

$$\Phi\left(\frac{0.741\ 937 - 0.872\ 025}{0.063\ 28}\right) = \Phi(-2.0557)$$

Note that $\Phi(x)$ is used to denote the standard normal distribution function. From standard tables, by interpolation, we evaluate this probability as 0.0199.

(ii) $S_{15} > 2.7 \Leftrightarrow \ln S_{15} > \ln 2.7 = 0.993\ 252$. The required probability is then

$$1 - \Phi\left(\frac{0.993\ 252 - 0.872\ 025}{0.063\ 28}\right)$$
$$= 1 - \Phi(1.9157) = 0.0277$$

they are produced by some well-defined numerical algorithm, rather than truly random. Truly random numbers are generated from physical processes, such as thermal noise and quantum phenomena that can be exploited in physical devices. (Although the use of such devices is clearly impractical in nearly all conceivable applications, in 1955 the RAND Corporation published a table of one million random digits obtained with such a device.) The essential point, however, about the sequence of values produced by an efficient pseudo-random number generator is that, from a statistical viewpoint, it is indistinguishable from a sequence of values produced by a genuinely random process. For practical purposes, therefore, the pseudo-random values generated by a computer or calculator may be considered as if they were, indeed, genuinely random numbers. In what follows we use the term "random numbers" although technically they may only be pseudo-random.

EXAMPLE 12.4.1

As a simple illustration of a simulation exercise, consider the model of independent annual rates of return (as described in Section 12.2) in relation to a fund for which the yield each year will be 2%, 4%, or 6%, each of these values being equally likely. Suppose that at the start of each year for 15 years, £1 is paid into the account. To how much will the account accumulate at the end of 15 years?

Solution

This question can be answered only with hindsight. At the start of the 15-year period, all that can be said with certainty is that the accumulated amount will lie between £17.6393

(i.e., $\ddot{s}_{\overline{15}|}$ at 2%) and £24.6725 (i.e., $\ddot{s}_{\overline{15}|}$ at 6%). The lower value will occur only if in every year of the period the annual yield is 2%. Likewise, the higher value will occur only if the yield every year is 6%. Since each year there is a probability of 1/3 that the yield will be 2% and (by assumption) the yields in different years are independent of one another, the probability that the yield will be 2% in *every* year of the period is $(1/3)^{15}$, which is less than 10^{-7}. This shows how unlikely it is that the accumulation will be the lowest possible value, for example.

The yield each year in Example 12.4.1 has expected value

$$j = E[i_t] = \frac{1}{3}(0.02 + 0.04 + 0.06) = 0.04 \qquad (12.4.1)$$

It follows from Eq. 12.2.12 that the accumulated amount of the fund (after 15 years) has mean value

$$\mu_{15} = E[A_{15}] = \ddot{s}_{\overline{15}|} \quad \text{at 4\%} \quad = 20.824533 \qquad (12.4.2)$$

The variance of the annual yield is

$$s^2 = E[(i_t - j^2)] = \frac{1}{3}[(-0.02)^2 + 0^2 + (0.02)^2] = \frac{8}{3}10^{-4} \qquad (12.4.3)$$

The variance of the accumulated amount may be found from Eqs 12.2.10, 12.2.13 (with $\mu_{n-1} = \ddot{s}_{\overline{n-1}|}$ at 4%), and 12.2.14. This gives the variance as

$$\text{var}[A_{15}] = 0.670\ 76 \qquad (12.4.4)$$

The standard deviation of the accumulated amount, σ_{15} say, is

$$\sigma_{15} = (\text{var}[A_{15}])^{1/2} = 0.819\ 00 \qquad (12.4.5)$$

How likely is it that the accumulated amount of the fund (after 15 years) will be less than £19 or greater than £22.50? The answers to these, and similar, questions are of practical importance. Theoretically, exact answers are difficult to obtain, but much valuable information may be obtained by *simulation*. This is illustrated in the following scenario:

Suppose that we generate a sequence of 15 random numbers from the interval $[0,1]$, say $\{x_i: i = 1, 2, \ldots, 15\}$. The value of x_t is used to determine i_t, the yield on the fund over year t, by the rule

$$i_t = \begin{cases} 0.02 & \text{if } 0 \le x_t < \frac{1}{3} \\ 0.04 & \text{if } \frac{1}{3} \le x_t < \frac{2}{3} \quad 1 \le t \le 15 \\ 0.06 & \text{if } \frac{2}{3} \le x_t \le 1 \end{cases} \qquad (12.4.6)$$

Since x_t is randomly chosen from the interval $[0,1]$, Eq. 12.4.6 ensures that each possible yield has probability 1/3 of being the actual yield in year t. The accumulated amount A_{15} (calculated by Eq. 12.2.3), corresponding to the sequence $\{x_i: i = 1, 2, \ldots, 15\}$ is then a fair realization of the experience of the fund over one particular period of 15 years.

We may repeat this procedure many times. For each simulation we generate a sequence of 15 random numbers from the interval $[0, 1]$ and then calculate the accumulated amount at the end of 15 years, using Eqs 12.4.6 and 12.2.3. By studying the distribution of the large number of resulting accumulations, we can gain valuable insights.

As an illustration, 10,000 simulations were carried out for the distribution of A_{15}. The sample mean of the resulting values was 20.82207, and the sample standard deviation was 0.81597. The lowest accumulation was 18.1394, and the greatest was 23.9657. The distribution of the values obtained by the simulations is indicated in Table 12.4.1.

Table 12.4.1 Distribution of $(A_{15} - \mu_{15})/\sigma_{15}$ 10,000 Simulations

Range	< -3	-3 to -2.5	-2.5 to -2.0	-2.0 to -1.5	-1.5 to -1.0	-1.0 to -0.5	-0.5 to 0.0	0.0 to 0.5	0.5 to 1.0	1.0 to 1.5	1.5 to 2.0	2.0 to 2.5	2.5 to 3.0	≥ 3
Number of Values in Range	1	22	165	462	930	1592	1914	1877	1463	880	446	171	59	18

In Table 12.4.1, each of the intervals used to define the groupings are closed on the left. The interval "−3 to −2.5" includes −3 but excludes −2.5.

EXAMPLE 12.4.2

Compare the sample mean and sample standard deviation resulting from the simulation to the values of μ_{15} and σ_{15}.

Solution

These values have already been calculated previously (Eqs 12.4.2 and 12.4.5) as $\mu_{15} = 20.824533$ and $\sigma_{15} = 0.81900$.

The simulated value of μ_{15} is therefore within 0.01% of the calculated value, and the simulation of σ_{15} is within 0.38%. The agreement between both values in each case is therefore very good and could be improved yet further with more simulations.

EXAMPLE 12.4.3

Using the data in Table 12.4.1, state how many of the 10,000 simulations will produce

(a) Accumulations which lie between μ_{15} and $\mu_{15} + 0.5\sigma_{15}$,
(b) Accumulations which are greater than or equal to $\mu_{15} + 3\sigma_{15}$,
(c) Accumulations which are less than $\mu_{15} - 3\sigma_{15}$,
(d) Accumulations which are less than μ_{15},
(e) Accumulations which are less than 19,
(f) Accumulations which are less than 22.5.

Solution

(a) 1,877 accumulations lay between μ_{15} and $\mu_{15} + 0.5\sigma_{15}$,
(b) 18 accumulations were greater than or equal to $\mu_{15} + 3\sigma_{15}$,
(c) 1 accumulation was less than $\mu_{15} - 3\sigma_{15}$,
(d) 5,086 accumulations were less than μ_{15},
(e) 80 of the accumulations were less than 19,
(f) 227 of the accumulations were greater than 22.5.

EXAMPLE 12.4.4

Using the data in Table 12.4.1, estimate a 95% confidence interval for

(a) The probability that any one 15-year period will produce an accumulated amount of less than £19,
(b) The probability that the accumulated fund will exceed £22.50.

Solution

(a) In this case, 80 accumulations from the 10,000 simulations produced an accumulated amount less than £19. This is equivalent to a probability of $\overline{p} = 80/10,000 = 0.008$. An estimate of the standard deviation is given by

$\sqrt{\overline{p}(1 - \overline{p})/n}$, where $n = 10,000$, and the confidence interval is

$$\left(\overline{p} - 1.96\sqrt{\overline{p}(1 - \overline{p})/n}, \overline{p} + 1.96\sqrt{\overline{p}(1 - \overline{p})/n} \right)$$

$$= (0.0063, 0.0097)$$

(b) In this case, 227 accumulations from the 10,000 simulations produced an accumulated amount greater than £22.50, leading to $\overline{p} = 0.0227$. The corresponding approximate 95% confidence interval is then calculated to be (0.0198, 0.0256).

Note that the standard deviation of the results will decrease with n, leading to a narrower confidence interval.

The preceding discussion and examples relate to one particular situation only, but the underlying ideas are readily extended to more general situations. In addition, more sophisticated simulation methods are available, for example to generate random samples drawn from a population with a normal, log-normal, or any other distribution.

12.5 RANDOM NUMBER GENERATION

The RND function on your calculator and RAND function in Excel, for example, generate uniformly distributed random numbers on the domain $[0,1]$, i.e., from the $U[0,1]$ distribution. At this stage it is worth briefly considering how we may simulate a more general random variable given only the ability to generate these uniformly distributed random numbers.

Only a small number of techniques are described here.

Inverse Transform Method

Suppose that the random variable X takes values on the interval $[a, b]$. (If $a = -\infty$ or $b = +\infty$, the interval is to be regarded as open or half-open, as appropriate.) Suppose that $f(x)$, the probability density function of X, is strictly positive on the interior of the interval. Let F be the distribution function for X. Therefore, for $a \leq x \leq b$,

$$F(x) = \text{probability } \{X \leq x\} \tag{12.5.1}$$

Clearly, $F(a) = 0$ and $F(b) = 1$. Since, by assumption, X has a strictly positive density function, F is a strictly increasing function over the interval $[a, b]$. The situation is illustrated in Figure 12.5.1.

If t is any real number in the interval $[0, 1]$, there is a unique real number in the interval $[a, b]$, $z(t)$ say, such that $F[z(t)] = t$. Suppose that α is a given value in the interval $[a, b]$ and that t is chosen randomly from the interval $[0, 1]$. What is the probability that $z(t)$ will not exceed α?

Consideration of Figure 12.5.2 shows that $z(t)$ will not exceed α if and only if $0 \leq t \leq F(\alpha)$. Since, by hypothesis, t is chosen at random from the interval $[0, 1]$,

$$\begin{aligned} \text{probability } \{z(t) \leq \alpha\} &= F(\alpha) \\ &= \text{probability } \{X \leq \alpha\} \end{aligned} \tag{12.5.2}$$

This equation shows that $z(t)$ may be regarded as a fair realization of the random variable X. Therefore, provided that we are able to "invert" the distribution function F (i.e., solve the equation $F(x) = t$) by generating random numbers from the interval $[0, 1]$, we may simulate the random variable with

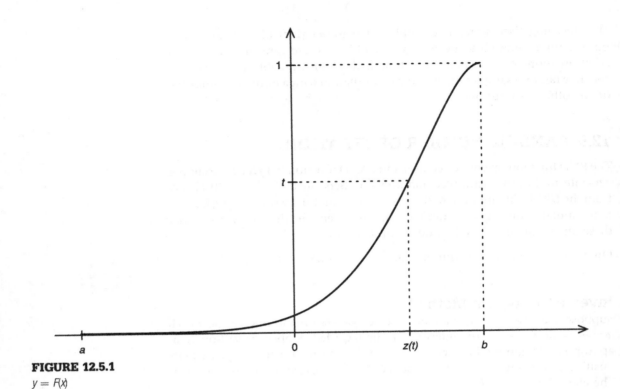

FIGURE 12.5.1
$y = F(x)$

distribution function $F(x)$. This is the *inverse transform* method for generating random numbers from a general distribution.

The inverse transform method for generating a sequence of random numbers $\{x_i\}$ from a distribution function F is easily stated in the following algorithm:

1. Generate a random number t_i from the uniform distribution $U[0, 1]$;
2. Return $x_i = F^{-1}(t_i)$;
3. $i = i + 1$;
4. Go to step 1.

The main disadvantage of the inverse transform method is the need for either an explicit expression for the inverse of the distribution function, $F^{-1}(y)$, or a numerical method to solve $y = F(x)$ for an unknown x. This means that the method cannot be used for particular distributions where the explicit expression does not exist, and the computation (e.g., using the Newton–Raphson method discussed in Appendix 2) is too computationally expensive. For example, to generate a random number from the standard normal distribution using the inverse transform method requires the inverse of the distribution function

EXAMPLE 12.5.1

Consider a non-negative random variable with exponential probability density function

$$f(x) = 2e^{-2x} \quad x \geq 0$$

Show how to simulate the drawing of a random sample from this distribution and generate five random numbers from it.

Solution

Since

$$F(x) = \int_0^x f(r)dr$$
$$= \int_0^x 2e^{-2r}dr$$
$$= 1 - e^{-2x}$$

it follows that, if

$$F[z(t)] = t$$

then

$$1 - e^{-2x(t)} = t$$

Hence,

$$z(t) = -\frac{1}{2}\ln(1 - t)$$

By generating a random number t from the interval [0, 1] and using the last equation to find $z(t)$, we may simulate the sampling procedure very simply.

Using the RAND function in Excel, we can generate five random numbers from $U[0,1]$, $\{t_i\}$, and apply the algorithm stated previously. For example,

i	t_i	z_i
1	0.5529	0.4025
2	0.4299	0.2810
3	0.4304	0.2814
4	0.7646	0.7232
5	0.4682	0.3157

$$F(x) = \frac{1}{\sqrt{2\pi}} \int_{-\infty}^{x} e^{-t^2/2}dt$$

Since no explicit solution to the equation can be found in this case, numerical methods must be incorporated into the inverse transform method.

Other methods for generating random samples from a general distribution exist which do not require the manipulation of the distribution function, for example the *acceptance-rejection method*. Using this method is one way to avoid the problems with the inverse transform method. However, a better alternative might be to use a method designed explicitly for generating normally distributed random numbers if such randomly distributed numbers are required. Examples of these methods include the **Box–Muller algorithm** and the **Polar algorithm**. A justification of these methods is beyond the scope of this book, but we state the algorithms for reference.

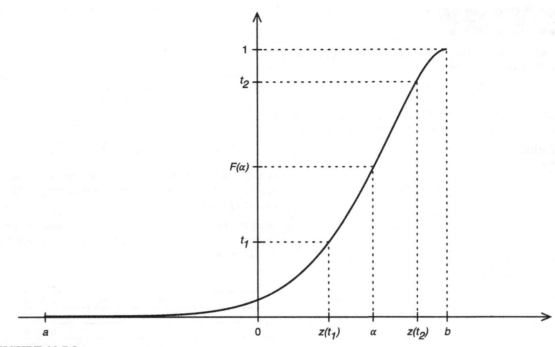

FIGURE 12.5.2
$y = F(x)$, showing $F(\alpha)$

The Box–Muller Algorithm

A pair of independent random numbers (z_1, z_2) from the standard normal distribution [i.e., $N(0, 1)$] can be generated from two uniformly distributed random numbers (t_1, t_2) (i.e., $U[0, 1]$).

1. Generate two random numbers t_i and t_{i+1} from the uniform distribution $U[0, 1]$.
2. Return $z_i = \sqrt{-2\ln t_i}\cos(2\pi t_{i+1})$ and $z_{i+1} = \sqrt{-2\ln t_i}\sin(2\pi t_{i+1})$.
3. $i = i + 2$.
4. Go to step 1.

The Box–Muller method is easy to incorporate into a computer code; however, it suffers from the disadvantage that the computation of sin and cos functions is time-consuming. For this reason an alternative formulation of this method is generally preferred when very large numbers of random numbers are required; this alternative is called the *Polar algorithm*.

The Polar Algorithm

The *polar algorithm* is very similar to the Box–Muller method but is modified through use of the acceptance-rejection method (not discussed here) to avoid computation of the trigonometric functions.

As with the Box–Muller method, the Polar method generates a pair of independent random numbers from the standard normal distribution from two uniformly distribution random numbers over [0,1].

1. Generate two random numbers t_i and t_{i+1} from the uniform distribution $U[0, 1]$.
2. Set $v_1 = 2t_i - 1$, $v_2 = 2t_{i+1} - 1$, and $s = v_1^2 + v_2^2$.
3. If $s > 1$, go to step 1.

   Otherwise, return $z_i = \sqrt{\dfrac{-2\ln s}{s}}\, v_1$ and $z_{i+1} = \sqrt{\dfrac{-2\ln s}{s}}\, v_2$.
4. $i = i + 2$.
5. Go to step 1.

EXAMPLE 12.5.2

Use the Box–Muller and Polar algorithms to generate random numbers from the $N(2,4)$ distribution. You should use the five pairs of $U(0, 1)$ random numbers.

Solution

For the purposes of demonstration, we use the same uniformly distributed random variables $\{t_i\}$ for both algorithms. These are generated from the RAND function in Excel. The two algorithms will then produce $\{z_i\}$ from the standard normal distribution from which we can generate $\{x_i = 2+2z_i\}$.

Note that in this particular instance, the Box–Muller method produced ten random numbers x_i (BM) but the Polar method produced only eight x_i (BM). This is a consequence of step 3 of the Polar algorithm.

i	t_i	t_{i+1}	z_i (BM)	z_{i+1} (BM)	x_i (BM)	x_{i+1} (BM)
1	0.3229	0.3858	−1.1329	0.9886	−0.2658	3.9773
3	0.8234	0.0221	0.6174	0.0864	3.2349	2.1728
5	0.3496	0.7289	−0.1913	−1.4371	1.6174	−0.8742
7	0.5124	0.8967	0.9213	−0.6991	3.8426	0.6019
9	0.7150	0.8118	0.3103	−0.7581	2.6205	0.4837
j	t_i	t_{i+1}	z_i (P)	z_{i+1} (P)	x_i (P)	x_{i+1} (P)
1	0.3229	0.3858	−1.5625	−1.0075	−1.1249	−0.0149
3	0.8234	0.0221	N/A	N/A	N/A	N/A
5	0.3496	0.7289	−0.8518	1.2967	0.2963	4.5935
7	0.5124	0.8967	0.0299	0.9607	2.0599	3.9214
9	0.7150	0.8118	0.5983	0.8678	3.1965	3.7356

EXAMPLE 12.5.3

By adopting a particular investment strategy, a company expects that on average the annual yield on its funds will be 8%. However, the investment policy is one of comparatively high risk, and it is anticipated that the standard deviation of the annual yield will be 7%. The yields in different years can be assumed to be independently distributed.

(a) Find the expected value and the standard deviation of the accumulated amount after 15 years of (i) a single premium of £1,000 and (ii) a series of 15 annual premiums, each of £1,000.

Assuming further that each year $1 + i$ (where i is the annual yield on the company's funds) has a log-normal distribution,

(b) Calculate the probability that a single premium accumulation will be (i) less than 60% of its expected value and (ii) more than 150% of its expected value.

(c) Using simulation methods, estimate the corresponding probabilities for an annual premium accumulation.

Solution

(a) In the notation previously used, we are given that

$$j = E[i] = 0.08$$

and that

$$s = (\mathrm{var}[i])^{1/2} = 0.07$$

(i) The single premium accumulation has expected value

$$1,000 E[S_{15}] = 1,000(1.08)^{15} = £3,172.17$$

and standard deviation

$$1,000(\mathrm{var}[S_{15}])^{1/2} = £808.14$$

(These values are given by Eqs 12.2.5 and 12.2.7, respectively.)

(ii) The annual premium accumulation has expected value

$$1,000 E[A_{15}] = 1,000 \ddot{s}_{\overline{15}} \text{ at } 8\% = £29,324.28$$

and standard deviation

$$1,000(\mathrm{var}[A_{15}])^{1/2} = £5,080.29$$

(The last value is obtained using Eq. 12.2.14.)

The relative magnitudes of these values should be noted. For example, in the case of a single premium accumulation, the standard deviation is some 25% of the expected value. This is a reflection of the wide range of possible yields each year and the corresponding spread of possible accumulations.

(b) We now assume that each year $\ln(1 + i_t)$ has a normal distribution. We are given that $E[i_t] = 0.08$ and var $[i_t] = 0.07^2$. It is easily shown that this implies that the normal distribution for $\ln(1 + i_t)$ has mean

$$\mu = 0.074865$$

and standard deviation

$$\sigma = 0.064747$$

Our remarks in Section 12.3 show that $\ln(S_{15})$ has a normal distribution with mean $\mu' = 15\,\mu = 1.1229\,75$ and standard deviation $\sigma' = \sigma\sqrt{15} = 0.250\,764$.

(i) Let Φ denote the distribution function of a standard normal random variable. Note that $0.6\,E[S_{15}] = 1.903\,30$. Since $S_{15} < 1.903\,30$ if and only if $\ln(S_{15}) < 0.643\,589$, it follows that

$$\text{probability } (S_{15} < 0.6E[S_{15}]) = \Phi\left(\frac{0.643\,589 - \mu'}{\sigma'}\right)$$
$$= \Phi(-1.91170)$$
$$= 0.028$$

(ii) Similarly, since $\ln(1.5\,E[S_{15}]) = \ln(4.758\,25) = 1.559880$, it follows that

$$\text{probability}(S_{15} > 1.5E[S_{15}]) = 1 - \Phi\left(\frac{1.559880 - \mu'}{\sigma'}\right)$$
$$= 1 - \Phi(1.742\ 29)$$
$$= 0.041$$

(c) In order to estimate the corresponding probabilities for the annual premium accumulation, we use simulation. Having found the values of μ and σ, we may readily simulate the accumulation using either the Box–Muller or Polar algorithm, for example. The yield in the tth year is determined by generating a value, x_t say, drawn randomly from a normal distribution with mean μ and standard deviation σ, and letting

Continued

> ## EXAMPLE 12.5.3   (cont'd)
>
> $$\ln(1 + i_t) = x_t$$
>
> or
>
> $$1 + i_t = e^{x_t}$$
>
> The outcome of the simulation of the annual premium accumulation is summarized in Table 12.5.1, in which we denote the expected value and standard deviation of the accumulation (as calculated previously) by μ_{15} and σ_{15}, respectively.
>
> The spread of the distribution should be noted. A total of 5,379 accumulations were less than the expected value. This is a reflection of the fact that more than one half of a log-normal distribution lies below its mean. Therefore, each year the yield was more likely to be less than 8% than to exceed 8%. The smallest accumulation was 16.0604, and the greatest was 54.1707. The smallest accumulation corresponds to an effective yield over the 15-year period of only 0.85% per annum. For the largest accumulation, the effective annual yield is 14.89%.
>
> Note that $0.6E[A_{15}] = 17.5946$ and $1.5E[A_{15}] = 43.9864$. Of the 10,000 simulations, 15 had $A_{15} < 17.5946$ and 69 had $A_{15} > 43.9864$. Accordingly, we estimate the probability that the accumulation will be less than 60% of its expected value to be approximately 0.0015. [A 95% confidence interval for the probability is (0.0007, 0.0023).] Similarly, we estimate the probability that the accumulation will exceed its expected value by more than 50% to be 0.0069 [with a corresponding 95% confidence interval (0.0053, 0.0085)].
>
> **Table 12.5.1** Distribution of $(A_{15} - \mu_{15})/\sigma_{15}$ for 10 000 Annual Premium Accumulations
>
> | Range | < -3 | -3 to -2.5 | -2.5 to -2.0 | -2.0 to -1.5 | -1.5 to -1.0 | -1.0 to -0.5 | -0.5 to 0.0 | 0.0 to 0.5 | 0.5 to 1.0 | 1.0 to 1.5 | 1.5 to 2.0 | 2.0 to 2.5 | 2.5 to 3.0 | ≥ 3 |
> |---|---|---|---|---|---|---|---|---|---|---|---|---|---|---|
> | Number of Values in Range | 0 | 4 | 77 | 401 | 1058 | 1839 | 2000 | 1797 | 1223 | 833 | 424 | 203 | 82 | 59 |

Note that the scaling property of the normal distribution means that any standard normally distributed random number generated from these methods, z_i [i.e., from $z \sim N(0, 1)$], is easily transformed to a *normally distributed* random number, x_i [i.e., from $X \sim N(\mu, \sigma^2)$], via $x_i = \mu + \sigma z_i$. This was used in Example 12.5.2.

12.6 DEPENDENT ANNUAL RATES OF RETURN

So far we have discussed stochastic interest rate models in which the yields in distinct years are independent of each other. In Section 12.4, using simulations, we gave illustrations of such models. Another area in which simulation techniques are particularly useful is models for which the yields in distinct years are *not* independent. In such models the yield in any one year, although containing a "random" component, depends in some clearly defined manner on the yields in previous years.

There exists a wide range of possible models which provide for some measure of interdependence of the annual yields in different years. A full discussion of such models is beyond the scope of this book, but, as the consequences of some form of dependence may be financially very significant, it is worth briefly considering an elementary example.

In the most simple form of dependent stochastic interest rate model, the yield in any one year is determined by (a) the yield in the preceding year and (b) a random component. For example, the model might anticipate that, if the yield in a particular year is relatively high, then (because the relevant underlying economic and market factors may continue) the yield in the following year is likely to be higher than would otherwise be the case.

For the independent log-normal model discussed in Section 12.3, we assume that, for each value of t, $\ln(1 + i_t)$ (where i_t is the yield in year t) has a normal distribution with specified mean μ and variance σ^2. One way of retaining a log-normal model while also allowing for some form of interdependence in the yields is provided by assuming that $\ln(1 + i_t)$ has a normal distribution with constant variance σ^2 and mean

$$\mu_t = \mu + k[\ln(1 + i_{t-1}) - \mu] \qquad (12.6.1)$$

where k is some positive constant.

By varying the value of k, we can allow for different levels of dependence between the yields in successive years. If $k = 0$, the model is simply the independent one. If $k = 1$, then $\mu_t = \ln(1 + i_{t-1})$ and the normal distribution for $\ln(1 + i_t)$ is centered on $\ln(1 + i_{t-1})$. For $0 < k < 1$ the model reflects an intermediate position. The greater the value of k, the greater the influence of the yield in 1 year on the determination of the yield in the succeeding year.

In the *independent uniform model*, we assume that each year the distribution of the yield is uniform over a specified interval $[\mu - d, \mu + d]$ (see Example 12.2.4). We may readily extend this model to allow for dependence in the annual yields by assuming that the distribution for i_t is uniform over the interval $[\mu_t - d, \mu_t + d]$, where

$$\mu_t = \mu + k(i_{t-1} - \mu) \qquad (12.6.2)$$

and μ, d, and k are specified constants, and $0 \leq k \leq 1$.

For this model the constant k is particularly simple to interpret. Each year the distribution for the yield is uniform over an interval of fixed length $2d$. The mean of the distribution for i_t is determined by linear interpolation (specified by the value of k) between i_{t-1} and μ. For example, if $k = 0.5$, the distribution for i_t is over an interval centred on a point midway between i_{t-1} and μ. When $k > 0$, the model implies that, if $i_{t-1} > \mu$, then i_t is more likely to exceed μ than to be less than μ and, if $i_{t-1} < \mu$, then i_t is more likely to be less than μ than to exceed μ.

As k increases from 0 to 1 in this **dependent uniform** model, there is greater dependence between the yields in successive years. This is reflected in an increasing spread in the distribution of both annual and single premium accumulations. This is illustrated by Example 12.6.1.

EXAMPLE 12.6.1

In order to assess the potential cost of certain minimum guarantees contained in its "special growth fund" savings policies, a company wishes to consider the likely future accumulation of its investments on the assumption that the distribution of i_t, the yield on the fund in year t, will be uniform over the interval $[\mu_t - 0.03, \mu_t + 0.03]$, where

$$\mu_t = 0.06 + k(i_{t-1} - 0.06) \qquad (1)$$

and k is a specified constant. For the year just ended, the yield on special growth fund was 6%.

Estimate by simulation, for $k = 0, 0.2, 0.4, 0.4, 0.8,$ and 1, the mean value, and the standard deviation of the accumulation in the special growth fund of 15 premiums of £1, payable annually in advance, the first premium being due now. Estimate also the upper and lower 5% and 10% points for the accumulation, and find the yields corresponding to these extreme values. Assume that expenses can be ignored.

Solution

Note that, since in the year just ended the yield on the special growth fund was 6% (i.e., $i_0 = 0.06$), it follows from Eq. 1 that for the first year of the accumulation period, the distribution of the yield will be uniform over the interval [0.03, 0.09]. The second year's yield will depend on i_1 and k. For example, if $k = 0.5$ and $i_1 = 0.04$, then the distribution of i_2 will be uniform over the interval [0.02, 0.08].

Note also that, if $k = 0$, then each year the distribution of the yield is uniform over the interval [0.03, 0.09], and we may apply the results of Section 12.3. The reader should verify that in this case the accumulation has mean value 24.673 and standard deviation 1.040.

For any given value of k, it is simple to simulate the accumulation by generating a sequence of 15 random numbers from the interval [0, 1], say $\{x_i : i = 1, 2, \ldots, 15\}$, and proceeding as indicated next.

Each year the distribution of the yield is uniform over an interval of length 0.06. The yield in the first year, i_1, is obtained as

$$i_1 = 0.03 + 0.06x_1$$

For $t \geq 2$, we determine i_t by (a) calculating μ_t from Eq. 1 and (b) defining

$$i = \mu_t - 0.03 + 0.06x_1$$

In this way the sequence of random numbers $\{x_i\}$ determines the yields for each year of the period. The final accumulation, A_{15}, is calculated from Eq. 12.2.3. For each of the given values of k, we carried out 10,000 simulations. Our results are summarized in Table 12.6.1, which gives the mean value and the standard deviation of the resulting accumulations and also the upper and lower 5% and 10% points for the resulting distributions.

Therefore, for example, when $k = 0.6$, the mean value of the accumulations was 24.774 and the standard deviation was 2.373. A total of 5% of the accumulations were less than 21.077, and 10% were less than 21.80; 10% of the accumulations exceeded 27.908, and 5% exceeded 28.901. Equivalently, we may say that the mean accumulation corresponded to an effective yield over the completed transaction of 6.05% per annum. Some 5% of the accumulations resulted in an effective yield over the completed transaction of less than 4.14% per annum, while 5% of the accumulations produced an overall effective yield of at least 7.83% per annum.

For each value of k, Table 12.6.2 shows the effective annual yield on the completed transaction for the mean accumulation and for each of the "tail" values given in Table 12.6.1. For values of k less than 0.6, the mean values of the accumulation are very similar. However, for larger values of k, there is a marked increase in the mean value. In the extreme case, when $k = 1$, the mean accumulation is some 9.6% greater

Continued

EXAMPLE 12.6.1   *(cont'd)*

than in the independent case, $k = 0$. Note also that, as k increases, the range of values of the accumulation becomes wider. When $k = 0$, some 90% of the values lie between 23.000 and 26.412, a relative difference of 14.8%. When $k = 0.6$, the corresponding quantiles are 21.077 and 28.901, with a relative difference of 37.1%.

Suppose that the special growth fund policies were such that the policyholder received the accumulated amount of his premiums, with the proviso that there would be a minimum return of 4% per annum effective. The simulation results enable the costs of such guarantees to be assessed. (Note

that a separate "guarantee fund" must be provided to meet these liabilities.) The simulation results show that, when $k = 0.6$, the expected amount to be paid from the guarantee fund is of the order of 1.5p per £1 annual premium. This figure is relatively small, since the probability that a policy will require supplementation from the guarantee fund is approximately 0.01. Of the 10,000 simulations, 114 produced accumulations less than $\ddot{s}_{\overline{15}|}$ at 4% (i.e., 20.8243). For these 114 policies, the average "shortfall" in the accumulation was £1.36 per £1 annual premium.

Table 12.6.1 Dependent Uniform Model; $\mu = 0.06$, $d = 0.03$, $i_0 = 0.06$; Term 15 years; Annual Premium £1; 10,000 Simulations for Each Value of the Dependency Constant k

Dependency Constant	Mean Accumulation	Standard Deviation of Accumulation	Lower Tail for Accumulation		Upper Tail for Accumulation	
			5%	10%	10%	5%
0.0	24.679	1.037	23.000	23.352	26.036	26.412
0.2	24.673	1.264	22.669	23.078	26.320	26.828
0.4	24.724	1.667	22.040	22.622	26.905	27.579
0.6	24.774	2.373	21.077	21.809	27.908	28.901
0.8	24.990	4.205	18.712	19.929	30.634	32.432
1.0	27.044	11.549	12.587	14.502	42.366	49.237

Table 12.6.2 Overall Effective Annual Yield Per Cent Corresponding to Values Given in Table 12.6.1

Dependency Constant	Yield for Mean Accumulation	Yield for Lower Tail		Yield for Upper Tail	
		5%	10%	10%	5%
0.0	6.00	5.18	5.36	6.63	6.79
0.2	6.00	5.01	5.22	6.75	6.97
0.4	6.02	4.67	4.98	7.01	7.29
0.6	6.05	4.14	4.55	7.43	7.83
0.8	6.15	2.72	3.47	8.50	9.15
1.0	7.07	−2.23	−0.42	12.16	13.83

12.7 AN INTRODUCTION TO THE APPLICATION OF BROWNIAN MOTION

In this section we briefly indicate how the theory of one-dimensional *random walks* with small steps may be used to help describe the progress of a single premium investment. As in the case of some other topics, the results of the theory may not always be directly applicable to practical problems, but they give a starting point for further discussion. We then briefly consider the applications of such walks to other financial problems.

Consider a *random walk* beginning at the origin and moving with independent steps of length δ at the rate of r steps per annum. At each of the times $0, 1/r, 2/r, \ldots,$ (measured in years), the probability of an upward movement is p and that of a downward movement is $1 - p$. Consider the distribution of the position $X(t)$ of the random walk at time t when δ becomes small in such a way that $(2p - 1)\delta r$ approaches a constant μ and $\delta^2 r$ approaches a constant σ^2. Note that these conditions imply that p tends to $1/2$ and r tends to infinity. $X(t)$ may then be considered to describe a *Brownian motion* with drift μ and variance σ^2.

Brownian motion is named after the nineteenth-century botanist Robert Brown, who observed the random movement of pollen particles in water. The link to Brown's work is made because the path of a two-dimensional Brownian motion process bears a resemblance to the track of a pollen particle.

The distribution of $X(t)$ may be shown to be approximately normal with mean μt and variance $\sigma^2 t$. Further, the probability that, at *some* time in the first n years, $X(t)$ falls below the value $-\xi$ (where $\xi > 0$) may be shown to be approximately equal to

$$
\begin{aligned}
P_n &= \frac{\xi}{\sigma\sqrt{2\pi}} \int_0^n \exp\left[-\frac{(\mu t + \xi)^2}{2\sigma^2 t} \right] t^{-3/2} dt \\
&= \Phi\left(\frac{-\xi - \mu n}{\sigma\sqrt{2\pi}} \right) + \exp\left(\frac{-2\mu\xi}{\sigma^2} \right) \Phi\left(\frac{-\xi + \mu n}{\sigma\sqrt{n}} \right)
\end{aligned}
\tag{12.7.1}
$$

where $\Phi(x)$ denotes the standard normal distribution function. Hence, the probability that $X(t)$ will *never* fall below $-\xi$ is

$$
1 - P_\infty = \begin{cases} 0 & \text{if } \mu \leq 0 \\ 1 - \exp\left(\dfrac{12\xi\mu}{\sigma^2} \right) & \text{if } \mu > 0 \end{cases}
\tag{12.7.2}
$$

Stochastic processes based on Brownian motion are fundamental to financial mathematics and can, in principle, be incorporated into the models for the prices of underlying assets. Indeed, Bachelier used Brownian motion to model the movements of the Paris stock exchange index around 1900. However, the disadvantage of using Brownian motion to model asset prices is that there is a significant probability that the pricing model will report negative values. No matter how successful Brownian motion can be modeling market prices in the short term, it is therefore of no use in the long run. For example, if the path shown in Figure 12.7.1 were to be used to model the price of coffee at particular discrete time intervals, the model clearly predicts a period of negative price. Negative prices have no meaning in most commodity and stock valuations.

With this in mind, one can build models within which the *log-price* is given by a Brownian motion process. Taking exponentials of a Brownian motion process leads to a **geometric Brownian motion** process that naturally takes non-negative values. Such models are clearly related to the log-normal distributions discussed in Section 12.3. In particular, geometric Brownian motion is log-normally distributed with parameters μ and σ^2. Such a process is used in the Black–Scholes formulation for option pricing, as discussed in Section 11.5.

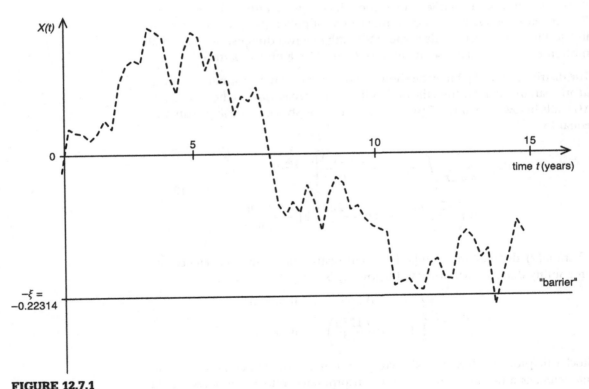

FIGURE 12.7.1
Possible path of *X(t)* showing Brownian motion

An application of these theoretical results to financial questions is as follows. Let $X(t)$ denote the logarithm of the proceeds at time t years of a single premium investment of 1 at time $t = 0$. Under idealized conditions, $X(t)$ may be considered to follow a Brownian motion with drift μ and variance σ^2, where μ and σ^2 are the mean and variance, respectively, of the logarithm of the growth rates in any given year. In addition, the logarithms of the growth rates in years 1, 2,..., n, respectively, are independent, identical, normally distributed variables, so that we obtain the mathematical model described in Section 12.3. However, Eq. 12.7.1 allows us to compute the probability that, at some time during the first n years, the proceeds of the investment will fall below a specified level. This may be important in that guarantees may be given by a life office issuing a policy that the proceeds of the investment will not fall below a certain figure. Also, the Brownian motion model may be used to help to determine the risk that certain share prices (which may sometimes be regarded as describing a Brownian motion) will, within a given time, fall below a fixed value (at which, perhaps, certain options may be exercised).

EXAMPLE 12.7.1

A single premium investment of £10,000 may be supposed to be such that the logarithm of the proceeds at time t per £1 initial investment follows a Brownian motion with $\mu = 0.045$ and $\sigma = 0.10$. Find the probability that, at some point in the next 15 years, the proceeds of this investment will be less than £8,000.

Solution

Let

$$X(t) = \ln[(\text{proceeds of investments at time } t)/10,000]$$

Note that the proceeds of the investment at time t are less than £8,000 if and only if $X(t) < \ln(0.8) = -0.223\ 14$. We are required to find the probability that at some point in the next 15 years $X(t) < -0.223\ 14$.

By Eq. 12.7.1, this probability is

$$P_{15} = \Phi\left[\frac{-0.223\ 14 - (0.045 \times 15)}{0.1\sqrt{15}}\right]$$

$$+0.134\ 22\Phi\left[\frac{-0.223\ 14 + (0.045 \times 15)}{0.1\sqrt{15}}\right]$$

$$= \Phi(-2.319) + 0.134\ 22\Phi(1.167)$$

$$= 0.0102 + 0.1179 = 0.1281$$

Note that $1 - P_\infty$, the probability that the proceeds of this investment will *never* fall below £8,000 is, by Eq. 12.7.2,

$$1 - \exp\left[-\frac{2 \times 0.223\ 14 \times 0.045}{(0.1)^2}\right] = 0.8658$$

A possible path taken by $X(t)$ is illustrated in Figure 12.7.1.

Similar calculations may be applied when the "barrier", or minimum level of proceeds, is of the form $A(1 + j)^t$ per unit invested, where j is a fixed escalation rate. In this case one may work in terms of "escalating" units of currency, the rate of escalation being j per annum. In terms of these units of currency, $X(t)$ has drift $\mu - \ln(1 + j)$ and variance σ^2, and the probability that the proceeds fall below the "barrier" level may be calculated as before. The technique is illustrated in Example 12.7.2.

EXAMPLE 12.7.2

Suppose that a single premium investment of £1 is such that the logarithm of the proceeds at time t describes a Brownian motion with $\mu = 0.07$ and $\sigma = 0.1395$. Find the probability that at some point within the next 10 years the proceeds will fall below $(0.75) \times (1.04)^t$, where t is the time in years from the present.

Solution

Work in monetary units which escalate at 4% per annum compound. Then $X(t)$, the logarithm of the proceeds at time t, describes a Brownian motion with drift $0.07 - \ln(1.04) = 0.030\ 78$ and variance $(0.1395)^2$. We require to find the probability that $X(t)$ falls below $\ln(0.75) = -0.287\ 68$ at some point in the next 10 years. By Eq. 12.7.1, this is

$$P_{10} = \Phi\left(\frac{-0.287\ 68 - 0.3078}{0.1395\sqrt{10}}\right)$$

$$+\exp\left[\frac{-2 \times 0.030\ 78 \times 0.287\ 68}{(0.1395)^2}\right]$$

$$\Phi\left(\frac{-0.287\ 68 + 0.3078}{0.1395\sqrt{10}}\right)$$

$$= \Phi(-1.350) + 0.4025 \times \Phi(0.0456) = 0.2971$$

SUMMARY

- In practice the assumption of deterministic interest rates used throughout this book is restrictive, and it may be necessary to introduce probabilistic distributions.
- A sensible first step towards a general *stochastic model* is to assume that interest rates in each subinterval of time (for example, a year) are *independent and identically distributed*. The advantage of this assumption is that explicit analytical expressions for the moments of the accumulation of a single and a series of payments can be obtained.
- The *log-normal distribution* is an important distribution for stochastic models of interest rates and other financial quantities. It has the advantages of being mathematically tractable and not permitting negative values.
- For more complicated stochastic models, analytical progress cannot be made, and computational *simulation techniques* are required. Fundamental to these are the generation of *pseudo-random numbers* from particular distributions. Various algorithms are available to generate these numbers.

EXERCISES

Institute and Faculty of Actuaries

12.1 The rate of interest is a random variable that is distributed with mean 0.07 and variance 0.016 in each of the next 10 years. The value taken by the rate of interest in any one year is independent of its value in any other year. Deriving all necessary formulae, calculate

  (a) The expected accumulation at the end of 10 years, if one unit is invested at the beginning of 10 years,

  (b) The variance of the accumulation at the end of 10 years, if one unit is invested at the beginning of 10 years,

(c) Explain how your answers in (a) and (b) would differ if 1,000 units had been invested.

12.2 A total of £80,000 is invested in a bank account which pays interest at the end of each year. Interest is always reinvested in the account. The rate of interest is determined at the beginning of each year and remains unchanged until the beginning of the next year. The rate of interest applicable in any one year is independent of the rate applicable in any other year.

Institute and Faculty of Actuaries

During the first year, the annual effective rate of interest will be one of 4%, 6%, or 8% with equal probability. During the second year, the annual effective rate of interest will be either 7% with probability 0.75 or 5% with probability 0.25. During the third year, the annual effective rate of interest will be either 6% with probability 0.7 or 4% with probability 0.3.

(a) Derive the expected accumulated amount in the bank account at the end of 3 years.

(b) Derive the variance of the accumulated amount in the bank account at the end of 3 years.

(c) Calculate the probability that the accumulated amount in the bank account is more than £97,000 at the end of 3 years.

12.3 The annual returns, i, on a fund are independent and identically distributed. Each year, the distribution of $1 + i$ is log-normal with parameters $\mu = 0.05$ and $\sigma^2 = 0.004$, where i denotes the annual return on the fund.

Institute and Faculty of Actuaries

(a) Calculate the expected accumulation in 25 years' time if £3,000 is invested in the fund at the beginning of each of the next 25 years.

(b) Calculate the probability that the accumulation of a single investment of £1 will be greater than its expected value 20 years later.

12.4 A capital redemption policy has just been issued with premiums payable annually in advance throughout the duration of the policy. The term of the policy is 20 years, and the sum assured is £10,000. The premiums will be invested in a fund which earns interest at a constant effective annual rate *throughout the duration of the policy*. This rate will be 3%, 6%, or 9%, each of these values being equally likely. There are no expenses.

(a) Find the expected value of the annual rate of interest. What would the annual premium be if it were calculated on the basis of this mean rate of interest?

(b) Let P be the annual premium.

(i) Find, in terms of P, the expected value of the accumulated profit on the policy at the maturity date. For what value of P is this expected value zero? What is the value of this expected value if P is as calculated in (a)?

(ii) Find, in terms of P, the expected value of the net present value of the policy immediately after it is effected. For what value of P is this expected value zero? What is the value of this expected value if P is as calculated in (a)?

12.5 The yields on a company's funds in different years are independently and identically distributed. Consider a single premium capital redemption policy with a term of 10 years and sum assured £1,000. Let P be the single premium for the policy and let i be the random variable denoting the yield in any given year. Expenses may be ignored.

(a) **(i)** Show that the expected value of the accumulated profit on the policy at the maturity date is

$$P(1 + E[i])^{10} - 1{,}000 \tag{1}$$

and find an expression (in terms of P) for the standard deviation of the accumulated profit.

(ii) Show that the expected value of the net present value of the policy immediately after it is effected is

$$P - 1{,}000\left(E\left[\frac{1}{1+i}\right]\right)^{10} \tag{2}$$

Show further that the standard deviation of the net present value is

$$1{,}000\left\{\left(E\left[\frac{1}{(1+i)^2}\right]\right)^{10} - \left(E\left[\frac{1}{1+i}\right]\right)^{20}\right\}^{1/2}$$

(b) For each of the three models for the distribution of i described below, find the value of P for which the expected value of the accumulated profit at the maturity date is zero and, using this value of P, calculate the standard deviation of the accumulated profit. For each model, find also the value of P for which the expected value of the net present value of the policy immediately after issue is zero, and calculate the standard deviation of the net present value.

Model I i takes each of the values 0.02, 0.04, and 0.06 with equal probability.
Model II i has a uniform distribution over the interval [0.02, 0.06].
Model III i has a triangular distribution over the interval [0.02, 0.06].

Compare your answers for the different models.

12.6 The yield i on a company's funds in any year is equally likely to be 3%, 6%, or 9%. Yields in different years are independent.

(a) Show that $E[i] = 0.06$ and that $\text{var}[i] = 0.0006$.

(b) Hence, or otherwise, derive expressions for $E[S_n]$ and $\text{var}[S_n]$, the expected value and the variance of the accumulation of a single premium of £1 over a period of n years. Find the mean and standard deviation of S_n for $n = 5, 10, 15,$ and 20.

(c) Let A_n be the random variable representing the accumulation over n years of n annual premiums of £1. *Write down* an expression for $E[A_n]$. Find, for $n = 1, 2, ..., 15$, the standard deviation of A_n.

(d) For $n = 5, 10,$ and 15, find the value of (standard deviation $[A_n])/E[A_n]$ and comment on your answers.

12.7 Each year the yield on a company's funds is either 2%, 4%, or 7% with corresponding probabilities 0.3, 0.5, and 0.2, respectively. Yields in different years are independent.

(a) Find the mean value and standard deviation of the accumulation of a single premium of £1,000 over a period of 15 years. Using simulation, estimate the probability that the accumulation will be (i) less than £1,600 and (ii) more than £2,000. Give 95% confidence intervals for each probability.

(b) Find the mean value and standard deviation of the accumulation of 15 annual premiums, each of £100. By simulation, or otherwise, estimate the probability that this accumulation will be (i) less than £1,900 and (ii) more than £2,300. Give 95% confidence intervals for each probability.

12.8 The yields on a company's fund in different years are independently and identically distributed. Each year the distribution of $(1 + i)$ is log-normal with parameters μ and σ^2. Let V_n be the random variable denoting the present value of £1 due at the end of n years.

(a) Show that V_n has a log-normal distribution and find the parameters of the distribution.

(b) Assuming further that each year the yield has mean value 0.08 and standard deviation 0.05, find, for $n = 5, 10, 15,$ and 20, the expected value and the standard deviation of the present value of £1,000 due at the end of n years.

12.9 Let

$$X = \frac{X_1 + X_2 + X_3}{3}$$

where X_1, X_2, and X_3 are independent random variables, each having a uniform distribution over the interval $[0, 1]$.

(a) Find the mean and standard deviation of X.

(b) By computationally generating three random numbers on the unit interval, estimate the probability that
(i) $X \leq 1/4$;

(ii) $X \geq 0.8$;

(iii) $1/3 \leq X \leq 2/3$;

(iv) $0.3 \leq X \leq 0.4$.

(c) Again, by simulation, estimate the value of t for which

(i) The probability that $X \leq t$ is 0.05;

(ii) The probability that $X \geq t$ is 0.01;

(iii) The probability that $X \leq t$ is 0.25.

Note that it is, in fact, quite simple to give exact answers to the preceding questions (The answers are included in the solutions for this exercise.) The purpose of the exercise is to give the reader an elementary introduction to simulation. The reader should carry out the simulation exercise twice, firstly with a relatively small number of simulations (e.g., 100), and secondly with a considerably greater number. Note the difference between the estimated values and the true values in each case.

Appendix 1: Theorem Proofs

PROOF OF THEOREM 2.4.1

Theorem

If $\delta(t)$ and $A(t_0, t)$ are continuous functions of t for $t \geq t_0$, and the principle of consistency holds, then, for $t_0 \leq t_1 \leq t_2$,

$$A(t_1, t_2) = \exp\left[\int_{t_1}^{t_2} \delta(t)dt\right]$$

Proof

Suppose that t_1 and t_2 are given with $t_0 \leq t_1 \leq t_2$. For $t \geq t_0$, let $f(t) = A(t_0, t)$. Note that f is continuous. For $t \geq t_0$, we have

$$\delta(t) = \lim_{h \to 0+} i_h(t)$$

$$= \lim_{h \to 0+} \frac{A(t, t+h) - 1}{h} \quad \text{(by definition)}$$

$$= \lim_{h \to 0+} \frac{A(t_0, t)A(t, t+h) - A(t_0, t)}{hA(t_0, t)}$$

$$= \frac{1}{A(t_0, t)} \lim_{h \to 0+} \frac{A(t_0, t+h) - A(t_0, t)}{h} \quad \text{(by principle of consistency)}$$

$$= \frac{1}{f(t)} \lim_{h \to 0+} \frac{f(t+h) - f(t)}{h}$$

$$= \frac{1}{f(t)} \left[f'^{+}(t)\right]$$

where f^+ denotes the right-sided derivative of f.

309

The last equation may be written as

$$f'^+(t) = f(t)\,\delta(t)$$

Since f and δ are both continuous functions, so too is f'^+. A continuous function which has a continuous right-sided derivative is in fact differentiable, so that we obtain

$$f'(t) = f(t)\delta(t).$$

By using the integrating factor $\exp\left\{-\int_{t_0}^t \delta(s)ds\right\}$, we immediately obtain

$$f(t) = c\,\exp\left\{\int_{t_0}^t \delta(s)ds\right\} \tag{A.1.1}$$

where c is some constant.

Now the principle of consistency implies that

$$A(t_1, t_2) = A(t_0, t_2) \,/\, A(t_0, t_1)$$
$$= f(t_2)\,/\,f(t_1) \qquad \text{(by definiton of } f(t))$$
$$= \exp\left\{\int_{t_1}^{t_2} \delta(s)ds\right\}$$

(from Eq. A.1.1), as required.

PROOF OF THEOREM 3.2.1

Theorem
For any transaction in which all the negative net cash flows precede all the positive net cash flows (or vice versa), the yield is well defined.

Proof
For such a transaction, we may assume without loss of generality that all the negative net cash flows precede all the positive net cash flows. (If the opposite holds, we simply multiply the equation of value by -1 to revert to this situation.) There is then an index l (i.e., time shift of length l) such that the equation of value Eq. 3.2.2 can be written as

$$-\left\{\alpha_1 e^{-\delta t_1} + \alpha_2 e^{-\delta t_2} + \cdots + \alpha_l e^{-\delta t_l}\right\}$$
$$+\left\{\alpha_{l+1} e^{-\delta t_{l+1}} + \alpha_{l+2} e^{-\delta t_{l+2}} + \cdots + \alpha_n e^{-\delta t_n}\right\} = 0 \tag{A.1.2}$$

where

$$t_1 < t_2 < \cdots < t_n \qquad\qquad (A.1.3)$$

and each $\alpha_i > 0 (1 \leq i \leq n)$.

After multiplication by $-e^{\delta t_i}$, Eq. A.1.2 becomes

$$g(\delta) \equiv g_1(\delta) - g_2(\delta) = 0$$

where

$$g_1(\delta) = \sum_{r=1}^{l} \alpha_r e^{\delta(t_i - t_r)}$$

$$g_2(\delta) = \sum_{r=l+1}^{n} \alpha_r e^{-\delta(t_r - t_i)}$$

Since each α_i is positive, the order condition of Eq. A.1.3 implies that g_1 is an increasing function of δ and that g_2 is a strictly decreasing function. Hence, g is a strictly increasing function. Since $\lim_{\delta \to \infty} g(\delta) = \infty$ and $\lim_{\delta \to -\infty} g(\delta) = -\infty$, it follows that the equation of value has a unique root (which may be positive, zero, or negative). This completes the proof of the theorem in the discrete case. The proof when continuous cash flows are present follows similarly.

Appendix 2: The Solution of Non-linear Equations

The problem of finding the solution (or solutions) of a non-linear equation occurs in many branches of mathematics. In compound interest problems, such an equation is often in a polynomial form (but this is not necessarily always the case).

Simple examples of non-linear equations are

$$6x^4 - 3x - 1 = 0 \qquad (A.2.1)$$

$$x - \ln x - 2 = 0 \qquad (A.2.2)$$

$$e^{-x} - \sin x = 0 \qquad (A.2.3)$$

Such an equation may be considered abstractly in its general form as

$$f(x) = 0$$

Although we may not be able to determine the *exact* value of a solution of the equation (i.e., the precise value of ξ such that $f(\xi) = 0$), in practice we may be able to estimate the value of ξ to any desired degree of accuracy. This means that, given any very small positive number ε (such as 10^{-8}), we can find a number η such that $|\xi - \eta| < \varepsilon$. For practical purposes, η is then regarded as a solution of the equation.

Special techniques exist for solving polynomial equations; however, the methods which are most suitable in the more general case can be applied easily to polynomial equations. Accordingly, we restrict our attention in this appendix to three of the principal methods which may be applied to solve the general non-linear equation.

Before one tries to find the root (or roots) of any equation, it is important to confirm by as elementary methods as possible that the equation does, indeed, have at least one root, thereby avoiding wasted time. In addition, one should try to determine the number of roots and find some information about their location.

In almost all practical problems, the function f will be continuous, and henceforth we shall assume that this is so. (If f is not continuous, the equation

CONTENTS

$f(x) = 0$ may require more careful analysis.) If $a < b$ and $f(a)$ is of opposite sign to $f(b)$, there exists ξ with $a < \xi < b$ such that $f(\xi) = 0$. This well-known property can be used in most cases to find an interval containing a solution of the equation and leads naturally to one method for finding it.

BISECTION METHOD

Suppose that $a_0 < b_0$ and $f(a_0) f(b_0) < 0$. Put $m = (a_0 + b_0)/2$. Then either

(a) $f(a_0) f(m) \leq 0$, in which case there is a solution in the interval $[a_0, m]$; or
(b) $f(a_0) f(m) > 0$, in which case $f(m) f(b_0) < 0$ so that there is a solution in the interval $[m, b_0]$.

We have therefore determined an interval, $[a_1, b_1]$ say, contained in the interval $[a_0, b_0]$ and one half its length, in which the solution lies. Repeating this procedure, we may construct a series of nested intervals $[a_i, b_i]$, each containing the root, such that $(b_i - a_i) = (b_0 - a_0)/2^i$. By making i sufficiently large, we can therefore find the solution to any desired degree of accuracy.

The principal advantage of the bisection method lies in the fact that, once an interval has been found which contains a solution, then the method guarantees that, provided we are willing to carry out a sufficient number of iterations, the value of the solution may be found to any specified degree of accuracy. The major drawback is that, if very high accuracy is required, the necessary number of iterations may be large. With a modern computer, this may not be a serious objection, but it is natural to seek faster methods.

In the following sections, we describe two further methods, both of which are generally satisfactory in nearly all practical situations.

SECANT METHOD (ALSO KNOWN AS THE MODIFIED REGULA FALSI METHOD)

A sequence $\{x_0, x_1, x_2, \ldots\}$ is constructed in such a way that as n increases x_n tends to the desired solution. The values of x_0 and x_1 are chosen arbitrarily, but in the neighbourhood of the solution. For $n \geq 1$, x_{n+1} is defined to be the x-coordinate of the point of intersection of the x-axis and the line joining the points $[x_{n-1}, f(x_{n-1})]$ and $[x_n, f(x_n)]$ (see Figure A.2.1).

It is easily verified that

$$x_{n+1} = x_n - f(x_n) \frac{x_n - x_{n-1}}{f(x_n) - f(x_{n-1})}$$

$$= \frac{f(x_n)x_{n-1} - f(x_{n-1})x_n}{f(x_n) - f(x_{n-1})}$$

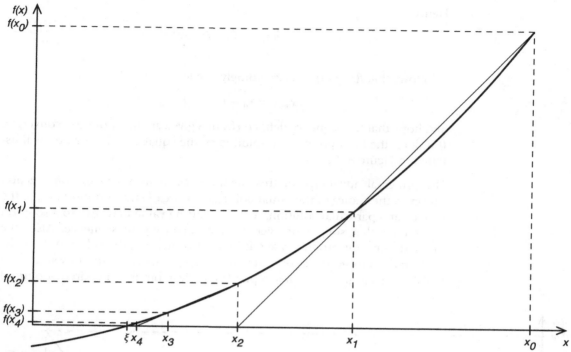

FIGURE A.2.1
Secant method

Given $\varepsilon > 0$, we continue the iteration until

$$f(x_n - \varepsilon)\, f(x_n + \varepsilon) \leq 0$$

at which stage we know that x_n is within ε of the solution.

NEWTON–RAPHSON METHOD

Finally, we consider a technique which gives theoretically faster convergence to a solution than either of the two preceding methods, but which requires that we are able to evaluate the derivative of f.

We try to construct a sequence $\{x_n\}$ which converges rapidly to the required solution. Intuitively, the rationale for the method may be described as follows.

Suppose that x_n is close to the solution so that $|f(x_n)|$ is small. We wish $(x_n + h)$ to be a better approximation. If h is small,

$$f(x_n + h) \simeq f(x_n) + h f'(x_n)$$

Hence, if

$$h = -f(x_n)/f'(x_n)$$

it follows that $f(x_n + h) \simeq 0$. Accordingly, we let

$$x_{n+1} = x_n - f(x_n)/f'(x_n)$$

and hope that the sequence defined (from a given starting point x_0) converges. *If it does, the limit point ξ is a solution of the equation.* The method is illustrated in Figure A.2.2.

The principal advantage of this method lies in its speed of convergence; however, the choice of the initial point x_0 may be of critical importance in the search for a particular solution. A "bad" starting value may lead to a solution other than that sought or even to a non-convergent sequence. Also, the method does not produce a small interval containing the solution. Usually, the iterative procedure is continued until $|x_{n+1} - x_n|$ is "sufficiently small". In practice, this is generally an acceptable criterion, but is not entirely foolproof

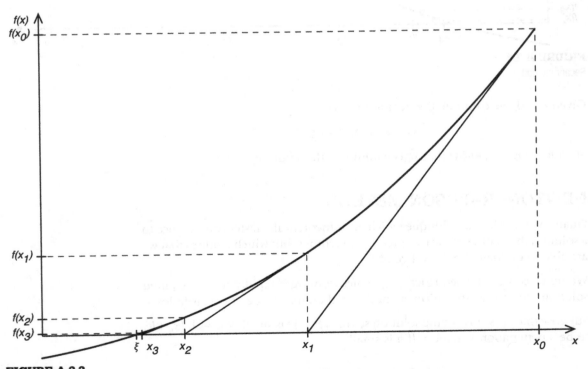

FIGURE A.2.2
Newton–Raphson method

EXAMPLE A.2.1

Find the positive solution of Eq. A.2.1 using the three methods discussed.

Solution

Here we have

$$f(x) = 6x^4 - 3x - 1$$

and

$$f'(x) = 24\left[x^3 - \frac{1}{8}\right]$$

Hence, f is a decreasing function for $x < 1/2$ and an increasing function for $x > 1/2$. Also $f(1/2) = -17/8$. Since

$f(x)$ is positive for large positive and large negative values of x, it follows that the equation has two real solutions. By our earlier remarks, we see that one solution lies in the interval $(-1, 0)$ and the other in the interval $(1/2, 1)$. For the bisection method, we put $a_0 = 0.5$, $b_0 = 1$; for the secant method, we let $x_0 = 0.5$ and $x_1 = 1$; for the Newton–Raphson method, we let $x_0 = 0.75$.

The number of iterations required by each method depends on the accuracy to which the solution is required. Our results are summarized in Table A.2.1. The speed of convergence of the three methods should be noted.

Table A.2.1 *Solution of $6x^4 - 3x - 1 = 0$*

| | Accuracy* | | | | |
| | $\varepsilon = 10^{-4}$ | | $\varepsilon = 10^{-8}$ | | |
Method of Solution	Solution	Number of Iterations	Solution	Number of Iterations
Bisection	0.8831	12	0.883 111 66	25
Secant	0.8831	5	0.883 111 66	7
Newton–Raphson	0.8831	3	0.883 111 67	4

*The true root lies within ε of the value given.

and, theoretically, an additional test (such as that described previously for the secant method) should be applied.

Although the calculation of the derivative may be somewhat laborious and extra care may be needed initially, the Newton–Raphson method may produce significant savings in time when many similar calculations are required.

EXERCISES

A.1 Find the negative solution of Eq. A.2.1.
A.2 Find all the real solutions of Eq. A.2.2.
A.3 How many real solutions does equation A.2.3 have? Find the two smallest solutions.

SOLUTIONS

(To five significant figures.)

A.1 −0.31391.
A.2 0.15859, 3.1462.
A.3 An infinite number. 0.58853, 3.0964.

Appendix 3: Solutions to Exercises

Below are detailed solutions to the exercises at the end of Chapters 2–12.

As explained in the text, there are often many ways to formulate the equation of value for a particular cash-flow stream. Care has been taken to present the most efficient formulation of each solution, although this may not be the most intuitively obvious formulation. For this reason the expressions stated here may not be the same as in your attempts. This does not necessarily mean that your attempts are incorrect, and you should compare the numerical solutions, which, of course, should be the same irrespective of the particular formulation used.

CHAPTER 2 EXERCISES

2.1 **(a)** Let the answer be t days; from the definition of simple interest, we have

$$1,500(1 + 0.05 \times t/365) = 1,550$$

which is solved for $t = 243.333$ days.

(b) Let the answer be t days, such that

$$1,500e^{0.05(t/365)} = 1,550$$

This is recast as $0.05\,(t/365) = \ln\,(1,550/1500)$, which is solved by $t = 239.366$ days.

2.2 **(a)**

$$A(0,5) = e^{\int_0^5 0.04dt} = e^{[0.04t]_0^5} = e^{0.2} = 1.22140$$

$$A(5,10) = e^{\int_5^{10} 0.008tdt} = e^{[0.004t^2]_5^{10}} = e^{0.3} = 1.34986$$

$$A(10,12) = e^{\int_{10}^{12}(0.005t^2+0.0003t^2)dt} = e^{[0.0025t^2+0.0001t^3]_{10}^{12}} = e^{0.1828} = 1.20057$$

The required present value is then

$$= \frac{1}{A(0,5)A(5,10)A(10,12)} = \frac{1}{1.22140 \times 1.34986 \times 1.20057} = \frac{1}{1.97941}$$
$$= 0.50520$$

(b) The equivalent effective annual rate is i, where $(1+i)^{12} = 1.97941$, so $i = 5.855\%$.

(c) The present value at time $t = 0$ is

$$= \int_2^5 e^{-0.05t} \left(e^{-\int_0^t 0.04 ds} \right) dt = \int_2^5 e^{-0.05t} \left(e^{-0.04t} \right) dt$$

$$= \int_2^5 e^{-0.09t} \, dt = \left[\frac{e^{-0.09t}}{-0.09} \right]_2^5 = \frac{e^{-0.18} - e^{-0.45}}{0.09} = 2.1960$$

2.3 Let time be measured in years from the beginning of the year.
$i_h(t_0) = $ the nominal rate of interest per annum

at time t_0 on transactions of term $h = \dfrac{\exp\left[\displaystyle\int_{t_0}^{t_0+h} \delta(t) dt \right] - 1}{h}$

Now

$$\delta(t) = 0.15 - 0.03t \quad \text{for } 0 \le t \le 1$$

so

$$\int_{t_0}^{t_0+h} \delta(t) dt = h\delta\left(t_0 + \frac{1}{2}h \right) = h\left[0.15 - 0.03\left(t_0 + \frac{1}{2}h \right) \right]$$

When $t_0 = 0$, we obtain the following answers:

$$\frac{\exp[h(0.15 - 0.015h)] - 1}{h} \quad \text{for } h = \frac{1}{4}, \frac{1}{12}, \frac{1}{365}$$

Which gives (a) 0.148957, (b) 0.149676, and (c) 0.149990.
When $t_0 = 1/2$, we obtain the following answers:

$$\frac{\exp[h(0.135 - 0.015h)] - 1}{h} \quad \text{for } h = \frac{1}{4}, \frac{1}{12}, \frac{1}{365}$$

which gives (a) 0.133427, (b) 0.134498, and (c) 0.134984. Note that $\delta(1/2) = 0.135$.

2.4 For $0 \leq t \leq 1$, let $F(t)$ denote the accumulation at time t of an investment of 1 at time 0. Note that $F(1/2) = 20{,}596.21/20{,}000$ and $F(1) = 21{,}183.70/20{,}000$. Hence, $\ln F(1/2) = 0.029\,375$ and $\ln F(1) = 0.057\,500$. By hypothesis, for $0 \leq t \leq 1$, $\delta(t) = a + bt$. Hence,

$$\int_0^t \delta(s)\,ds = at + \frac{1}{2}bt^2$$

and so

$$\ln F(t) = at + \frac{1}{2}bt^2$$

Letting $t = 0.5$ in the last equation, we obtain

$$0.029375 = \frac{1}{2}a + \frac{1}{8}b$$

and, letting $t = 1$, we have

$$0.057500 = a + \frac{1}{2}b$$

These last two equations imply that $a = 0.06$ and $b = -0.005$. Therefore,

$$\delta(t) = 0.06 - 0.005t$$

Also,

$$\ln F\left(\frac{3}{4}\right) = \frac{3}{4}a + \frac{9}{32}b = 0.043594$$

from which it follows that $F(3/4) = 1.044558$. The accumulated amount of the account at time 3/4 is $20{,}000\,F(3/4)$, which equals £20,891.16.

2.5 Measure time in years from the present. We have

$$v(t) = v^t = (1.08)^{-t} \quad \text{for all } t \geq 0$$

(a) Let X be the single payment at time 5. We must have

$$6{,}280v^4 + 8{,}460v^7 + 7{,}350v^{13} = Xv^5$$

from which we obtain $X = £18{,}006$.

(b) Let t be the appropriate future time. We find t from the equation

$$6{,}280v^4 + 8{,}460v^7 + 7{,}350v^{13} = 22{,}090v^t$$

from which we obtain $t = 7.66$ years.

2.6 **(a)** We begin with

$$v(t) = \exp\left[-\int_0^t \delta(s)ds\right]$$

Now

$$\int_0^t \delta(s)ds = \begin{cases} 0.08t & \text{for} & 0 \le t \le 5 \\ 0.1 + 0.06t & \text{for} & 5 \le t \le 10 \\ 0.3 + 0.04t & \text{for} & t \ge 10 \end{cases}$$

Hence,

$$v(t) = \begin{cases} \exp(-0.08t) & \text{for} & 0 \le t \le 5 \\ \exp(-0.1 - 0.06t) & \text{for} & 5 \le t \le 10 \\ \exp(-0.3 - 0.04t) & \text{for} & t \ge 10 \end{cases}$$

(b) **(i)** Let the single payment be denoted by S. The equation of value, at the present time, is

$$600[v(0) + v(1) + \cdots + v(14)] = Sv(15)$$

Hence, we obtain

$$S = \frac{600(1 + e^{-0.08} + \cdots + e^{-0.86})}{e^{-0.9}} \quad \text{(using (a))}$$

$$S = £14,119.$$

(ii) Let A be the annual payment. The equation of value, at the present time, is

$$600[v(0) + v(1) + \cdots + v(14)] = A[v(15) + v(16) + \cdots + v(22)]$$

Hence,

$$A = \frac{600(1 + e^{-0.08} + \cdots + e^{-0.86})}{(e^{-0.9} + e^{-0.94} + \cdots + e^{-1.18})} \quad \text{(using (a))}$$

$$= £2,022$$

2.7 (a) We have

$$v(t) = \exp\left[-\int_0^t \delta(s)\,ds\right]$$

$$= \exp\left[-\int_0^t ae^{-bs}\,ds\right]$$

$$= \exp\left[\frac{a}{b}\left(e^{-bt}-1\right)\right]$$

(b) (i) Since $\delta(0) = 0.1$, $a = 0.1$, and $\delta(10)/\delta(0) = e^{-10b} = 0.5$, we find that $b = 0.069315$. Hence, the present value of the payments is

$$1.000[v(1) + v(2) + v(3) + v(4)] = \text{£3,205.43}$$

(ii) We find δ by solving the equation

$$1{,}000\left(e^{-\delta} + e^{-2\delta} + e^{-3\delta} + e^{-4\delta}\right) = 3{,}205.43$$

that is,

$$e^{-\delta} + e^{-2\delta} + e^{-3\delta} + e^{-4\delta} = 3.20543$$

By numerical methods (see Appendix 2), we obtain $\delta = 0.09063$.

2.8 (a) We have

$$v(t) = \exp\left[-\int_0^t \delta(y)\,dy\right]$$

$$= \exp\left[-\int_0^t (r + se^{-ry})\,dy\right]$$

$$= \exp\left[-rt + \frac{s}{r}\left(e^{-rt}-1\right)\right]$$

$$= \exp\left(-\frac{s}{r}\right)\exp(-rt)\exp\left(\frac{s}{r}\,e^{-rt}\right)$$

(b) **(i)** The present value of the payment stream is

$$1{,}000 \int_0^n v(t)\,\mathrm{d}t \;=\; 1{,}000\exp\left(-\frac{s}{r}\right) \int_0^n \exp(-rt)\exp\left(\frac{s}{r}e^{-rt}\right)\,\mathrm{d}t$$

$$= 1{,}000\exp\left(-\frac{s}{r}\right)\left[-\frac{1}{s}\exp\left(\frac{s}{r}e^{-rt}\right)\right]_{t=0}^{t=n}$$

$$= 1{,}000\exp\left(-\frac{s}{r}\right)\left(-\frac{1}{s}\right)\left[\exp\left(\frac{s}{r}e^{-m}\right) - \exp\left(\frac{s}{r}\right)\right]$$

$$= \frac{1{,}000}{s}\left\{1 - \exp\left[\frac{s}{r}\left(e^{-m} - 1\right)\right]\right\}$$

(ii) When we substitute $n = 50$, $r = \ln 1.01$, and $s = 0.03$, the present value is found to be £23,109.

CHAPTER 3 EXERCISES

3.1 We recognize that we have to compare the present values of a lump sum and an annuity.
The present value of the annuity is

$$50{,}000\,a_{\overline{20}|} = 50{,}000\left(\frac{1 - 1.06^{-20}}{0.06}\right) = €573{,}496$$

The annuity plan is better by €73,496 in present value terms.

3.2 Recognize the problem as a deferred annuity of $10,000 per annum paid for 4 years with the first payment at time $t = 3$. The present value can then be expressed as

$$10{,}000\,{}_{2|}a_{\overline{4}|} = 10{,}000 \times v\,{}^{2}a_{\overline{4}|} = 10{,}000 \times (1.08)^{-2}\left(\frac{1 - 1.08^{-4}}{0.08}\right)$$

$$= \$28{,}396.15$$

You should therefore set aside $28,396.15 today to cover the payments. Note that an alternative, but equivalent, expression for the present value is $10{,}000\,{}_{3|}\ddot{a}_{\overline{4}|}$.

3.3 Working in millions:

$$PV \text{ of liabilities} = 9 + 12v\,\bar{a}_{\overline{1}|} \text{ at } 9\%$$

$$= 9 + 12v.\frac{i}{\delta}v$$

$$= 9 + 12 \times 0.91743^2 \times 1.044354$$

$$= 19.54811$$

3.4 (a) Working in thousands:
Outlay

$$PV = 500 + 90a_{\overline{5}|} + 10(Ia)_{\overline{5}|} \text{ at } 11\%$$

$$PV = 500 + 90 \times 3.695897 + 10 \times 10.3199$$

$$= 935.8297$$

Income

$$PV = 80\left(\bar{a}_{\overline{1}|} + 1.04v\,\bar{a}_{\overline{1}|} + (1.04)^2 v^2\,\bar{a}_{\overline{1}|} + \cdots \cdots + (1.04)^{24} v^{24}\,\bar{a}_{\overline{1}|}\right)$$

$$= 80\bar{a}_{\overline{1}|} \times \left[\frac{1 - (1.04v)^{25}}{1 - (1.04v)}\right]$$

where $\bar{a}_{\overline{1}|} = \dfrac{i}{\delta} \cdot v = \dfrac{0.11}{\ln 1.11} \cdot \dfrac{1}{1.11} = 0.949589$

and so $PV = 80 \times 0.949589 \times 12.74554 = 968.2421$
The present value of the cost of further investment

$$= 300v^{15} = 62.7013$$

$$PV \text{ of sale} = 700v^{25} = 51.5257$$

Hence, $NPV = 968.2421 + 51.5257 - 935.8297 - 62.7013 = 21.2368$ (i.e., 21,237).

(b) If interest > 11%, then v decreases. The present value of both income and outgo would then go down. However, the present value of the outgo is dominated by initial outlay of £500k at time 0, which is unaffected. Therefore, PV of income decreases by more than decrease in PV of outgo, and the net present value would reduce (and possibly become negative).

3.5 (a) The equations of value for each of the three savings plans, and the corresponding yield per annum (i.e., solution of the equation of value), are as follows:

    (i) $-1,330 + 1,000(1+i)^3 = 0$, $i \simeq 0.0997$ or 9.97%;
    (ii) $-1,550 + 1,000(1+i)^5 = 0$, $i \simeq 0.0916$ or 9.16%;
    (iii) $-1,000(1+i)^4 + 425a_{\overline{4}|} = 0$, $i \simeq 0.0858$ or 8.58%.

(b) Let this rate of interest per annum be i. We must have
$$1,330(1+i)^2 \geq 1,550$$

Hence, $i \geq 0.0795$, i.e., at least 7.95%.

(c) Let the rate of interest per annum used to calculate the amount of the annuity-certain be i. Then i must be such that
$$1{,}550 \geq 425\ddot{a}_{\overline{4}|i}$$

i.e.,
$$\ddot{a}_{\overline{4}|i} \leq 3.64706$$

i.e.,
$$a_{\overline{3}|i} \leq 2.64706$$

By interpolation between 6% and 7%, we find that $a_{\overline{3}|i} = 2.647\,06$ when $i \simeq 6.53\%$, so we require i to be at least 6.53%.

3.6 (a) The equation of value for plan (i) is
$$100\ddot{s}_{\overline{10}|i} = 1{,}700$$

where i is the yield per annum. We therefore have
$$\ddot{s}_{\overline{10}|i} = 17$$

i.e.,
$$s_{\overline{11}|i} = 18$$

By linear interpolation between 9% and 10%, we obtain $i \simeq 9.45\%$. The equation of value for plan (ii) is
$$100\ddot{s}_{\overline{15}|i} = 3{,}200$$

where i is yield per annum. We therefore have
$$\ddot{s}_{\overline{15}|} = 32$$

i.e.,
$$s_{\overline{16}|} = 33$$

By linear interpolation between 8% and 9%, we obtain $i \simeq 9.00\%$.

(b) Let i be the minimum fixed rate of interest which will give the desired proceeds. The equation of value, 15 years after the date on which the first premium was paid, is
$$1{,}700(1+i)^5 + 100\ddot{s}_{\overline{5}|i} - 3{,}200 = 0.$$

By linear interpolation between 8% and 9%, we obtain $i \simeq 8.50\%$. The fixed rate of interest must therefore be at least 8.50%.

3.7 (a) Consider the purchase of an article with retail price £100. The cash price is £70 or one may pay £75 in 6 months' time. The effective rate of discount per annum d is found by solving the equation

$$70 = 75(1-d)^{1/2}$$

and so $d = 0.1289$ or 12.89%.
The effective annual rate of interest is

$$i = \frac{d}{1-d} = 0.1480 \text{ or } 14.80\%$$

(b) One may pay £72.50 in 3 months' time instead of £70 now, so the effective annual rate of discount d is found from the equation

$$70 = 72.50(1-d)^{1/4}$$

which gives $d = 0.1310$ or 13.10%.
The new arrangement therefore offers a greater effective annual rate of discount to cash purchasers.

3.8 (a)

$$_{5|}a_{\overline{32}|} = a_{\overline{37}|} - a_{\overline{5}|} = 14.6908$$

$$\ddot{a}_{\overline{62}|} = \frac{1 - v^{62}}{d} = 23.7149$$

$$\bar{a}_{\overline{62}|} = \frac{1 - v^{62}}{\delta} = \frac{d}{\delta}\ddot{a}_{\overline{62}|} = 23.2560$$

$$_{12|}\ddot{a}_{\overline{50}|} = v^{12}\ddot{a}_{\overline{50}|} = v^{12}(1 + a_{\overline{49}|}) = 13.9544$$

$$s_{\overline{62}|} = \frac{(1+i)^{62} - 1}{i} = 259.451$$

$$\ddot{s}_{\overline{61}|} = s_{\overline{62}|} - 1 = 258.451$$

$$(I\ddot{a})_{\overline{62}|} = \frac{\ddot{a}_{\overline{62}|} - 62v^{62}}{d} = 474.905$$

$$_{5|}(Ia)_{\overline{20}|} = v^5(Ia)_{\overline{20}|} = 102.867$$

$$(I\bar{a})_{\overline{25}|} = \frac{\ddot{a}_{\overline{25}|} - 25v^{25}}{\delta} = 175.136$$

$$(\bar{I}\bar{a})_{\overline{25}|} = \frac{\bar{a}_{\overline{25}|} - 25v^{25}}{\delta} = 167.117.$$

(b) Since $\ddot{a}_{\overline{2n}|} = \ddot{a}_{\overline{n}|} + v^n \ddot{a}_{\overline{n}|} = \ddot{a}_{\overline{n}|}(1 + v^n)$
we have

$$1 + v^n = \frac{10.934563}{7.029584}$$

So

$$v^n = 0.555506$$

Hence,

$$\ddot{a}_{\overline{n}|} = \frac{1 - v^n}{d} = \frac{1 - 0.555506}{d} = 7.029584$$

Hence $d = 0.063232$ and $i = d/(1 - d) = 0.0675$ or 6.75%.

Therefore,

$$n = \frac{\ln v^n}{\ln v} = \frac{\ln(0.555\,506)}{\ln(1 - 0.063\,232)} = 9.$$

3.9 (a) It is clear at once that the value of the annuity is

$$5\ddot{a}_{\overline{6}|} + 7_{6|}\ddot{a}_{\overline{9}|} + 10_{15|}\ddot{a}_{\overline{5}|} \tag{i}$$

$$= 5\ddot{a}_{\overline{6}|} + 7(\ddot{a}_{\overline{15}|} - \ddot{a}_{\overline{6}|}) + 10(\ddot{a}_{\overline{20}|} - \ddot{a}_{\overline{15}|})$$

$$= 10\ddot{a}_{\overline{20}|} - 3\ddot{a}_{\overline{15}|} - 2\ddot{a}_{\overline{6}|} \tag{ii}$$

$$= 10(1 + a_{\overline{19}|}) - 3(1 + a_{\overline{14}|}) - 2(1 + a_{\overline{5}|})$$

$$= 5 + 10a_{\overline{19}|} - 3a_{\overline{14}|} - 2a_{\overline{5}|} \tag{iii}$$

(b) The accumulated amount of the annuity payments is

$$5(1 + i)^{14}s_{\overline{6}|} + 7(1 + i)^5 s_{\overline{9}|} + 10s_{\overline{5}|} \tag{i}$$

$$= 5(s_{\overline{20}|} - s_{\overline{14}|}) + 7(s_{\overline{14}|} - s_{\overline{5}|}) + 10s_{\overline{5}|}$$

$$= 5s_{\overline{20}|} + 2s_{\overline{14}|} + 3s_{\overline{5}|} \tag{ii}$$

3.10 The accumulation of the premiums until time 5 years, just before the payment of the premium then due, is

$$100\ddot{s}_{\overline{5}|0.08}$$

Therefore, the accumulation of the premiums until just before the payment of the premium due at time $t = 12$ is

$$100\ddot{s}_{\overline{5}|0.08}(1.06)^7 + 100\ddot{s}_{\overline{7}|0.06}$$

The accumulation until time 20 is therefore

$$\left[100\ddot{s}_{\overline{5}|0.08}(1.06)^7 + 100\ddot{s}_{\overline{7}|0.06}\right](1.05)^8 + 100\ddot{s}_{\overline{8}|0.05}$$

$$= 100\left[\ddot{s}_{\overline{5}|0.08}(1.06)^7(1.05)^8 + \ddot{s}_{\overline{7}|0.06}(1.05)^8 + \ddot{s}_{\overline{8}|0.05}\right]$$

$$= £3,724.77$$

The yield per annum i obtained by the investor on the completed transaction is found by solving the equation of value

$$100\ddot{s}_{\overline{20}|} = 3,724.77$$

i.e.,

$$s_{\overline{21}|} = 38.2477$$

By interpolation between 5% and 6%, we obtain $i = 5.59\%$.

CHAPTER 4 EXERCISES

4.1 **(a)**   **(i)** $100 \times (1 + 0.05/12)^{-12 \times 10} = £60.716$;
     (ii) $100 \times (1 - 0.05/12)^{12 \times 10} = £60.590$;
    (iii) $100 \times e^{-10\delta} = £60.6531$.

  (b) $98.91 = 100(1+i)^{-91/365}$
    $\ln(1+i) = (-365/91) \times \ln(98.91/100) = 0.04396$
    therefore $i = 0.04494$

4.2 **(a)**

$$\left[1 + \frac{i^{(4)}}{4}\right]^4 = 1 + i$$

and so $i^{(4)} = 0.061\ 086$.

$$e^{\delta} = 1 + i$$

and so $\delta = \ln(1+i) = 0.060\ 625$.

$$\left[1 - \frac{d^{(2)}}{2}\right]^2 = 1 - d = v = (1+i)^{-1}$$

and so $d^{(2)} = 0.059\ 715$.

(b) Since
$$\left[1+\frac{i^{(12)}}{12}\right]^{12} = \left[1+\frac{i^{(2)}}{2}\right]^{2} = 1+i$$

we obtain $i^{(12)} = 0.061701$.

Since
$$e^{\delta} = 1+i = \left[1+\frac{i^{(2)}}{2}\right]^{2}$$

from which we obtain $\delta = 0.061543$.

Since
$$\left[1-\frac{d^{(4)}}{4}\right]^{4} = 1-d = (1+i)^{-1} = \left[1+\frac{i^{(2)}}{2}\right]^{-2}$$

which gives $d^{(4)} = 0.061\ 072$.

(c) Since
$$\left[1+\frac{i^{(2)}}{2}\right]^{2} = e^{\delta} = (1-d)^{-1} = \left[1-\frac{d^{(12)}}{12}\right]^{-12}$$

we obtain $i^{(2)} = 0.063\ 655$, $\delta = 0.062\ 663$, and $d = 0.060\ 740$.

(d) Since
$$\left[1+\frac{i^{(4)}}{4}\right]^{4} = \left[1-\frac{d^{(2)}}{2}\right]^{-2} = e^{\delta}$$

we obtain $i^{(4)} = 0.062\ 991$ and $d^{(2)} = 0.061534$

4.3

$$a^{(4)}_{\overline{67}|} = \frac{1-v^{67}}{i^{(4)}} = 23.5391$$

$$\ddot{s}^{(12)}_{\overline{18}|} = \frac{i}{d^{(12)}} s_{\overline{18}|} = 26.1977$$

$$14|\ddot{a}^{(2)}_{\overline{10}|} = \ddot{a}^{(2)}_{\overline{24}|} - \ddot{a}^{(2)}_{\overline{14}|} = \frac{i}{d^{(2)}}\left(a_{\overline{24}|} - a_{\overline{14}|}\right) = 4.8239$$

$$\bar{s}_{\overline{56}|} = \frac{(1+i)^{56} - 1}{\delta} = 203.7755$$

$$a^{(4)}_{\overline{16.5}|} = \frac{1-v^{16.5}}{i^{(4)}} = 12.0887$$

$$\ddot{s}^{(12)}_{\overline{15.25}|} = \frac{(1+i)^{15.25} - 1}{d^{(12)}} = 20.9079$$

$$4.25|\bar{a}^{(4)}_{\overline{3.75}|} = v^{4.25}\left(\frac{1-v^{3.75}}{i^{(4)}}\right) = 2.9374$$

$$\bar{a}_{\overline{26/3}|} = \frac{1-v^{26/3}}{\delta} = 7.3473$$

4.4 (a) (i) We work in time units of 1 year. The accumulation at time 15, just before the payment then due is made, is

$$\frac{100}{a_{\overline{3}|}}s_{\overline{15}|} \quad \text{at } 10\% = £1{,}277.59, \text{ say } £1{,}278.$$

(ii) We now work in units of a half-year. The accumulation at time 30, just before the payment then due is made, is

$$\frac{100}{a_{\overline{6}|}}s_{\overline{30}|} \quad \text{at } 5\% = £1{,}308.96, \text{ say } £1{,}309.$$

(b) Let us proceed from first principles. Let time be measured in years, and let $i = 0.08$. The present value is

$$240\left(v + v^4 + v^7 + \ldots + v^{46}\right)$$

$$= 240v\left[1 + (v^3) + (v^3)^2 + \ldots + (v^3)^{15}\right]$$

$$= 240v\frac{\left[1 - (v^3)^{16}\right]}{1 - v^3}$$

$$= 240v\frac{a_{\overline{48}|}}{a_{\overline{3}|}}$$

$$= £1{,}051.07, \text{ say } £1{,}051$$

Note that the value of the payments 2 years before the present time is easily seen to be $(240/s_{\overline{3}|})a_{\overline{48}|}$ at 8%. Multiplying this value by 1.08^2, we obtain the same present value.

4.5 (a) Work in time units of 1 year. The present values are as follows:

(i) $600\,a_{\overline{20}|}$ at 12% $= £4{,}482$

(ii) $600\,a_{\overline{20}|}^{(4)}$ at 12% $= 600[i/i^{(4)}]a_{\overline{20}|}$ at 12%
$= £4{,}679$

(iii) $600\,a_{\overline{20}|}^{(12)}$ at 12% $= 600[i/i^{(12)}]a_{\overline{20}|}$ at 12%
$= £4{,}723$

(iv) $600\,\ddot{a}_{\overline{20}|}$ at 12% $= 600(i/\delta)a_{\overline{20}|}$ at 12%
$= £4{,}745$

(b) We now work in time units of a quarter-year with an interest rate of 3%. The present values are as follows:

(i) $(600/s_{\overline{4}|})\, a_{\overline{80}|}$ at 3% $= £4{,}331$

(ii) $150\, a_{\overline{80}|}$ at 3% $= £4{,}530$

(iii) $150\, a_{80}^{(3)}$ at 3% $= 150[i/i^{(3)}]a_{\overline{80}|} = £4{,}575$

(iv) $150\, \bar{a}_{\overline{80}|}$ at 3% $= £4{,}598$

4.6 Let X be the initial annual amount of the annuity. The present value of the annuity is found by summing the present values of the payments in each of the periods 1–4 years, year 5, 6–10 years, 11–12 years, and 13–15 years. This gives the equation of value

$$2049 = \frac{X}{2}\left(\frac{a_{\overline{16}|0.02}}{s_{\overline{2}|0.02}}\right) + v\frac{16}{0.02}\left(\frac{X}{2}\right)a_{\overline{2}|0.04}$$

$$+ Xv_{0.02}^{16}v_{0.04}^{2}a_{\overline{10}|0.04}^{(2)} + 2\,Xv_{0.02}^{16}v_{0.04}^{12}a_{\overline{4}|0.04}^{(6)}$$

$$+ 4\,Xv_{0.02}^{16}v_{0.04}^{16}a_{\overline{3}|0.08}^{(12)} = 17.0763X$$

Hence $X = £119.99$, say £120. $\left(\text{Note that at } 4\%,\, i^{(6)} = 0.039\,349.\right)$

4.7 (a) Let the present value of this annuity be A. We have

$$A = 1 + 4v + 9v^2 + \cdots + 361v^{18} + 400v^{19}$$

So

$$vA = v + 4v^2 + 9v^3 + \cdots + 361v^{19} + 400v^{20}$$

and

$$A - vA = 1 + 3v + 5v^2 + \cdots + 39v^{19} - 400v^{20}$$

$$= 1 + v + v^2 + \cdots + v^{19} + 2(v + 2v^2 + 3v^3 + \cdots + 19v^{19}) - 400v^{20}$$

$$= 1 + a_{\overline{19}|} + 2(Ia)_{\overline{19}|} - 400v^{20}$$

Hence,

$$A = \frac{1 + a_{\overline{19}|} + 2(Ia)_{\overline{19}|} - 400v^{20}}{1 - v} \quad \text{at } 5\% = £1{,}452.26$$

(b) The present value of the annuity is $[d/d^{(4)}]$ A at 5%, i.e., £1,426.06.

(c) The present value of the annuity is $[d/i^{(2)}]$ A at 5%, i.e., £1,400.18.

(d) The present value of the annuity is (d/δ) A at 5%, i.e., £1,417.40.

4.8 (a) Work in time units of 1 year. The accumulation is

$$200\ddot{s}^{(4)}_{\overline{13.5}|} \text{ at } 12\% = 200\left[\frac{(1+i)^{13.5}-1}{d^{(4)}}\right] \text{ at } 12\%$$

$$= £6,475.64$$

(b) Work in time units of a half-year. The accumulation is

$$100\ddot{s}^{(2)}_{\overline{27}|} \text{ at } 6\% = £6,655.86$$

(c) Work in time units of a quarter-year. The accumulation is

$$50\ddot{s}_{\overline{54}|} \text{ at } 3\% = £6,753.58$$

(d) Work in time units of 1 month. The accumulation is

$$\frac{50s_{\overline{162}|}}{a_{\overline{3}|}} \text{ at } 1\% = £6,821.85$$

4.9 Let X be the amount of the revised annuity, and consider the position at the date of the request. The equation of value is

$$X(1+i)^{1/4}a^{(2)}_{\overline{21.5}|} = 200v^{1/4}\ddot{a}_{\overline{22}|} + 320(1+i)^{1/12}a^{(4)}_{\overline{16.25}|}$$

$$+180a^{(12)}_{\overline{18.75}|} \text{ at } 8\%$$

from which we obtain

$$X = \frac{6,899.89}{10.5091} = £656.56$$

4.10 (a) **(i)** Let X be the monthly annuity payment. Working in time units of 1 year, we have the equation of value

$$400\ddot{s}^{(2)}_{\overline{20}|} = 12X\ddot{a}^{(12)}_{\overline{15}|} \text{ at } 12\%$$

and so $X = £361.01$, say £361.

(ii) Let Y be the monthly annuity payment. Working in time units of a half-year, we obtain the equation

$$200\ddot{s}_{\overline{40}|} = 6Y\,\ddot{a}^{(6)}_{\overline{30}|} \text{ at } 6\%$$

and so (on calculating $i^{(6)} = 6[(1+i)^{1/6} - 1]$ at $6\% = 0.058\,553$) $Y = £383.93$, say £384.

(iii) Let Z be the monthly annuity payment. Working in time units of 1 month, we have

$$200\left(\frac{s_{\overline{240}|}}{a_{\overline{6}|}}\right) = Z\ddot{a}_{\overline{180}|} \text{ at } 1\%$$

and so $Z = £405.67$, say £406.

(b) The monthly annuity payments are now as follows:
(i) $X\,v^{5/12}$ at $12\% = £344.36$, say £344;
(ii) $Yv^{5/6}$ at $6\% = £365.73$, say £366;
(iii) Zv^5 at $1\% = £385.98$, say £386.

CHAPTER 5 EXERCISES

5.1 If we work in months, the relevant interest rate $i = 7\%/12$ per month effective.
(a) The schedule is as follows:

Month	Beginning of Month Balance	Payment	Interest	Capital Repayment	End of Month Balance
1	200,000.00	1,797.66	1,166.67	630.99	199,369.01
2	199,369.01	1,797.66	1,162.99	634.67	198,734.33
3	198,734.33	1,797.66	1,159.28	638.38	198,095.96
4	198,095.96	1,797.66	1,155.56	642.10	197,453.86
5	197,453.86	1,797.66	1,151.81	645.85	196,808.01
6	196,808.01	1,797.66	1,148.05	649.61	196,158.40

(b) In the last month (month 180), after the final monthly mortgage payment is made, the ending mortgage balance will be zero. That is, the mortgage will be fully paid.
(c) The cash flow is unknown even if the borrower does not default. The reason is that the borrower has the right to prepay in whole or in part the mortgage balance at any time.

5.2 (a) Let X_A, X_B be the monthly repayments under Loans A and B, respectively.
For loan A:

$$\text{Flat rate of interest} = 10.715\% = \frac{60X_A - L_A}{5L_A}$$

$$= \frac{60X_A - 10,000}{50,000}, \text{ and so } X_A = £255.96$$

For loan B:

$$L_B = 15{,}000 = 12X_B\left(a^{(12)}_{\overline{2}|12\%} + v^2_{12\%}a^{(12)}_{\overline{3}|10\%}\right)$$

$$X_B = \frac{1{,}250}{\left(\dfrac{i}{i^{(12)}}a_{\overline{2}|}\right)_{12\%} + v^2_{12\%}\left(\dfrac{i}{i^{(12)}}a_{\overline{3}|}\right)_{10\%}}$$

$$= \frac{1{,}250}{(1.053875 \times 1.6901) + 0.79719(1.045045 \times 2.4869)}$$

$$X_B = £324.43$$

Hence, the student's overall surplus $= 600 - X_A - X_B = £19.61$.

(b) Effective rate of interest under loan A is $i\,\%$ where

$$12 \times 225.96a^{(12)}_{\overline{5}|} = 10{,}000$$

Try $i = 20\%$: $a^{(12)}_{\overline{5}|} = 3.2557$, as required.
So capital outstanding after 24 months is $12 \times 255.96a^{(12)}_{\overline{3}|}$ at 20%

$$= 12 \times 255.96 \times 1.088651 \times 2.1065 = 7{,}043.74$$

Capital outstanding under B is $12 \times 324.43\ a^{(12)}_{\overline{3}|}$ at 10%

$$= 12 \times 324.43 \times 1.045045 \times 2.4869 = 10{,}118.02$$

So interest paid in month 25 under loans A and B

$$= 7{,}043.74\frac{i^{(12)}_{20\%}}{12} + 10{,}118.02\frac{i^{(12)}_{10\%}}{12} = 107.84 + 80.68 = £188.52$$

and capital repaid

$$= (255.96 - 107.84) + (324.43 - 80.68)$$

$$= 148.12 + 243.75 = £391.87$$

(c) Under the new loan, the capital outstanding is the same as under the original arrangement $= 17{,}161.76$.

The monthly repayment $= \left(\dfrac{255.96 + 324.43}{2}\right) = £290.20$

The effective rate of interest on the new loan A is i where

$$= 12 \times 290.20 a_{\overline{10|}}^{(12)} = 17,167.76 \implies a_{\overline{10|}}^{(12)} = 4.9281$$

Try $i = 20\%$: $a_{\overline{10|}}^{(12)} = 4.5642$

Try $i = 15\%$: $a_{\overline{10|}}^{(12)} = 5.3551$

By interpolation $i = 15\% + \left(\dfrac{5.3551 - 4.9281}{5.3551 - 4.5642} \right)$

$$\times (20\% - 15\%) \approx 17.7\%$$

Hence, interest paid in month 25

$$= 17161.76 \frac{i_{17.7\%}^{(12)}}{12} = 234.66$$

and capital repaid is £290.20 $-$ £234.66 $=$ £55.54.

(d) The new strategy reduces the monthly payments but repays the capital more slowly. The student could consider the following options:
  ■ keeping loan B and taking out a smaller new loan to repay loan A (which has the highest effective interest rate);
  ■ taking out the new loan for a shorter term to repay the capital more quickly.

5.3 (a) (i) The flat rate of interest is $(2 \times 2{,}400 - 2{,}000)/(2 \times 2{,}000) = 70\%$.
   (ii) The flat rate of interest is not a good measure of the cost of borrowing because it takes no account of the timing of payments and the timing of repayment of capital.

(b) If the consumers' association is correct, then the present value of the repayments is greater than the loan at 200%,

i.e., $2{,}000 < 2{,}400 \dfrac{i}{d^{(12)}} a_{\overline{2|}}$

$i = 2$; $a_{\overline{2|}} = 0.44444$; $d^{(12)} = 1.04982$ gives RHS $= 2{,}032$.

The consumers' association is correct.

If the banks are correct, then the present value of the payments received by the bank, after expenses, is less than the amount of the loan at a nominal (before inflation) rate of interest of $(1.01463 \times 1.025 - 1)$ per annum effective $= 0.04$,

i.e., $2{,}000 > 720 \dfrac{i}{d^{(12)}} a_{\overline{2|}} + 720 \dfrac{i}{d^{(2)}} a_{\overline{1.5|}} + 960 \dfrac{i}{d^{(12)}} a_{\overline{1|}} - 0.3$

$\times 2{,}400 \dfrac{i}{d^{(12)}} a_{\overline{2|}}$

$$\frac{i}{d^{(12)}} = 1.021529; \ a_{\overline{2}|} = 1.8861; \ a_{\overline{1}|} = 0.9615; \ a_{\overline{1.5}|}$$

$$= \frac{1 - 1.04^{-1.5}}{0.04} = 1.4283$$

So RHS $= 72 \times 1.021529 \times 1.8861 + 720 \times 1.021529 \times 1.4283$
$+ 960 \times 1.021529 \times 0.9615 - 0.3 \times 2,400 \times 1.021529 \times 1.8861$
$= 1,387.23 + 1,050.52 + 942.91 - 1,387.23 = 1,993.43$
Therefore, the banks are also correct.

5.4 (a) **(i)** Let the annual repayment be X. We have

$$Xa_{\overline{25}|} = 3,000 \text{ at } 12\%$$

So

$$X = \frac{300}{a_{\overline{25}|}} = £382.50$$

(ii) (1) The loan outstanding just after the payment at time 9 years is

$$382.50a_{\overline{16}|} = 2,667.56$$

and just after the payment at time 10 years it is

$$382.50a_{\overline{15}|} = 2,605.17$$

Hence, the capital repaid at the end of the tenth year is

$$2,667.56 - 2,605.17 = £62.39$$

and the interest paid at this time is

$$382.50 - 62.39 = £320.11$$

(2) The capital outstanding just after the payment at time 24 years is

$$382.50a_{\overline{1}|} = £341.52$$

The capital repaid at the end of the 25th year is therefore £341.52, and the interest paid at this time is

$$382.50 - 341.52 = £40.98$$

(iii) The loan outstanding after the tth repayment is $382.50a_{\overline{25-t}|}$. This first falls below 1,800 when $a_{\overline{25-t}|}$ (at 12%) first falls below 4.7059. By the compound interest tables, the smallest value of t for which $a_{\overline{25-t}|} < 4.7059$ is 18, so the answer is the 18th repayment.

(iv) Using the loan schedule, the capital content exceeds the interest content of the tth installment when

$$1 - v^{26-t} < v^{26-t}$$

i.e., when

$$v^{26-t} > 0.5$$

This first occurs when $t = 20$, i.e., for the 20th payment.

(b) The loan outstanding just after the 15th annual payment has been made is

$$382.50a_{\overline{10}|} \text{ at } 12\% = 2,161.20$$

Let Y be the revised annual payment. The equation of value is

$$Ya_{\overline{16}|} = 2,161.20 \text{ at } 12\%$$

which gives $Y = £309.89$.

5.5 (a) Let X be the initial annual repayment. The equation of value is

$$16,000 = Xa_{\overline{10}|} \text{ at } 8\%$$

from which we obtain $X = £2,384.46$.

(b) The loan outstanding just after the fourth payment is made is

$$2,384.46a_{\overline{6}|} \text{ at } 8\% = 11,023.14$$

Let Y be the revised annual installment. The equation of value is

$$Ya_{\overline{6}|} = 11,023.14 \text{ at } 10\%$$

which gives $Y = £2,530.99$.

(c) The loan outstanding just after the seventh payment is made is

$$2,530.99a_{\overline{3}|} \text{ at } 10\% = 6,294.32$$

Let Z be the revised annual installment; the equation of value is

$$Za_{\overline{3}|} = 6,294.32 \text{ at } 9\%$$

and so $Z = £2,486.60$.
The equation of value for the entire transaction is

$$16,000 = 2,384.46a_{\overline{4}|} + 2,530.99_4|a_{\overline{3}|} + 2,486.60_7|a_{\overline{3}|}$$

By interpolation between 8% and 9%, the yield, or effective rate of interest, per annum is 8.60%.

5.6 (a) (i) Let X be the annual repayment. The amount lent is the present value, on the stated interest basis, of the repayments. Therefore,

$$X\left[a_{\overline{6}|0.10} + v_{0.10}^6 a_{\overline{12}|0.09}\right] = 2,000$$

from which we obtain $X = £238.17$.

(ii) The loan outstanding just after payment of the third payment is

$$2,000(1.1)^3 - 238.17 s_{\overline{3}|0.10} = 1,873.65$$

The loan outstanding just after payment of the fourth payment is

$$2000(1.1)^4 - 238.17 s_{\overline{4}|0.10} = 1,822.84$$

The capital repaid at time $t = 4$ is therefore

$$1,873.65 - 1,822.84 = £50.81 \tag{1}$$

The loan outstanding just after payment of the 11th payment is

$$238.17 a_{\overline{7}|} \text{ at } 9\% = 1,198.71$$

The loan outstanding just after payment of the 12th payment is

$$238.17 a_{\overline{6}|} \text{ at } 9\% = 1,068.42$$

Hence, the capital repaid at time $t = 12$ years is

$$1198.71 - 1068.42 = £130.29 \tag{2}$$

(b) After the special payment, the capital outstanding is £968.42. The revised annual repayment, Y, is found from the equation of value

$$Y a_{\overline{6}|0.09} = 968.42$$

which gives

$$Y = £215.88$$

5.7 The total cash payable is

$$\int_0^n t\,\mathrm{d}t = \frac{1}{2}n^2$$

We find n from the equation

$$(\bar{I}\bar{a})_{\overline{n}|} = \frac{1}{4}n^2 \text{ at } 5\%$$

i.e.,

$$\frac{\bar{a}_{\overline{n}|} - nv^n}{\delta} = \frac{1}{4}n^2$$

i.e.,

$$\frac{1 - v^n}{\delta} - nv^n - \frac{1}{4}\delta n^2 = 0 \text{ at } 5\%$$

By trials and interpolation, we find that $n = 22.37$.

5.8 **(a)** Let $X/12$ be the monthly repayment. We solve the equation of value

$$Xa^{(12)}_{\overline{25}|} = 9{,}880 \text{ at } 7\%$$

to obtain $X = 821.76$. Hence, the monthly repayment is £68.48.

(b) The loan outstanding is $Xa^{(12)}_{\overline{34/3}|}$ at $7\% = £6{,}486$.

(c) The loan outstanding just after the repayment on 10 September 2009 is made is $Xa^{(12)}_{\overline{83/6}|}$ at 7%.

The loan outstanding just after the repayment on 10 October 2009 is made is $Xa^{(12)}_{\overline{13.75}|}$ at 7%.

The capital repaid on 10 October 2009 is therefore

$$X\left(a^{(12)}_{\overline{83/6}|}\right) - a^{(12)}_{\overline{13.75}|} \text{ at } 7\%$$

$$= X\left(\frac{v^{13.75} - v^{83/6}}{i^{(12)}}\right) \text{ at } 7\% = £26.86$$

(d) **(i)** The capital repayment contained in these 12 installments is

$$X\left(a^{(12)}_{\overline{22/3}|} - a^{(12)}_{\overline{19/3}|}\right) \text{ at } 7\% = £516.20$$

(ii) The total interest in these 12 installments is [by (i)]

$$X - 516.20 = £305.56$$

(e) The capital and interest contents of each installment may be found by means of the loan schedule, using the rate of interest $j = i^{(12)}_{0.07}/12$ per month. The capital and interest contents of the tth repayment are therefore

$$\frac{9,880}{a_{\overline{300}|}}v^{301-t} \text{ at rate } j$$

and

$$\frac{9,880}{a_{\overline{300}|}}(1 - v^{301-t}) \text{ at rate } j$$

respectively. Note that $j = 0.005654$. We therefore require the smallest t such that

$$v^{301-t} > \frac{1}{2}(1 - v^{301-t}) \text{ at rate } j$$

i.e.,

$$(1 - j)^t > \frac{(1+j)^{301}}{3} \text{ at rate } j$$

i.e.,

$$t > \frac{-\ln 3 + 301 \ln(1+j)}{\ln(1+j)} = 106.15$$

The required value of t is then 107. The capital repaid first exceeds one-half of the interest content at the 107th installment, payable on 10 June 2017. (The capital content of this installment is £22.94, and the interest content is £45.54.)

5.9 (a) Let $X/12$ be the original monthly repayment, and let $Y/12$ be the monthly payment payable if the borrower had elected not to extend the term of the loan. We have

$$X a^{(12)}_{\overline{20}|} = 19,750 \text{ at } 9\%$$

and so $X = 2,079.12$ and $X/12 = £173.26$.
Also, the loan outstanding just after the 87th monthly payment had been made is

$$X a^{(12)}_{\overline{12.75}|} \text{ at } 9\% = 16,027.52$$

and hence

$$Y a^{(12)}_{\overline{12.75}|} = 16,027.52 \text{ at } 10\%$$

This gives $Y = 2{,}180.52$ and $Y/12 = £181.71$.
The increase in the monthly installment would then have been $(Y - X)/12 = £8.45$.

(b) Let the revised outstanding term of the loan be n months. We find n from the inequalities

$$2{,}079.12a^{(12)}_{\overline{(n-1)/12}|} < 16{,}027.52 \le 2{,}079.12a^{(12)}_{\overline{n/12}|} \quad \text{at } 10\%$$

(Notice that n must be an integer, and the final repayment may be of reduced amount.) By trials, we obtain $n = 169$, so that the revised outstanding term is 14 years and 1 month. The loan outstanding just after the penultimate installment is paid is

$$[16{,}027.52 - 2{,}079.12a^{(12)}_{\overline{14}|}](1+i)^{14} \quad \text{at } 10\% = 81.13.$$

The final installment must comprise this sum plus interest on it for 1 month; the final payment is therefore

$$81.13(1+i)^{1/12} \quad \text{at } 10\% = £81.78$$

Note that an alternative method of finding this reduced final payment is from the equation of value

$$2{,}079.12a^{(12)}_{\overline{14}|} + (\text{final payment})v^{169/12} = 16{,}027.52 \quad \text{at } 10\%$$

5.10 (a) Let X be the initial quarterly payment. We have the equation of value

$$4Xa^{(4)}_{\overline{15}|} + 160_5|a^{(4)}_{\overline{10}|} + 160_{10}|a^{(4)}_{\overline{5}|} = 11{,}820 \quad \text{at } 12\%$$

This gives $X = 389.96$.

(b) (i) The initial annual amount of the annuity is $4X$, i.e., $£1{,}559.84$. The loan outstanding just after payment of the eighth installment (at the end of the second year) is

$$11{,}820(1+i)^2 - 1{,}559.84s^{(4)}_{\overline{2}|} \quad \text{at } 12\% = 11{,}374.85$$

and, just after the 12th installment is paid, the loan outstanding is

$$11{,}820(1+i)^3 - 1{,}559.84s^{(4)}_{\overline{3}|} \quad \text{at } 12\% = 11{,}111.45$$

The capital repaid during the third year of the loan is therefore

$$11{,}374.85 - 11{,}111.45 = £263.40$$

(ii) In the final 5 years it is easier to calculate the outstanding loan by valuing the *future* annuity installments than by accumulating the original loan less past installments of the annuity. Therefore, we obtain the expressions

$$1,879.84a_{\overline{3}|}^{(4)} \text{ at } 12\%$$

$$1,879.84a_{\overline{2}|}^{(4)} \text{ at } 12\%$$

for the loan outstanding at the end of the 12th and 13th years, respectively (after payment of the annuity installments due at these times). The capital repaid in the 13th year is then

$$1,879.84\left(a_{\overline{3}|}^{(4)} - a_{\overline{2}|}^{(4)}\right) \text{ at } 12\% = \pounds1,369.82$$

(c) The loan outstanding just after the 33rd installment is paid is (on valuing future payments)

$$1,879.84a_{\overline{6.75}|}^{(4)} - 160a_{\overline{1.75}|}^{(4)} \text{ at } 12\% = 8,493.09$$

Let the revised quarterly annuity payment be Y. We have the equation of value

$$4Ya_{\overline{6.75}|}^{(4)} = 8,493.09 \text{ at } 12\%$$

which gives $Y = 456.50$.

5.11 Let X_t denote the gross annuity payment at time t; let c_t denote the capital repayment at time t; and let L_t denote the loan outstanding just after the payment at time t is made. We have the relations

$$X_t = 0.1 L_{t-1} + c_t \text{ with } 1 \le t \le 10 \tag{1}$$

(where L_0 denotes the amount of the original loan), and

$$(0.7 \times 0.1L_{t-1}) + c_t = 5,000 \text{ for } 1 \le t \le 10 \tag{2}$$

We also have, for $t \ge 1$,

$$\begin{aligned} L_t &= L_{t-1} - c_t \\ &= L_{t-1} - 5,000 + 0.07L_{t-1} \quad \text{(by Eq. 2) (3)} \\ &= 1.07L_{t-1} - 5,000 \end{aligned}$$

This shows that

$$\begin{aligned} L_1 &= 1.07L_0 - 5,000 \\ L_2 &= 1.07^2L_0 - 5,000(1 + 1.07) \\ &= 1.07^2L_0 - 5,000s_{\overline{2}|0.07} \end{aligned}$$

and so on, giving

$$L_t = 1.07^t L_0 - 5{,}000 s_{\overline{t}|0.07} \quad \text{for } 1 \le t \le 10 \tag{4}$$

(This result may be rigorously established from Eq. 3 by finite induction or by solving a first-order difference equation.) But $L_{10} = 0$, so

$$1.07^{10} L_0 - 5{,}000 s_{\overline{10}|0.07} = 0$$

and so

$$L_0 = 5{,}000 a_{\overline{10}|0.07} = 35{,}117.91$$

Equations 1, 2, and 4 now show that for $1 \le t \le 10$,

$$X_t = 0.1 L_{t-1} + (5{,}000 - 0.07 L_{t-1})$$
$$= 5{,}000 + 0.03 L_{t-1}$$
$$= 5{,}000 + 0.03\left\{ (1.07)^{t-1} L_0 - 5{,}000 \left[\frac{(1.07)^{t-1} - 1}{0.07} \right] \right\}$$
$$= 7{,}142.86 - 1{,}089.32(1.07)^{t-1}$$

Therefore, the price that the investor should pay to obtain a yield of 8% effective per annum is

$$\sum_{t=1}^{10} X_t v^t \quad \text{at } 8\%$$

$$= 7142.86 a_{\overline{10}|0.08} - (1{,}089.32/1.07) \sum_{t=1}^{10} \left(\frac{1.07}{1.08} \right)^t$$

$$= 7142.86 a_{\overline{10}|0.08} - 1{,}089.32 \left[\frac{1 - \left(\frac{1.07}{1.08} \right)^{10}}{1.08 - 1.07} \right]$$

$$= \pounds 38{,}253$$

CHAPTER 6 EXERCISES

6.1 We have $i = 12\%$ per annum.

(a) *Company A:* The annual payment under the perpetuity is £300,000 (assume in arrears), and the price is £2,000,000. The NPV is then obtained from

$$\text{NPV}_A = -2,000,000 + \frac{300,000}{0.12} = £500,000$$

Company B: The annual payment under the perpetuity is £435,000 (assume in arrears), and the price is £3,000,000. The NPV is then obtained from

$$\text{NPV}_B = -3,000,000 + \frac{435,000}{0.12} = £625,000$$

Both Company A and Company B would be positive NPV acquisitions, but Westcott-Smith cannot purchase both because the total purchase price of £5 million exceeds the budgeted amount of £4 million. Because Company B's NPV of £625,000 is higher than Company A's NPV of £500,000, Westcott-Smith should purchase Company B according to the NPV rule.

(b) *Company A:* The equation of value is

$$-2,000,000 + \frac{300,000}{\text{IRR}} = 0$$

which can be solved to give the IRR as 15% per annum.
Company B: The equation of value is

$$-3,000,000 + \frac{435,000}{\text{IRR}} = 0$$

which can be solved to give the IRR as 14.5% per annum.
Both Company A and Company B have IRRs that exceed Westcott-Smith's opportunity cost of 12%, but Westcott-Smith cannot purchase both because of its budget constraint. According to the IRR rule, Westcott-Smith should purchase Company A because its IRR of 15% is higher than Company B's IRR of 14.5%.

(c) Westcott-Smith should purchase Company B. When the NPV and IRR rules conflict in ranking mutually exclusive investments, we should follow the NPV rule because it directly relates to shareholder wealth maximization.

6.2 (a) The money-weighted rate of return, r, is the discount rate that equates the present value of inflows to the present value of outflows.
Outflows:
At $t = 0$ (1 January 2002):
150 shares purchased $\times$ \$156.30 per share $= \$23,445$.
Inflows:
At $t = 1$ (1 January 2003):
150 shares $\times$ \$10 dividend per share $= \$1,500$
100 shares sold $\times$ \$165 per share $= \$16,500$

At $t = 2$ (1 January 2004):

50 shares remaining $\times$ \$15 dividend per share $=$ \$750

50 shares sold $\times$ \$165 per share $=$ \$16,500

Therefore, PV (outflows) $=$ PV (inflows) becomes

$$23{,}455 = \frac{1{,}500 + 16{,}500}{1 + r} + \frac{750 + 8{,}500}{(1 + r)^2} = \frac{18{,}000}{1 + r} + \frac{9{,}250}{(1 + r)^2}$$

(b) In this case, we can solve the preceding expression for the money-weighted rate of return as the real root of the quadratic equation

$$18{,}000x + 9{,}250x^2 - 23{,}445 = 0$$

where $x = 1/(1 + r)$. The solution is $r = 0.120017$, or approximately 12% per annum.

(c) The time-weighted rate of return is the solution to

$$(1 + \text{time-weighted rate})^2 = (1 + r_1)(1 + r_2)$$

where r_1 and r_2 are the holding-period returns in the first and second years, respectively. The value of the portfolio at $t = 0$ is \$23,445. At $t = 1$, there are inflows of sale proceeds of \$16,500 and \$1,500 in dividends, or \$18,000 in total. The balance of 50 shares is worth \$8,250. So at $t = 1$ the valuation is \$26,250. Therefore,

$$r_1 = \frac{(\$26{,}250 - \$23{,}445)}{\$23{,}445} = 0.119642$$

The amount invested at $t = 1$ is \$8,250. At $t = 2$, \$750 in dividends are received, as well as sale proceeds of \$8,500. So at $t = 2$, the valuation is \$9,250. Therefore,

$$r_2 = \frac{(\$9{,}250 - \$8{,}250)}{\$8{,}250} = 0.121212$$

The time-weighted rate of return $= \sqrt{1.119642 \times 1.121212} - 1 = 0.1204$, or approximately 12% per annum.

(d) If Wilson is a private investor with full discretionary control over the timing and amount of withdrawals and additions to his portfolios, then the money-weighted rate of return is an appropriate measure of portfolio returns.

(e) If Wilson is an investment manager whose clients exercise discretionary control over the timing and amount of withdrawals and additions to the portfolio, then the time-weighted rate of return is the appropriate measure of portfolio returns. Time-weighted rate of return is standard in the investment management industry.

6.3 Choice C is correct. When one is valuing mutually exclusive projects, the decision should be made with the NPV method because this method uses the most realistic discount rate, namely the opportunity cost of funds. In this example, the reinvestment rate for the NPV method (here 10%) is more realistic than the reinvestment rate for the IRR method (here 21.86% or 18.92%).

6.4 Choice B is correct. The required value of i can be found from the expression

$$-100 + 36\, a_{\overline{4}|} = -100 + 175v^4$$

which reduces to

$$\frac{(1+i)^4 - 1}{i} - \frac{175}{36} = 0$$

The left side of this expression changes sign between $i = 10\%$ and 15.02%. (In fact, it is solved by $i = 13.16\%$.)

6.5 (a) Working in millions:

$$PV\text{ of liabilities} = 9 + 12v\,\bar{a}_{\overline{1}|} \text{ at 9%}$$

$$= 9 + 12v.\frac{i}{\delta}v$$

$$= 9 + 12 \times 0.91743^2 \times 1.044354$$

$$= 19.54811$$

The assets up to $(k+2)$ years from 1 January 2006 have

$$PV = 5v^2 a_{\overline{k}|}^{(2)} = 5v^2\, \frac{i}{i^{(2)}} a_{\overline{k}|}$$

$$= 5 \times 0.84168 \times 1.022015 \times a_{\overline{k}|}$$

$$= 4.301048 a_{\overline{k}|}$$

with $k = 6, PV = 4.301048 \times 4.4859$

$$= 19.2941$$

The next payment of 2.5 million at $k = 6.5$ is made at time 8.5 and has present value $= 2.5 \times v^{8.5} = 1.2018$.
This would make PV of assets (20.5m) > PV of liabilities (19.5m). This discounted payback period is then 8.5 years.

(b) The income of the development is received later than the costs are incurred. Hence, an increase in the rate of interest will reduce the present value of the income more than the present value of the outgo. Hence, the DPP will increase.

6.6 **(a)** Net present value of costs

$$= 5{,}000{,}000 + 3{,}500{,}000\bar{a}_{\overline{2}|} = 5{,}000{,}000 + 3{,}500{,}000\frac{i}{\delta}a_{\overline{2}|}$$

$$= 5{,}000{,}000 + 3{,}500{,}000 \times 1.073254 \times 1.6257 = 11{,}106{,}762$$

Net present value of benefits

$$= 450{,}000v^2\ddot{a}^{(4)}_{\overline{n-2}|} + 50{,}000v^2(I\ddot{a})^{(4)}_{\overline{n-2}|} + S_n v^n$$

$$= 450{,}000v^2\frac{i}{d^{(4)}}a_{\overline{n-2}|} + 50{,}000v^2\frac{i}{d^{(4)}}(Ia)_{\overline{n-2}|} + S_n v^n$$

where n is the year of sale and S_n are the sale proceeds if the sale is made in year n.

If $n = 3$, the NPV of benefits

$$= (450{,}000 \times 0.75614 \times 1.092113 \times 0.86957)$$
$$+(50{,}000 \times 0.75614 \times 1.092113 \times 0.86957)$$
$$+(16{,}500{,}000 \times 0.65752)$$
$$= 323{,}137 + 35{,}904 + 10{,}849{,}080 = 11{,}208{,}121$$

Hence, the net present value of the project is $11{,}208{,}121 - 11{,}106{,}762 = 101{,}359$.

Note that if $n = 4$, the extra benefits in year 4 consist of an extra £1.5 million on the sale proceeds and an extra £650,000 rental income. This is clearly less than the amount that could have been obtained if the sale had been made at the end of year 3 and the proceeds invested at 15% per annum. Hence, selling in year 4 is not an optimum strategy.

If $n = 5$, the NPV of benefits

$$= (450{,}000 \times 0.75614 \times 1.092113 \times 2.2832)$$
$$+(50{,}000 \times 0.75614 \times 1.092113 \times 4.3544)$$
$$+(20{,}500{,}000 \times 0.49718)$$
$$= 848{,}450 + 179{,}791 + 10{,}192{,}190 = 11{,}220{,}431$$

Hence, the net present value of the project is $11{,}220{,}431 - 11{,}106{,}762 = 113{,}669$.

Hence, the optimum strategy if net present value is used as the criterion is to sell the housing after 5 years.

(b) If the discounted payback period is used as the criterion, the optimum strategy is that which minimises the first time when the net present value is positive. By inspection, this is when the housing is sold after 3 years.

(c) We require

$$5,000,000 + 3,500,000 \frac{i}{\delta} a_{\overline{2}|} = 450,000 v^2 \ddot{a}^{(4)}_{\overline{n-2}|}$$

$$+ 50,000 v^2 (I\ddot{a})^{(4)}_{\overline{n-2}|} + S_n v^n \text{ at } 17.5\%$$

$$\text{LHS} = 5,000,000 + 3,500,000 \left(\frac{1 - v^2_{0.175}}{\delta_{0.175}} \right)$$

$$= 5,000,000 + 3,500,000 \left(\frac{1 - 0.72431}{0.16127} \right) = 10,983,227$$

$$\text{RHS} = 450,000 v^2_{0.175} \left(\frac{1 - v^4_{0.175}}{d^{(4)}} \right)$$

$$+ 50,000 v^2_{0.175} \left(\frac{\ddot{a}_{\overline{4}|} - 4v^4_{0.175}}{d^{(4)}} \right) + S_6 v^6_{0.175}$$

$$d^{(4)}_{0.175} = 4 \left(1 - v^{\frac{1}{4}} \right) = 0.15806$$

$$\ddot{a}_{\overline{4}|} = \frac{1 - v^4}{d} = 3.1918$$

Therefore, we have on the RHS

$$450,000 \times 0.72431 \times 3.0076 + 50,000 \times 0.72431$$

$$\times \left(\frac{3.1918 - 2.0985}{0.15806} \right) + 0.37999 S_6$$

$$= 980,296 + 250,502 + 0.37999 S_6$$

For equality, $S_6 = \dfrac{10,983,227 - 1,230,798}{0.37999} = £25,665,000.$

(d) Reasons investors may not achieve the internal rate of return:
- allowance for expenses when buying/selling may be significant;
- there may be periods when the property is unoccupied and no rental income is received;
- rental income may be reduced by maintenance expenses;
- there is tax on rental income and/or sale proceeds.

6.7 (a) Working in units of £1,000:

Money-weighted rate of return (MWRR) is i such that

$$21(1 + i)^3 + 5(1 + i)^2 + 8(1 + i) = 38$$

Try $i = 5\%$, LHS $= 38.223$

$i = 4\%$, LHS $= 37.350$

By interpolation, $i = 4.74\%$ per annum.

(b) Time-weighted rate of return (TWRR) is i such that

$$(1+i)^3 = \frac{24}{21} \times \frac{32}{29} \times \frac{38}{40} \text{ and so } i = 6.21\% \text{ per annum}$$

(c) MWRR is lower than TWRR because of the large cash flow on 1 July 2005; the overall return in the final year is much lower than in the first 2 years, and the payment at 1 July 2005 gives this final year more weight in the MWRR but does not affect the TWRR.

6.8 (a) *Project A:* The internal rate of return i is the solution of the equation

$$-1{,}000{,}000 + 270{,}000a_{\overline{8}|} = 0 \quad \text{at rate } i$$

which gives $i = 0.2120$ or 21.20%.
Project B: The internal rate of return i is the solution of the equation

$$-1{,}200{,}000 - 20{,}000a_{\overline{3}|} + 1{,}350{,}000{}_5|a_{\overline{3}|} = 0 \quad \text{at rate } i$$

which gives $i = 0.1854$ or 18.54%.

(b) We have

$$NPV_A(0.15) = 1{,}000[-1{,}000 + 270a_{\overline{8}|}] \text{ at } 15\% = £211{,}577$$
$$NPV_B(0.15) = 1{,}000[-1{,}200 - 20a_{\overline{3}|} + 1{,}350{}_5|a_{\overline{3}|}] \text{ at } 15\%$$
$$= £286{,}814$$

Comment: Project A is viable when money can be borrowed at up to 21.2% per annum interest, but Project B is viable only for interest rates up to 18.5% per annum. At 15% per annum interest, however, Project B has the greater net present value.

6.9 (a) *Project A:* The internal rate of return i is the solution of the equation

$$-160 - 80v + 60(a_{\overline{11}|} - a_{\overline{4}|}) = 0 \quad \text{at rate } i$$

(where we have worked in money units of £1,000). This gives $i = 0.0772$, or 7.72%.
Project B: The internal rate of return i is the solution of the equation

$$-193 - 80v + 70(a_{\overline{11}|} - a_{\overline{4}|}) = 0 \quad \text{at rate } i$$

(where we again work in money units of £1,000). This gives $i = 0.0804$, or 8.04%.

(b) The purchase money of £33,000 must buy at least £10,000 per annum for 7 years, beginning at time 5 years, for the combined transaction to be more favourable to the businessman than Project B. The minimum rate of interest per annum i must therefore satisfy

$$-33{,}000 + 10{,}000(a_{\overline{11}|} - a_{\overline{4}|}) = 0 \quad \text{at rate } i$$
$$\text{i.e., } a_{\overline{11}|} - a_{\overline{4}|} = 3.3 \quad \text{at rate } i$$

This gives $i = 0.1011$, or 10.11%

6.10 **(a)** Working in money units of £1,000, we have

$$-10 - 3v + \left(\frac{1}{2}v^2 + \frac{2}{2}v^3 + \frac{3}{2}v^4 + \ldots + \frac{8}{2}v^9\right) = 0 \quad \text{at rate } i$$

where i is the internal rate of return. That is,

$$-10 - 3v + \frac{1}{2}v(Ia)_{\overline{8}|} = 0 \quad \text{at rate } i$$

Hence, $i = 0.0527$, or 5.27%

(b) The investor should proceed with the venture, since its internal rate of return exceeds the rate of interest he must pay on borrowed money. The profit in 9 years' time is found by noting that the investor's bank account, just *before* receiving the final income (but *after* paying interest on his borrowing), is

$$-10,000(1.05)^9 - 3,000(1.05)^8 + 500(1.05)^8(Ia)_{\overline{7}|0.05} = -3,680$$

so the final income payment (of £4,000) extinguishes the loan and leaves a profit of £320.

6.11 **(a)** *Sheep rearing:* The internal rate of return i is the solution of the equation of value

$$20,000 = 1,100a_{\overline{20}|} + 20,000v^{20}$$

which gives $i = 5.5\%$ exactly. This is also clear from the fact that income is sufficient to pay interest at 5.5% per annum on the initial cost, and the capital is repaid at the end of 20 years.
Goat breeding: The internal rate of return i is found by solving

$$20,000 = 900a_{\overline{20}|} + 25,000v^{20} \quad \text{at rate } i$$

which gives $i = 5.24\%$.
Forestry: The internal rate of return i is found by solving

$$20,000(1+i)^{20} = 57,300$$

which gives $i = 5.40\%$.

(b) We calculate the profit after 20 years from each of these three ventures.
Sheep rearing: The "surplus" income is £100 per annum, so the profit is $100s_{\overline{20}|}$ at 4% = £2,978.
Goat breeding: A further £100 must be borrowed annually in arrears for 20 years, and these loans grow at 5% per annum compound. The total indebtedness at time 20 years is then

$$20,000 + 100s_{\overline{20}|} \quad \text{at } 5\% = 23,307$$

Hence, the profit to the investor at the end of the project is $25{,}000 - 23{,}307 = £1{,}693$.

Forestry: The original debt of £20,000 will have grown to $20{,}000 \times (1.05)^{20} = 53{,}066$ after 20 years, so the profit is $57{,}300 - 53{,}066 = £4{,}234$.

Hence, the largest profit is obtained from forestry.

6.12 Let us work in money units of £1,000.

(a) We must find any values of i, the rate of interest per annum, satisfying

$$1 - 2v + 2v^2 = 0 \text{ for Project } A$$

and

$$-4 + 7v - 1.5v^2 = 0 \text{ for Project } B$$

Now the equation $1 - 2v + 2v^2 = 0$ has no real solution, so there is no value of i satisfying the given condition for Project A. But

$$-4 + 7v - 1.5v^2 = (-1.5v + 1)(v - 4)$$

which equals 0 for $v = 2/3$ or 4. Since $i = v^{-1} - 1$, the required values of i for Project B are 0.5 and -0.75, i.e., 50% and -75%.

(b)

$$NPV_A(i) - NPV_B(i) = \left(1 - 2v + 2v^2\right) - \left(-4 + 7v - 1.5v^2\right)$$
$$= 5 - 9v + 3.5v^2$$
$$= v^2\left(5x^2 - 9x + 3.5\right) \quad \text{(where } x = 1 + i\text{)}.$$

By the properties of quadratics,

$$5x^2 - 9x + 3.5 > 0 \text{ for } x > 1.23166 \text{ and } x < 0.56833$$

So

$$NPV_A(i) - NPV_B(i) > 0 \quad \text{for } i > 23.166\% \text{ and } i < -43.167\%$$

The range of positive interest rates for which $NPV_A(i) > NPV_B(i)$ is therefore $i > 23.166\%$.

(c) It follows from (b) that Project B is the more profitable in case (i) and Project A is the more profitable in case (ii). The accumulated profits, at an interest rate of i per annum, of projects A and B are (in units of £1,000)

$$(1+i)^2 - 2(1+i) + 2$$

and

$$-4(1+i)^2 + 7(1+i) - 1.5$$

respectively. This gives the accumulations

(i) £1,040 for A, £1,140 for B;
(ii) £1,062.5 for A, £1,000 for B.

6.13 (a) The internal rate of return i is the solution of the equation of value

$$1{,}000\left[-10 + 6(1+i)^{-1} + 6.6(1+i)^{-2}\right] = 0$$

which gives $i = 0.1660$, say 16.6%.

(b) (i) Since the internal rate of return exceeds 16%, the person should proceed with the investment. The balance at the end of 1 year (just after receipt of the payment then due) is

$$-10{,}000(1.16) + 6{,}000 = -5{,}600$$

So the balance at the end of 2 years (just after receipt of the payment then due) is

$$-5{,}600(1.16) + 6{,}600 = £104$$

(ii) Under the conditions of the loan, the borrower must pay £1,600 in interest at time 1 year, and £11,600 in interest and capital repayment at time 2 years. He will therefore have £4,400 to invest at time 1 year, which gives £4,400(1.13) = £4,972 at time 2 years. Together with the second payment of £6,600, he will therefore have £11,572 available at time 2 years. But this is less than £11,600, so the investor should *not* proceed.

6.14 (a) Let us work in money units of £1,000. The net cash flows associated with the project are shown in the Figure S.6.1.

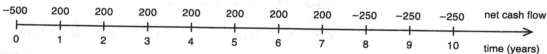

FIGURE S.6.1
Cash flow diagram for Solution 6.14

Until the initial loan is repaid the chemical company's balance just after the transaction at time t years is

$$-500(1+i)^t + 200s_t \quad \text{at } 15\%$$

This is negative if $a_{\overline{t}|} < 2.5$ at 15%, i.e., if $t \leq 3$. The bank loan is therefore paid off after 4 years.

(b) The balance at time $t = 4$ years, just after repaying the bank loan, is

$$-500(1+i)^4 + 200s_{\overline{4}|} \quad \text{at } 15\% = 124.172$$

The balance at time 10 years, when the project ends, is therefore

$$124.172(1+i)^6 + 200(s_{\overline{6}|} - s_{\overline{3}|}) - 250s_{\overline{3}|} \quad \text{at } 12\%$$
$$= 349.651 \text{ or } £349,651$$

6.15 (a) (i) Let us work in money units of £1,000. The DPP is the value of t such that

$$-80(1+i)^t - 5(1+i)^{t-1} + 10s_{\overline{t-2}|} \quad \text{at } 7\% = 0$$

By interpolation, we obtain $t = 17.767$, say 17.77 years.

(ii) The accumulated profit after 22 years is found by accumulating the income, after the DPP has elapsed, at 6% per annum. This gives

$$10s_{\overline{4.233}|} \quad \text{at } 6\% = 48.00$$

The accumulated amount is then £48,000.

(b) (i) The businessman may accumulate his rental income for each year at 6%, giving $10s_{\overline{1}|}$ at 6% = 10.2971 at the end of the year. The DPP is then the smallest integer t such that

$$-80(1+i)^t - 5(1+i)^{t-1} + 10.2971s_{\overline{t-2}|} \quad \text{at } 7\% \geq 0$$

By trials, this gives $t = 18$ years.

(ii) The balance in the businessman's account just after the transaction at time 18 years is

$$-80(1+i)^{18} - 5(1+i)^{17} + 10.2971s_{\overline{16}|} \quad \text{at } 7\% = 0.97714.$$

Hence, the accumulated amount in his account after 22 years is

$$0.97714(1+i)^4 + 10\bar{s}_{\overline{4}|} \quad \text{at } 6\%$$

$$= 46.279, \quad \text{i.e.,} \quad \pounds 46,279.$$

6.16 (a) The time-weighted rates of return are

$$\frac{164}{124} - 1 = 0.3226, \quad \text{or } 32.26\% \text{ for the property fund}$$

and

$$\frac{155}{121} - 1 = 0.2810, \quad \text{or } 28.10\% \text{ for the equity fund.}$$

(b) We first consider the property fund.

(i) Assuming that the investor buys 100 units per quarter, we obtain the equation of value

$$124(1+i) + 131(1+i)^{3/4} + 148(1+i)^{1/2} + 158(1+i)^{1/4}$$

$$= 4 \times 164$$

where i is the yield per annum. This gives $i = 0.2932$, or 29.32%.

(ii) Now assume that the investor purchases $\pounds 1$ worth of units each quarter. The equation of value is

$$4\ddot{s}^{(4)}_{\overline{1}|} = 1.64 \left(\frac{1}{1.24} + \frac{1}{1.31} + \frac{1}{1.48} + \frac{1}{1.58} \right)$$

i.e. $\ddot{s}^{(4)}_{\overline{1}|} = 1.1801$. This gives $i = 0.2979$, or 29.79%

(c) We now consider the equity fund.

(i) The equation of value is

$$121(1+i) + 92(1+i)^{3/4} + 103(1+i)^{1/2} + 131(1+i)^{1/4} = 4 \times 155,$$

which gives $i = 0.6731$, or 67.31%.

(ii) The equation of value is

$$4\ddot{s}^{(4)}_{\overline{1}|} = 1.55 \left(\frac{1}{1.21} + \frac{1}{0.92} + \frac{1}{1.03} + \frac{1}{1.31} \right)$$

i.e., $\ddot{s}^{(4)}_{\overline{1}|} = 1.4135$. This yield is therefore 0.7088, or 70.88%.

Comment: For each fund, the yield to the investor depends only slightly on whether a fixed sum was invested, or a fixed number of units was purchased each quarter. The equity fund has the lower time-weighted rate of return, but much higher yields. The reason is that the price of the equity fund fluctuates more than that of the property fund and the equity fund price at the end of the year was relatively high.

6.17 (a) (i) The time-weighted rate of return is

$$\left(\frac{352}{186}\right)^{1/6} - 1 = 0.1122, \text{ or } 11.22\%$$

(ii) Let U_r be the unit price on 1 April 1999 + r.
The yield per annum i is found from the equation

$$200 \sum_{r=0}^{5} U_r(1+i)^{6-r} = 1{,}200U_6$$

i.e., $186(1+i)^6 + 211(1+i)^5 + 255(1+i)^4 + 249(1+i)^3$
$+ 288(1+i)^2 + 318(1+i) = 2{,}112$

which gives $i = 0.1060$, or 10.60%.

(iii) The yield per annum i is found from the equation

$$500\ddot{s}_{\overline{6}|} = 500\left(\sum_{r=0}^{5} \frac{1}{U_r}\right)U_6 \text{ at rate } i$$

i.e., $\ddot{s}_{\overline{6}|} = 8.6839$. Hence, $i = 0.1067$, or 10.67%.

(b) The revised answer to (a) (ii) is

$$200 \sum_{r=0}^{5} (1.02U_r)(1+i)^{6-r} = 1{,}200 \times 0.98U_6$$

which gives

$$186(1+i)^6 + 211(1+i)^5 + 255(1+i)^4 + 249(1+i)^3$$
$$+ 288(1+i)^2 + 318(1+i) = 2{,}029.18$$

Hence, $i = 0.0933$, or 9.33%.
The revised answer to (a) (iii) is

$$500\ddot{s}_{\overline{6}|} = 500\left[\sum_{r=0}^{5}\left(\frac{1}{1.02U_r}\right)\right](0.98U_6)$$

i.e., $\ddot{s}_{\overline{6}|} = 8.3434$, which gives $i = 0.0950$, or 9.50%.

CHAPTER 7 EXERCISES

7.1 (a) Below nominal value since the coupon rate is less than the yield required by the market.

   (b) Below nominal value since the coupon rate is less than the yield required by the market.

   (c) Below nominal value since the coupon rate is less than the yield required by the market.

   (d) Above nominal value since the coupon rate is greater than the yield required by the market.

   (e) At nominal value since the coupon rate is equal to that the yield required by the market.

7.2 We work in units of 6 months and assume $100 nominal of the bond.

Coupon payment $= \$3.70$

$i = 2.8\%$

Number of time units $= 10$

Present value of cash flows $= 3.70^a\overline{10|} + 100v^{10} = 31.8864 + 75.8698$

The value is therefore 107.7562%.

7.3 We work in units of 6 months and assume $100 nominal of the bond.

$i = 3.8\%$

Number of time units $= 40$

Present value of cash flows $= 1,000,000v^{40} = \dfrac{1,000,000}{1.038^{40}}$

The value is therefore $\$224,960.29$.

7.4

$$i^{(2)} = 0.049390$$

$$g(1 - t_1) = 0.0625 \times 0.80 = 0.05$$

and we see that $i^{(2)} < (1 - t_1)g$.

There is therefore a capital loss on contract, and we assume it is redeemed as early as possible (i.e., after 10 years) to obtain minimum yield.

Price of stock per £100 nominal, P:

$$P = 100 \times 0.0625 \times 0.80 \times a_{\overline{a|}}^{(2)} + 100v^{10} \quad \text{at } 5\%$$

$$\Rightarrow P = 5a_{\overline{10|}}^{(2)} + 100v^{10}$$

$$= (5 \times 1.012348 \times 7.7217) + (100 \times 0.61391)$$

$$= 39.0852 + 61.3910 = £100.4762$$

7.5 (a) • Payments are guaranteed by the issuing government.

   • There can be various different indexation provisions but, in general, protection is given against a fall in the purchasing power of money.

- It is fairly liquid (i.e., large issue size and ability to deal in large quantities) compared with corporate issues, but not compared with conventional issues.
- Typically coupon and capital payments are both indexed to increases in a given price index with a lag.
- It has low volatility of return and low expected real return.
- It has more or less guaranteed real return if held to maturity (although can vary due to indexation lag).
- Nominal return is not guaranteed.

(b) The first coupon the investor will receive is on 31 December 2003. The net coupon per £100 nominal will be

$0.8 \times 1 \times$ (Index May 2003/Index November 2001)

$$= 0.8 \times 1 \times \frac{113.8}{110}$$

In real present value terms, this is $0.8 \dfrac{113.8}{110} \dfrac{v}{(1+r)^{0.5}}$ where

$r = 2.5\%$ per annum and v is calculated at 1.5% (per half-year). The second coupon on 30 June 2004 per £100 nominal is

$0.8 \times 1 \times \dfrac{113.8}{110}(1+r)^{0.5}$.

In real present value terms, this is $0.8(1+r)^{0.5}\dfrac{113.8}{110}\dfrac{v^2}{(1+r)}$.

The third coupon on 31 December 2004 per £100 nominal is

$$0.8 \times 1 \times \frac{113.8}{110}(1+r)$$

In real present value terms, this is $0.8(1+r)\dfrac{113.8}{110}\dfrac{v^2}{(1+r)^{1.5}}$

Continuing in this way, the last coupon payment on 30 June 2009 per £100 nominal is $0.8 \times 1 \times \dfrac{113.8}{110}(1+r)^{5.5}$.

In real present value terms, this is $0.8(1+r)^{5.5}\dfrac{113}{110}\dfrac{v^{12}}{(1+r)^6}$

By similar reasoning, the real present value of the redemption payment is

$$100(1+r)^{5.5}\frac{113.8}{110}\frac{v^{12}}{(1+r)^6}$$

The present value of the succession of coupon payments and the capital payment can be written as

$$P = \frac{1}{(1+r)^{0.5}} \frac{113.8}{110} \left(0.8\left(v + v^2 + \ldots + v^{12}\right) + 100v^{12}\right)$$

$$= \frac{1}{1.0124224} \frac{113.8}{110} \left(0.8a_{\overline{12}|1.5\%} + 100v^{12}_{1.5\%}\right)$$

$$= 1.02185 \times (0.8 \times 10.9075 + 100 \times 0.83639)$$

$$= 94.3833$$

7.6 We convert all cash flow to amounts in time 0 values.

Dividend paid at $t = 1$: $10{,}000 \times 0.041 \times \dfrac{147.7}{153.4} = 394.77$

Dividend paid at $t = 2$: $10{,}000 \times 0.046 \times \dfrac{147.7}{158.6} = 428.39$

Dividend paid at $t = 3$: $10{,}000 \times 0.051 \times \dfrac{147.7}{165.1} = 456.25$

Sale proceeds at $t = 3$: $10{,}000 \times 0.93 \times \dfrac{147.7}{165.1} = 8{,}319.87$

Therefore, the equation of value involving v and r ($=$ real rate of return) is

$$7{,}800 = 394.77v + 428.39v^2 + 8{,}776.17v^3 \qquad (1)$$

To estimate r:
Approximate nominal rate of return is

$$\left(4.6 + \frac{(93 - 78)}{3}\right)\Big/ 78 = 12.3\% \text{ per annum.}$$

Average inflation over a 3-year period comes from

$$\left(\frac{165.1}{147.1}\right)^{1/3} - 1 = 3.8\% \text{ per annum.}$$

Therefore, the approximate real return: $\dfrac{1.123}{1.038} - 1 = 8.2\%$ per annum
Try $r = 8\%$ RHS of Eq. (1) $= 7{,}699.61$

$$r = 7\%, \text{ RHS of Eq. (1)} = 7{,}907.09$$

$$r = 7\% + \frac{7{,}907.09 - 7{,}800}{7{,}907.69 - 7{,}699.61} \times 1\%$$

$$= 7.52\% \text{ per annum}$$

7.7 Working in half-years:

The present value of the security on 1 June would have been $\dfrac{3.5}{i^{(2)}}$

On 20 August, 80 days later, the present value is $\dfrac{3.5}{i^{(2)}}(1+r)^{80/365}$

Hence, the price per £100 nominal is $\dfrac{3.5}{0.097618}(1.1)^{80/365} = £36.611$

7.8 Let us work in time units of a half-year. We note that the period from 14 August 2012 to 1 December 2012 is 109 days, and the period from 1 June 2012 to 1 December 2012 is 183 days. Define $t = 109/183$. The yield per half-year, i, is the solution of the equation of value

$$35.125 = 1.75(1+i)^{-t}\left(\frac{1+i}{i}\right)$$

which gives $i = 0.05083$. The yield per annum is therefore $(1+i)^2 - 1 = 0.1042$, or 10.42%. Expressed as a nominal rate of interest convertible half-yearly, the yield per annum is $2(0.050\,83)$, i.e., $0.101\,66$ or 10.17%.

7.9 Let us work in time units of 1 year, measured from 31 July 2006, and in money units of £500. In the notation of this chapter, we have $C = 525$, $p = 2$, and $g = 0.1/1.05$.

(a) We first find the value of the loan just after payment of the interest on 31 July 2006. We have for K, the present value of the capital repayments,

$$K = 105(a_{\overline{34}|} - a_{\overline{29}|}) \quad \text{at 9%}$$

$$= 33.553$$

and, by Makeham's formula, the value of the loan on 31 July 2006 is

$$K + \frac{0.1(1 - 0.4)}{1.05 \times 0.09^{(2)}}(525 - K) = 352.45$$

Hence, the "ex dividend" value on 19 June 2006 is

$$352.45v^{42/365} \quad \text{at 9%} \quad = 348.975, \quad \text{or} \quad £69.79\%$$

(b) The "cum dividend" value on 19 June 2006 is

$$[352.452 + (0.6 \times 25)]v^{42/365} = 363.826, \quad \text{or} \quad £72.77\%$$

7.10 Let us work in time units of 1 year and in money units of £1,000. Assume first that redemption is always at par: an adjustment to allow for the actual redemption prices will be added later. In the notation of this chapter, we have $C = 300$ and

$$K = 10a_{\overline{30}|} \quad \text{at } 7\% \quad = 124.0904$$

By Makeham's formula, the value of the loan is

$$K + \frac{0.08 \times 0.6}{0.07^{(4)}}(300 - K) = 247.836$$

We now add the value of the premiums on redemption, namely,

$$2(a_{\overline{30}|} - a_{\overline{15}|}) \quad \text{at } 7\% \quad = 6.602$$

which gives a total value of 254.438, or £84.81%.

7.11 (a) Let $I_0 = 187.52$ (the index value for August 2005) and let $I_1 = 192.10$ (the index value for February 2006). Let r denote the assumed rate of increase per annum in the index. Therefore, the index value increases by the factor $(1 + r)^{1/2}$ every 6 months. Measure time in half-years from the issue date. Let Q_0 be the index value at the issue date. (Note that Q_0 is the value at the *precise* date of issue and *not* the index value "for April 2006".)

Consider a stock of term n. (We have $n = 40$ or 60.) Let the issue price per cent nominal be A. Consider a purchase of £100 nominal of stock. The purchaser has an outlay of A at time $t = 0$ (when the index value is Q_0).

On 7 October 2006 [i.e., at time $t = 1$, when the index value will be $Q_0(1 + r)^{1/2}$], the purchaser will receive an indexed interest payment of amount

$$\frac{3}{2}\frac{(\text{index value for February 2006})}{(\text{index value for August 2005})} = \frac{3}{2}\frac{I_1}{I_0}$$

On 7 April 2007 [i.e., at time $t = 2$, when the index value will be $Q_0(1 + r)$], the purchaser will receive interest of

$$\frac{3}{2}\frac{(\text{index value for August 2006})}{(\text{index value for August 2005})}$$

$$= \frac{3}{2}\frac{(1 + r)^{1/2}(\text{index value for February 2006})}{(\text{index value for August 2005})} = \frac{3}{2}(1 + r)^{1/2}\frac{I_1}{I_0}$$

On 7 October 2007 [i.e., at time $t = 3$, when the index value will be $Q_0(1 + r)^{3/2}$], the purchaser will receive interest of

$$\frac{3}{2}\frac{(\text{index value for February 2007})}{(\text{index value for August 2005})}$$

$$=\frac{3}{2}\frac{(1+r)(\text{index value for February 2006})}{(\text{index value for August 2005})}$$

$$=\frac{3}{2}(1+r)\frac{I_1}{I_0}$$

More generally, for $1 \leq t \leq n$, at time t [when the index value will be $Q_0(1+r)^{t/2}$], the purchaser will receive an indexed interest payment of amount

$$\frac{3}{2}(1+r)^{(t-1)/2}\frac{I_1}{I_0}$$

The indexed amount of the capital repayment, at time n, is

$$100(1+r)^{(n-1)/2}\frac{I_1}{I_0}$$

Let i be the effective real yield per half-year. By combining the preceding remarks, we see that i is obtained from the equation

$$-\frac{A}{Q_0}+\sum_{t=1}^{n}\left[\frac{3}{2}(1+r)^{(t-1)/2}\frac{I_1}{I_0}\right]\left[\frac{1}{Q_0(1+r)^{t/2}}\right](1+i)^{-t}$$

$$+\left[100(1+r)^{(n-1)/2}\frac{I_1}{I_0}\right]\frac{1}{Q_0(1+r)^{n/2}}(1+i)^{-n}=0$$

which gives

$$A=\frac{I_1}{(1+r)^{1/2}I_0}\left(\frac{3}{2}a_{\overline{n}|}+100v^n\right)\quad\text{at rate }i \tag{1}$$

For a real yield of 3% per annum convertible half-yearly $i = 0.015$, and it follows from Eq. 1 that

$$A=\frac{I_1}{(1+r)^{1/2}I_0}100$$

for *any* value of n. Since, on the assumption that $r = 0.06$, there is an effective real yield of $1\frac{1}{2}$% per half-year, by letting $r = 0.06$ in the last expression, we find the issue price of either stock to be

$$\frac{(192.10)(100)}{1.06^{1/2}(187.52)}=99.50\%$$

(b) It follows from Eq. 1 that, if the index increases at the rate r per annum, the real yield for a stock with term n and issue price 99.50% is i per half-year, where

$$\frac{3}{2}\,a_{\overline{n}|} + 100v^n = \frac{99.50(1+r)^{1/2}\,187.52}{192.10} \quad \text{at rate } i$$

When $r = 0.04$, the right side of this last equation equals 99.05, which is *less* than 100. Hence, the shorter term gives the higher yield. When $r = 0.08$, the right side of the last equation equals 100.94, which is *greater* than 100. In this case, therefore, the shorter term gives the lower yield.

(When $r = 0.04$, the values of i for $n = 40$ and for $n = 60$ are 1.5319% and 1.5242%, respectively. The corresponding values when $r = 0.08$ are 1.4688% and 1.4763%.)

7.12 Let us work in time units of a quarter-year and let L be the original loan. The capital repaid in 2005 is, according to the original schedule,

$$\frac{L}{a_{\overline{48}|}}(a_{\overline{36}|} - a_{\overline{32}|}) \quad \text{at } 3\%$$

and hence, the interest paid in 2005 is

$$\frac{4L}{a_{\overline{48}|}} - \frac{L}{a_{\overline{48}|}}(a_{\overline{36}|} - a_{\overline{32}|}) \quad \text{at } 3\%, \quad \text{i.e.} \quad 0.101181L$$

Equating this expression to 6,374.41 gives $L = £63,000$. Hence, the loan outstanding at the date of purchase was, according to the original schedule,

$$\left[\frac{L}{a_{\overline{48}|}}\right]a_{\overline{32}|} \quad \text{at } 3\%, \quad \text{i.e.} \quad £50,837.35$$

and each quarterly installment is of amount

$$\frac{L}{a_{\overline{48}|}} \quad \text{at } 3\% = 2,493.40$$

(a) The net yield per quarter is 2%. Accordingly, we value the outstanding installments, less tax, at this rate of interest. If there were no income tax, we would have the value of outstanding installments as

$$A = 2493.40\,a_{\overline{32}|} \quad \text{at } 2\% \quad = 58,515.95$$

and hence, we may find K, the value of the capital repayments, by Makeham's formula, i.e.,

$$58{,}515.95 = K + \frac{0.03}{0.02}(50{,}837.35 - K)$$

(Note that £50,837.35 is the outstanding loan at the date of purchase.) Hence, $K = 35{,}480.16$, and the purchase price paid by the investor is (again by Makeham's formula)

$$K + \frac{0.03(0.6)}{0.02}(50{,}837.35 - K) = \pounds 49{,}302$$

(b) The net yield per quarter is now $0.08^{(4)}/4$. Ignoring income tax, the value of the outstanding installments is

$$2{,}493.40 a_{\overline{32}|} \quad \text{at rate } 0.08^{(4)}/4$$

$$= (4 \times 2{,}493.40) a_{\overline{8}|}^{(4)} \quad \text{at rate } 0.08$$

$$= 59{,}006.55$$

We may find K (the value of the capital repayments) by Makeham's formula, i.e.,

$$59{,}006.55 = K + \frac{0.03}{0.08^{(4)}/4}(50{,}837.35 - K)$$

This gives $K = 35{,}828.12$, and the price paid by the investor is (again using Makeham's formula)

$$K + \frac{0.03 \times 0.6}{0.08^{(4)}/4}(50{,}837.35 - K) = \pounds 49{,}735$$

Note that the value may also be obtained as $K + 0.6$ $(59{,}006.55 - K)$.

7.13 Let i be the net annual yield. The value of the capital repayments is

$$K = 10{,}000 \left(v^8 + v^{16} + v^{24} \right) \quad \text{at rate } i$$

The value of the net interest payments is found, by first principles, to be

$$I = 30{,}000 \times 0.06 \times 0.7 \, a_{\overline{8}|}$$

$$+ 20{,}000 \times 0.04 \times 0.7 (a_{\overline{21}|} - a_{\overline{8}|})$$

$$+ 20{,}000 \times 0.04 \times 0.5 (a_{\overline{16}|} - a_{\overline{12}|})$$

$$+ 10{,}000 \times 0.02 \times 0.5 (a_{\overline{24}|} - a_{\overline{16}|}) \quad \text{at rate } i$$

$$= 700 a_{\overline{8}|} + 160 a_{\overline{12}|} + 300 a_{\overline{16}|} + 100 a_{\overline{24}|} \quad \text{at rate } i$$

We find i by solving the equation of value

$$26{.}000 = K + I$$

A rough solution gives $i \approx 4.5\%$, and by successive trials and interpolation, we find that $i \approx 4.63\%$.

7.14 **(a)** Let the amount of the monthly installment be X. Working in time units of a quarter-year, we have $3Xa_{\overline{60}|}^{(3)} = 100,000$ at 4%. This gives $X = £1,454.17$.

(b) Work in time units of 1 month. Let $j = 0.04^{(3)}/3 = 0.013159$. Note that the interest payable at the end of each month is j times the loan outstanding at the start of the month.

The net yield is somewhat less than 5% per annum. Let us value the loan to obtain a net yield of $4\frac{1}{2}\%$ per annum.

On this basis, if income tax is ignored, the value of the annuity payments is

$$Xa_{\overline{180}|} \text{ at rate } \frac{0 \cdot 045^{(12)}}{12} = 12Xa_{\overline{15}|}^{(12)} \text{ at } 4\frac{1}{2}\% = 191,240$$

By Makeham's formula, the value of the capital repayments K is such that

$$191,240 = K + \frac{0.013159}{0.045^{(12)}/12}(100,000 - K)$$

from which we obtain $K = 64,645$. Hence, the value of the annuity payments, net of tax, is (again using Makeham's formula)

$$K + \frac{0.2 \times 0.013159}{0 \cdot 045^{(12)}/12}(100,000 - K) = £89,964$$

Note that this value may also be obtained as $K + 0.2(191\,240 - K)$. The corresponding value if the net annual yield were 5% is given as £86,467. By interpolation between 4.5% and 5%, we find that the net annual yield i is approximately equal to 4.78%.

7.15 Let the redemption prices per cent be $100 + \lambda$, $100 + 2\lambda$, $100 + 3\lambda$, and $100 + 4\lambda$ at times 2, 4, 6, and 8 years, respectively. We shall first value the loan on the assumptions that redemption is at par and that income tax is at 50%, and later apply "corrections" to allow for the premiums on redemption and the "excess" tax. On these assumptions, we have

$$K = 2,000\left(v^2 + v^4 + v^6 + v^8\right) \quad \text{at } 7\%$$

$$= \frac{2,000a_{\overline{8}|}}{s_{\overline{2}|}} \quad \text{at } 7\% \quad = 5,769.37$$

By Makeham's formula, the net value of the loan is

$$K + \frac{0.1 \times 0.5}{0.07^{(4)}}(8{,}000 - K) = 7{,}403.91$$

The value of the "excess" income tax in the first 5 years is (by first principles)

$$20\left[a_{\overline{2}|}^{(4)} + a_{\overline{4}|}^{(4)} + 2a_{\overline{5}|}^{(4)}\right] \quad \text{at } 7\% = 274.85$$

and the premiums on redemption have value

$$20\lambda(v^2 + 2v^4 + 3v^6 + 4v^8) \quad \text{at } 7\% \quad = 134.526\lambda$$

Hence, we have the equation

$$7{,}403.91 + 274.85 + 134.526\lambda = 7{,}880.55$$

whence $\lambda = 1.5$. The redemption prices are therefore 101.5%, 103%, 104.5%, and 106%, respectively.

7.16 **(a)** We see at once that the amount of each annuity payment is

$$\frac{100{,}000}{a_{\overline{15}|}} \quad \text{at } 8\% \quad = 11{,}682.95$$

If we ignore income tax, the value of the loan repayments is

$$11{,}682.95 a_{\overline{15}|} \quad \text{at } 7\% \quad = 106{,}407$$

By Makeham's formula, if the value of the capital payments is K, then

$$K + \frac{0.08}{0.07}(100{,}000 - K) = 106{,}407$$

Hence, we may find K, which is 55,149. The price to be paid may be found, by Makeham's formula, to be

$$K + \frac{0.08 \times 0.6}{0.07}(100{,}000 - K) = £85{,}904$$

Note that this value may also be found as $K + 0.6(106{,}407 - K)$.

(b) Since $g = i$, we cannot use Makeham's formula to find K, the value of the capital. We must find K by the direct method, which gives

$$K = 11{,}682.95\left(v_{0.08}^{15}v_{0.08} + v_{0.08}^{14}v_{0.08}^2 + \cdots + v_{0.08}v_{0.08}^{15}\right)$$

$$= 11{,}682.95 v_{0.08}^{16} \times 15$$

$$= 51{,}152.08$$

Hence, by Makeham's formula, the price to be paid for the loan is

$$K + \frac{0.6(0.08)}{0.08}(100,000 - K) = £80,461$$

CHAPTER 8 EXERCISES

8.1

$$\left(1 + \frac{i^{(4)}}{4}\right)^4 = 1.04 \implies i^{(4)} = 0.039414$$

$$g(1 - t_1) = \frac{0.05}{1.03} \times 0.80 = 0.038835$$

Therefore, $i^{(4)} > (1 - t_1)g$ and there is a capital gain on contract.
We assume redeemed as late as possible (i.e., after 20 years) to obtain minimum yield.
Price of stock,

$$P = 100,000 \times 0.05 \times 0.80 \times a_{\overline{20}|}^{(4)} + (103,000$$
$$- 0.25(103,000 - P))v^{20} \quad \text{at } 4\%$$

And we have

$$P = \frac{4,000\ a_{\overline{20}|}^{(4)} + 77,250v^{20}}{1 - 0.25v^{20}}$$

$$= \frac{4,000 \times 1.014877 \times 13.5903 + 77,250 \times 0.45639}{1 - 0.25 \times 0.45639}$$

$$= 102,072.25$$

8.2

$$\left(1 + \frac{i^{(4)}}{4}\right)^4 = 1.05 \implies i^{(4)} = 0.049089$$

$$g(1 - t_1) = \frac{0.07}{1.08} \times 0.75 = 0.04861$$

Therefore, there is a capital gain on contract, and we assume the loan is redeemed as late as possible (i.e., after 20 years) to obtain minimum yield.
Let price of stock $= P$, such that

$$P = 0.07 \times 100,000 \times 0.75 \times a_{\overline{20}|}^{(4)} + (108,000$$
$$- 0.35(108,000 - P))v^{20} \text{at } 5\%$$

$$P = \frac{5{,}250a_{\overline{20}|}^{(4)} + 70{,}200v^{20}}{1 - 0.35v^{20}}$$

$$= \frac{5{,}250 \times 1.018559 \times 12.4622 + 70{,}200 \times 0.37689}{1 - 0.35 \times 0.37689} = 107{,}245.38$$

The equation of value for the investor is

$$12(1+i)^5 = 34.9206$$

$$i = 0.23817 \text{ or } 23.817\%$$

(b) $$12(1+i)^5 = 34.9206 - (34.9206 - 12) \times 0.25$$

where i is the net rate of return.

$$12(1+i)^5 = 29.1905$$

$$i = 0.1946 \text{ or } 19.46\%$$

(c) The cash flow received in nominal terms is still the same: 29.190495. The equation of value expressed in real terms is

$$12 = \frac{29.1905}{(1+f)^5} v^5 \text{ where } f = 0.04$$

$$v^5 = \frac{12 \times (1.04)^5}{29.1905} = 0.50016$$

and so $v = 0.50016^{\frac{1}{5}} = 0.87061$

$$i = 14.86\%$$

8.3 (a) Expected price of the shares in 5 years is

$$X = 2v + 2.5v^2 + 2.5 \times 1.01 \times v^3 + 2.5 \times 1.01^2 v^4 + \dots$$

$$= 2v + 2.5v^2 + 2.5v^2 \left(1.01v + 1.01^2 v^2 + \dots \right)$$

$$1.01v + 1.01^2 v^2 + \dots \quad \text{at } 8\% = \frac{1}{i'}$$

where $i' = \frac{1.08}{1.01} - 1 = 0.069307$

$$X = 2 \times 0.92593 + 2.5 \times 0.85734 + \frac{2.5 \times 0.85734}{0.069307}$$

$$= 3.9952 + 30.9254 = 34.9206$$

8.4 **(a)** Let us work in terms of £100 nominal of loan. We have $C = 110$, and the value of capital repayments (ignoring tax) is

$$K = 110v^{15} \quad \text{at } 8\% \quad = 34.6766$$

By Makeham's formula, with allowance for capital gains tax, the price A must satisfy the relationship

$$A = K + \frac{0.09(0 \cdot 55)}{1.1 \times 0.08^{(2)}}(110 - K) - 0.3\left(\frac{110 - A}{110}\right)K$$

which gives $A = £74.52$ per £100 nominal.

(b) We note that the capital gain, per £100 nominal, is £30, of which £9 must be paid in tax. The net proceeds on redemption are therefore £101 so that the net yield per annum is the solution of the equation

$$80 = 0.55 \times 9a^{(2)}_{\overline{15|}} + 101v^{15} \quad \text{at rate } i.$$

This gives $i = 7.32\%$.

8.5 Let us work in terms of £100 nominal. We have $C = 120$ and

$$K = \frac{120}{15}\left(a_{\overline{19|}} - a_{\overline{4|}}\right) \quad \text{at } 7\% \quad = 55.5871$$

By a straightforward modification of Makeham's formula, to allow for the fact that interest is payable in advance rather than in arrears, the value of the loan, ignoring capital gains tax, is

$$K + \frac{0.7(0.07)}{d^{(2)}_{0.07}}(C - K)$$

Let the price per cent, allowing for capital gains tax, be A. We have

$$A = K + \frac{0.7(0.07)}{d^{(2)}_{0.07}}(C - K) - 0.25\left(\frac{C - A}{C}\right)K$$

from which we obtain $A = £100.81\%$.

8.6 We first calculate A_0, the price paid by investor A for the bond. This is given by the equation of value

$$A_0 = 60a^{(2)}_{\overline{10|}} + 1{,}000v^{10} - 0.4(1{,}000 - A_0)v^{10} \quad \text{at } 10\%$$

and so $A_0 = 720.04$.

(a) Let A_1 be the price paid by investor B. We

$$A_1 = 60a^{(2)}_{\overline{5|}} + 1{,}000v^5 - 0.4(1{,}000 - A_1)v^5 \quad \text{at } 10\%$$

which gives $A_1 = 805.65$. Hence, investor A makes a capital gain of $805.65 - 720.04 = 85.61$ on selling the bond, and pays capital gains tax of $0.4(85.61) = 34.24$ at the date of sale. His net proceeds on the sale of the bond are therefore $805.65 - 34.24 = 771.41$, and his net annual yield is the solution of the equation

$$720.04 = 60a_{\overline{5}|}^{(2)} + 771.41v^5 \quad \text{at rate } i.$$

By trials and interpolation, $i = 9.71\%$.

(b) Let X be the *net* proceeds to investor A on the sale of the bond. We find X by means of the equation of value

$$720.04 = 60a_{\overline{5}|}^{(2)} + Xv^5 \quad \text{at 10\%}$$

which gives $X = 784.39$. The price paid by B, A_2, must therefore satisfy

$$A_2 - 0.4(A_2 - 720.04) = 784.39$$

which gives $A_2 = 827.29$.

Hence, the net proceeds to investor B on redemption of the bond are $1,000 - 0.4(1,000 - 827.29) = 930.92$, so that his net annual yield is the solution of

$$827.29 = 60a_{\overline{5}|}^{(2)} + 930.92v^5 \quad \text{at rate } i$$

By trials and interpolation, $i = 9.50\%$.

8.7 We first find the number of capital repayments n by solving the equation

$$115,000 = 1,000\{15 + (15 + 2) + (15 + 6) + \cdots + [15 + n(n-1)]\}$$

By trials, $n = 5$. Hence, there are five repayments of capital, the cash amounts being 15,000, 17,000, 21,000, 27,000, and 35,000, respectively. The value of the capital repayments is

$$\begin{aligned} K &= 15,000v^{10} + 17,000v^{13} + 21,000 v^{16} + 27,000v^{19} \\ &\quad + 35,000v^{22} \quad \text{at 7\%} \\ &= 37,158.5 \end{aligned}$$

Let the price paid be P per unit nominal. We have $g = 0.06/1.15$, and the equation of value is

$$100,000P = K + 0.7\frac{(0.06/1.15)}{0.07^{(2)}}\left(115,000 - K\right)$$

$$-0\cdot 4\left(\frac{1.15 - P}{1.15}\right)K$$

which gives $P = 0.73049$. Hence, the value of the whole loan is £73,049.

8.8 The price paid by the investor is

$$100v^{15\cdot5} + 3a^{(2)}_{\overline{15\cdot5}|} + a^{(2)}_{\overline{5\cdot5}|} \quad \text{at } 6\% \quad = 75\cdot34$$

The net yield per annum obtained by the investor is the solution of the equation

$$75.34 = 0.6\left[3a^{(2)}_{\overline{15\cdot5}|} + a^{(2)}_{\overline{5\cdot5}|}\right] + \left[100 - 0.2(100 - 75.34)\right]v^{15\cdot5} \quad \text{at rate } i$$

By trials, $i = 4.00\%$.

8.9 (a) Divide the loan into two sections.

Section 1: First 7 years:

$C_1 = \text{cash repaid} = 400 \times 7 \times 100 = 280,000$
$K_1 = \text{value of capital repayments} = 400 \times 100a_{\overline{7}|0\cdot06}$
$\quad = 223,296$
$g_1 = 0.04$

Section 2: Next 12 years:

$$C_2 = 600 \times 12 \times 115 = 828,000$$
$$K_2 = 600 \times 115(a_{\overline{19}|0\cdot06} - a_{\overline{7}|0\cdot06}) = 384,723$$
$$g_2 = 4/115 = 0\cdot034\,78$$

Let P be the price per bond. The equation of value is

$$10,000P = K_1 + \frac{0.65g_1}{i^{(2)}}(C_1 - K_1) - 0.3\frac{(C_1 - 2,800P)}{C_1}K_1$$

$$+K_2 + \frac{0.65g_2}{i^{(2)}}(C_2 - K_2) - 0.3\frac{(C_2 - 7,200P)}{C_2}K_2$$

i.e.,

$$10,000P = 223,296 + 24,934 - 66,989(1 - 0.01P)$$

$$+384,723 + 169,488 - 115,417(1 - 0.008\,6956P)$$

i.e.,

$$(10,000 - 669.89 - 1,003.62)P = 620,035$$

which gives $P = £74.47$.

(b) During years 1–7 and also years 8–19, yields decrease (because one must wait longer for the capital gain). The yield on a bond redeemed at time n years is found by solving the equation

$$74.47 = 100v^n + \left[0.65 \times 4a_{\overline{m}|}^{(2)}\right] - \frac{0.3(100 - 74.47)}{100}100v^n$$

$$= 92.34v^n + 2.6a_{\overline{m}|}^{(2)} \quad \text{for} 1 \le n \le 7$$

or

$$74.47 = 115v^n + \left[0.65 \times 4a_{\overline{m}|}^{(2)}\right] - 0.3\left(\frac{115 - 74.47}{115}\right)115v^n$$

$$= 102.84v^n + 2.6a_{\overline{m}|}^{(2)} \quad \text{for } 8 \le n \le 19$$

For $n = 7$, the RHS at 6% is 76.14, so yield > 6% for $1 \le n \le 7$.
For $n = 11$, the RHS at 6% is 74.98.
For $n = 12$, the RHS at 6% is 73.23.
Hence, if $n \le 11$, the yield is greater than 6%, and if $n \ge 12$, the yield is less than 6%. The number of bonds yielding less than 6% is then $8 \times 600 = 4,800$.

8.10 Assuming that redemption is always at par, the value of the capital repayments is

$$25,000a_{\overline{20}|} \quad \text{at } 8\% \quad = 245,454$$

Hence, the value of the net interest payments is, by Makeham's formula,

$$(1 - 0.4)\frac{0.12}{0.08^{(12)}}(500,000 - 245,454) = 237,375$$

The value of the premiums on redemption is

$$25,000 \times 0.1(a_{\overline{20}|} - a_{\overline{10}|}) \quad \text{at } 8\% \quad = 7,770$$

so the value of the capital and net interest repayments, ignoring capital gains tax, is

$$245,454 + 7,770 + 237,375 = 490,599$$

Let A be the price paid for the loan. We have

$$A = 490,599-\text{value of capital gains tax payments}$$

It is clear from this equation that the price must be below par, so capital gains tax will be paid on redemption of all the stock, not merely on the part redeemed at 110%. The value of the capital gains tax payments is

$$0 \cdot 4 \left[\left(25{,}000 - \frac{A}{20} \right) v + \left(25{,}000 - \frac{A}{20} \right) v^2 + \cdots + \left(25{,}000 - \frac{A}{20} \right) v^5 \right]$$

$$+ 0.3 \left[\left(25{,}000 - \frac{A}{20} \right) v^6 + \left(25{,}000 - \frac{A}{20} \right) v^7 + \cdots + \left(25{,}000 - \frac{A}{20} \right) v^{10} \right]$$

$$+ 0.3 \left\{ \left[\left(25{,}000 \times 1.1 \right) - \frac{A}{20} \right] v^{11} + \left[\left(25{,}000 \times 1.1 \right) - \frac{A}{20} \right] v^{12} + \cdots \right.$$

$$+ \left. \left[\left(25{,}000 \times 1.1 \right) - \frac{A}{20} \right] v^{20} \right\} \quad \text{at } 8\%$$

$$= 10{,}000 a_{\overline{5}|} + 7{,}500 (a_{\overline{10}|} - a_{\overline{5}|}) + 8{,}250 (a_{\overline{20}|} - a_{\overline{10}|})$$

$$- \frac{A}{20} \left[0.4 a_{\overline{5}|} + 0.3 \left(a_{\overline{10}|} - a_{\overline{5}|} \right) + 0 \cdot 3 \left(a_{\overline{20}|} - a_{\overline{10}|} \right) \right] \quad \text{at } 8\%$$

$$= 2{,}500 a_{\overline{5}|} - 750 a_{\overline{10}|} + 8{,}250 a_{\overline{20}|} - \frac{A}{20} \left(0.1 a_{\overline{5}|} + 0.3 a_{\overline{20}|} \right) \quad \text{at } 8\%$$

$$= 85{,}949 - 0.167236 A$$

Hence, we have
$$A = 490{,}599 - (85{,}949 - 0.167236 A)$$

from which we obtain $A = £485{,}912$.

8.11 Let us work in terms of £100 nominal of loan. The value of the capital repayments, ignoring capital gains tax, is

$$K = \frac{5}{s_{\overline{2}|}} (a_{\overline{48}|} - a_{\overline{8}|}) \quad \text{at } 8\% \quad = 15.486\ 77$$

Hence, if interest were at 6% throughout, the value of the gross interest payments would be

$$\frac{0.06}{0.08} (100 - K) = 63.38492$$

We now find the value of the "extra" gross interest in the final 18 years. The value, at time 30 years, of the outstanding capital repayments is

$$K' = \frac{5}{s_{\overline{2}|}} a_{\overline{18}|} \quad \text{at } 8\% \quad = 22.528\ 54$$

Note that the loan outstanding at time 30 years is £45 nominal. The value at time 30 of the "extra" gross interest is then

$$\frac{0.01}{0.08} (45 - K') = 2.808\ 93$$

Hence, the value at time $t = 0$ of the "extra" gross interest is $2.808\ 93 v^{30}$ at $8\% = 0.279\ 85$. The price per cent A therefore satisfies the equation

$$A = 15.486\ 77 + 0.65(63.384\ 92 + 0.279\ 85) - 0.3(100 - A)\frac{15.486\ 77}{100}$$

This gives $A = £54.77$.

8.12 We first find the term of the loan, n years. The capital repaid at time $t + 4$ is $44{,}000 + 6{,}000t$ $(t = 1, 2, \ldots, n)$. Hence, the capital repaid by time n is

$$44{,}000n + 6{,}000 \sum_{t=1}^{n} t = 44{,}000n + 3{,}000n(n+1)$$

$$= 880{,}000 \qquad \text{for} \qquad n = 11.$$

Hence, the loan has term 15 years, and the value of capital is

$$K = 50{,}000\ v^5 + 56{,}000\ v^6 + \ldots + 110{,}000\ v^{15} \qquad \text{at rate } i$$

(where i is the net annual yield)

$$= 44{,}000\left(a_{\overline{15}|} - a_{\overline{4}|}\right) + 6{,}000v^4(Ia)_{\overline{11}|}$$

We now solve the equation of value

$$831{,}600 = 0.945 \times 880{,}000$$

$$= K + \frac{0.625 \times 0.055}{i^{(2)}}(880{,}000 - K)$$

$$-0.3 \times \frac{880{,}000 - (0.945 \times 880{,}000)}{880{,}000}K \qquad \text{at rate } i$$

(1)

A rough solution is

$$i \approx (0.055 \times 0.625) + \frac{(100 - 94.5) \times 0.7}{94.5 \times 11} = 3.8\%$$

When $i = 4\%$, $K = 581{,}513$ and the right side of Eq. 1 is 830,969. The right side of Eq. 1 at 3½% is £867,387, and hence $i = 4.0\%$ (to the nearest 0.1%).

8.13 Let us work in terms of £300 nominal, so £10 nominal is redeemed each year, and let us consider the loan to be in three sections, referring to years 1−10, 11−15, and 16−30, respectively. Assuming that redemption is at par throughout, we have

$$K_1 = 10a_{\overline{10}|} \quad \text{at } 7\% = 70.235\ 82$$
$$K_2 = 10(a_{\overline{15}|} - a_{\overline{10}|}) \quad \text{at } 7\% = 20.843\ 32$$
$$K_3 = 10(a_{\overline{30}|} - a_{\overline{15}|}) \quad \text{at } 7\% = 33.011\ 27$$

Let
$$K = K_1 + K_2 + K_3 = 10a_{\overline{30}|} \quad \text{at } 7\% = 124.090\ 41$$

Then the value of the net interest is (by Makeham's formula)

$$\frac{0.08}{0.07^{(4)}}(0.6)(300 - 124.090\ 41) = 123.7455$$

The value of the loan, ignoring capital gains tax (CGT), is therefore

$$K_1 + K_2 + 1.2K_3 + 123.7455 = 254.438\ 16$$

Let A be the value of the loan allowing for CGT. We have

$$A = 254.438\ 16 - \text{value of CGT payments.}$$

This shows that $A \leq 254.438\ 16$, so CGT is payable on all redemptions. The value of the CGT payments is

$$0.4 \sum_{t=1}^{10}\left(10 - \frac{A}{30}\right)v^t + 0.25\sum_{t=11}^{15}\left(10 - \frac{A}{30}\right)v^t$$

$$+0.25\sum_{t=16}^{30}\left(12 - \frac{A}{30}\right)v^t \quad \text{at 7\%}$$

$$= 0.4\frac{300 - A}{300}K_1 + 0.25\frac{300 - A}{300}K_2$$

$$+0.25\frac{360 - A}{360}1.2K_3$$

Hence, we have

$$A = 254.438\ 16 - 0.4\frac{300 - A}{300}K_1 - 0.25\frac{300 - A}{300}K_2$$

$$-0.25\frac{360 - A}{360}1.2K_3$$

which gives $A = 245.196$. The price paid by the investor is therefore £81.73 per bond.

CHAPTER 9 EXERCISES

9.1 All other factors constant, the longer the maturity, the greater the volatility and the greater the price change when interest rates change. So, Bond B is the answer.

9.2 The information we have for estimating the effective duration:
Price if yields decline by 30 basis points = 83.50
Price if yields rise by 30 basis points = 80.75
Initial price = 82.00
Change in yield from current = 0.0030
Then,

$$\text{Effective duration} \approx \frac{1}{28} \times \frac{83.50 - 80.75}{2 \times 0.0030} = 5.59$$

9.3 According to the expectations theory, a humped yield curve means that short-term interest rates are expected to rise for a time and then begin to fall.

9.4 Let f_{tn} be the n-year forward rate at time t; we need to link this to n-spot rates, y_n. We use

$$f_{tn} = \left(\frac{(1+y_{t+n})^{t+n}}{(1+y_t)^t}\right)^{1/n} - 1$$

(a) $f_{0.5,0.5} = \left(\frac{(1+y_1)^1}{(1+y_{0.5})^{0.5}}\right)^2 - 1 = \left(\frac{1.054}{1.05^{0.5}}\right)^2 - 1$

$= 5.8\%$ per annum.

(b) $f_{1,0.5} = \left(\frac{(1+y_{1.5})^{1.5}}{(1+y_1)^1}\right)^2 - 1 = \left(\frac{1.058^{1.5}}{1.054}\right)^2 - 1$

$= 6.6\%$ per annum.

(c) $f_{3,0.5} = \left(\frac{(1+y_{3.5})^{3.5}}{(1+y_3)^3}\right)^2 - 1 = \left(\frac{1.074^{3.5}}{1.072^3}\right)^2 - 1$

$= 8.6\%$ per annum.

(d) $f_{1,2} = \left(\frac{(1+y_3)^3}{(1+y_1)^1}\right)^{1/2} - 1 = \left(\frac{1.072^3}{1.054}\right)^{\frac{1}{2}} - 1$

$= 8.1\%$ per annum.

(e) $f_{2,1} = \left(\frac{(1+y_3)^3}{(1+y_2)^2}\right)^1 - 1 = \left(\frac{1.072^3}{1.064^2}\right)^1 - 1 = 8.8\%$ per annum.

9.5 (a) From the 2-year stock information:

Price $= 3a_{\overline{2}|} + 102v^2$ at 5.5%

$= 3 \times 1.84632 + 102 \times 0.89845$

$= 97.1811$

Therefore, from the 1-year forward rate information,

$$97.1811 = \frac{3}{1+i_1} + \frac{3+102}{\left(1+i_1\right)\left(1+f_{1,1}\right)}$$

where $i_1 =$ 1-year spot rate and $f_{1,1} =$ 1-year forward rate from $t = 1$

$$97.1811 = \frac{3}{1+i_1} + \frac{105}{(1+i_1)1.05}$$

and so $i_1 = 5.9877\%$ per annum

(b) From the 3-year stock information:

$$108.9 = \frac{10}{1+i_1} + \frac{10}{(1+i_1)1.05} + \frac{110}{\left(1+i_1\right)1.05\left(1+f_{2,1}\right)}$$

where $f_{2,1} = 1$-year forward rate from $t = 2$
Hence,

$$108.9 = \frac{10}{1.059877} + \frac{10}{1.059877 \times 1.05} + \frac{110}{1.059877 \times 1.05 \times \left(1+f_{2,1}\right)}$$

and so $f_{2,1} = 9.245\%$ per annum.

(c) Let $y_2\%$ per annum be the 2-year par yield

then $$100 = y_2\left(\frac{1}{1+i_1} + \frac{1}{(1+i_1)(1+f_{1,1})}\right) + \frac{100}{(1+i_1)(1+f_{1,1})}$$

$$100 = y_2\left(\frac{1}{1.059877} + \frac{1}{1.059877 \times (1.05)}\right) + \frac{100}{1.059877 \times 1.05}$$

and so $y_2 = 5.506\%$ per annum.

9.6 (a) Bond yields are determined by investors' expectations of future short-term interest rates, so that returns from longer-term bonds reflect the returns from making an equivalent series of short-term investments.

(b) **(i)** Let i_t be the spot yield over t years.
One year: yield is 8%; therefore, $i_1 = 0.08$.
Two years: $(1+i_2)^2 = 1.08 \times 1.07$; therefore, $i_2 = 0.074988$.
Three years: $(1+i_3)^2 = 1.08 \times 1.07 \times 1.06$; therefore, $i_3 = 0.06997$.
Four years: $(1+i_4)^4 = 1.08 \times 1.07 \times 1.06 \times 1.05$; therefore, $i_4 = 0.06494$.

(ii) Price of the bond is
$$5[(1.08)^{-1} + (1.074988)^{-2} + (1.06997)^{-3}]$$
$$+105 \times (1.06494)^{-4} = 13.0822 + 81.6373 = 94.67552.$$

We find the gross redemption yield from
$94.67552 = 5a_{\overline{2}|} + 100v^4$.
Try 7%; $a_{\overline{2}|} = 3.3872$; $v^4 = 0.76290$ gives RHS $= 93.226$.
GRY must be lower, so try 6%; $a_{\overline{2}|} = 3.4651$; $v^4 = 0.79209$ gives RHS $= 96.5345$.
Interpolating between 6% and 7% gives
$i = 0.06562$.

9.7 **(a)** The present value of the assets is equal to the present value of the liabilities at the starting rate of interest.

The duration/discounted mean term/volatility of the assets is equal to that of the liabilities.

The convexity of the assets (or the spread of the timings of the asset cash flows) around the discounted mean term is greater than that of the liabilities.

(b) **(i)** PV of liabilities is £100m $a_{\overline{40|}}$ at 4%.

$$= £100\text{m} \times 19.7928$$
$$= £1{,}979.28\text{m}$$

(ii) The duration of the liabilities is

$$\sum_{t=1}^{t=40} 100t\ v^t / \sum_{t=1}^{t=40} 100v^t \text{ (working in £m)}$$

$$= \frac{100 \sum_{t=1}^{t=40} t\ v^t}{1{,}979.28} = \frac{100(Ia)_{\overline{40|}}}{1{,}979.28} \text{ at 4\%}$$

$$= \frac{100 \times 306.3231}{1{,}979.28} = 15.4765 \text{ years}$$

(c) Let x = nominal amount of 5-year bond;
y = nominal amount of 40-year bond.
Working in £m

$$1{,}979.28 = xv^5 + yv^{40} \tag{1}$$

$$30{,}632.31 = 5xv^5 + 40yv^{40} \tag{2}$$

multiply Eq. 1 by 5:

$$9{,}896.4 = 5xv^5 + 5yv^{40} \tag{1a}$$

subtract Eq. 1a from Eq. 2 to give

$$20{,}735.91 = 35yv^{40}$$

$$\frac{20{,}735.91}{35 \times v^{40}} = y$$

with $v^{40} = 0.20829$

$$y = 2{,}844.38$$

Substitute into Eq. 1 to give

$$1{,}979.28 = Xv^5 + 2{,}844.38 \times 0.20829$$

$$v^5 = 0.82193$$

$$\frac{1{,}979.28 - 2{,}844.38 \times 0.20829}{0.82193} = x = 1{,}687.28$$

Therefore, £1,687.28m nominal of the 5-year bond and £2,844.38m nominal of the 40-year bond should be purchased.

(d) **(i)** The duration of the liabilities is 15.4765.
The value of the liabilities would therefore change by

$$1.5 \times \frac{0.154765}{1.04} \times 1{,}979.28m = £441.81m$$

and the revised present value of the liabilities will be £2,421.09m.

(ii) PV of liabilities is £100m $a_{\overline{40}|}$ at 2.5%

$$= £100m \times \frac{1 - 1.025^{-40}}{0.025}$$
$$= £2{,}510.28m.$$

(iii) The PV of liabilities has increased by £531m. This is significantly greater than that estimated in (d) (i). This estimation will be less valid for large changes in interest rates, as in this case.

9.8 **(a)** Measure time in years throughout. The DMT is

$$\frac{\left(100 \times 0v^0\right) + \left(230 \times 5v^5\right) + \left(600 \times 13v^{13}\right)}{100v^0 + 230v^5 + 600v^{13}}$$

which equals (i) 8.42 years when $i = 5\%$ and (ii) 5.90 years when $i = 15\%$.

(b) **(i)** The DMT is

$$\frac{1{,}000\left(v + 2v^2 + 3v^3 + \cdots + 20v^{20}\right)}{1{,}000(v + v^2 + \cdots + v^{20})} \quad \text{at } 8\%$$

$$= \frac{(Ia)_{\overline{20}|}}{a_{\overline{20}|}} \quad \text{at } 8\% \quad = 8.04 \text{ years}$$

(ii) The DMT is

$$\frac{1{,}000\left[v + \left(1.1 \times 2v^2\right) + \left(1.2 \times 3v^3\right) + \cdots + \left(2.9 \times 20v^{20}\right)\right]}{1{,}000\left(v + 1.1v^2 + 1.2v^3 + \cdots + 2.9v^{20}\right)} \quad \text{at } 8\%$$

$$= \frac{0.9(Ia)_{\overline{20}|} + 0.1 \sum_{t=1}^{20} t^2 v^t}{0.9a_{\overline{20}|} + 0.1(Ia)_{\overline{20}|}} \quad \text{at } 8\%$$

Let

$$X = \sum_{t=1}^{n} t^2 v^t = v + 4v^2 + 9v^3 + \cdots + (n^2 - 2n + 1)v^{n-1} + n^2 v^n$$

Note that

$$vX = v^2 + 4v^3 + \cdots + (n^2 - 2n + 1)v^n + n^2 v^{n+1}$$

and so

$$(1 - v)X = v + 3v^2 + 5v^3 + \cdots + (2n - 1)v^n - n^2 v^{n+1}$$
$$= 2(Ia)_{\overline{n}|} - a_{\overline{n}|} - n^2 v^{n+1}$$

It follows that

$$X = \sum_{t=1}^{n} t^2 v^t = \frac{2(Ia)_{\overline{n}|} - a_{\overline{n}|} - n^2 v^{n+1}}{d}$$

hence, we may calculate the DMT, which is found to be 9.78 years.

(iii) The DMT is

$$\frac{1{,}000\left[v + (1.08 \times 2v^2) + (1.08^2 \times 3v^3) + \cdots + (1.08^{19} \times 20v^{20})\right]}{1{,}000(v + 1.08v^2 + 1.08^2 v^3 + \cdots + 1.08^{19} v^{20})} \quad \text{at } 8\%$$

$$= \frac{v(1 + 2 + 3 + \cdots + 20)}{20v} \quad \text{at } 8\%$$

$$= \frac{20 \times 21}{2 \times 20} = 10.5 \text{ years}$$

(iv) The DMT is

$$\frac{\sum_{t=1}^{20} 1{,}000 t (1.1)^{t-1} v^t}{\sum_{t=1}^{20} 1{,}000 (1.1)^{t-1} v^t} \quad \text{at } 8\%$$

$$= \left(\sum_{t=1}^{20} t\theta^{t-1} \right) \bigg/ \left(\sum_{t=1}^{20} \theta^{t-1} \right) \quad \text{(where } \theta = 1.1/1.08\text{)}$$

$$= \frac{(Ia)_{\overline{20}|}}{a_{\overline{20}|}} \quad \text{at rate } j \text{ (where } \theta = 1/1 + j)$$

$$= \frac{\ddot{a}_{\overline{20}|} - 20v^{20}}{1 - v^{20}} \quad \text{at rate } j = \frac{\left(\dfrac{1 - \theta^{20}}{1 - \theta}\right) - 20\theta^{20}}{1 - \theta^{20}}$$

$$\text{(since } \theta = v \text{ at rate } j) = 11.11 \text{ years}$$

9.9 **(a)** Let the cash flows occur at times $r_1, r_2, \ldots, r_k$, where $0 \le r_1 \le r_2 \le \ldots \le r_k$. Assume that for the first series of payments, the cash flow at time r_i is x_i and that the cash flow for the second series at this time is y_i. (Note that some of the $\{x_i\}$ and $\{y_i\}$ may be zero.) The discounted mean term for the first series is

$$t_1 = \frac{\sum x_i r_i v^{r_i}}{\sum x_i v^{r_i}} = \frac{\sum x_i r_i v^{r_i}}{V_1} \quad (\text{say, defining } V_1)$$

The discounted mean term for the second series is

$$t_2 = \frac{\sum y_i r_i v^{r_i}}{\sum y_i v^{r_i}} = \frac{\sum y_i r_i v^{r_i}}{V_2} \quad (\text{say, defining } V_2)$$

These equations imply that

$$V_1 t_1 + V_2 t_2 = \sum (x_i + y_i) r_i v^{r_i}$$

so that, if $V_1 + V_2$ is non-zero,

$$\frac{V_1 t_1 + V_2 t_2}{V_1 + V_2} = \frac{\sum (x_i + y_i) r_i v^{r_i}}{\sum (x_i + y_i) v^{r_i}}$$

which is the discounted mean term for the combined series. For the more general result, relating to n series of payments, let V_j and t_j denote the present value and discounted mean term of the jth series. Let t be the discounted mean term of the cash flow obtained by combining all n series. Then, if $\sum_{j=1}^{n} V_j \ne 0$,

$$t = \sum_{j=1}^{n} V_j t_j \Big/ \sum_{j=1}^{n} V_j$$

(b) **(i)** (1) Receipts by investor: $V_1 = 10{,}000 a_{\overline{10}|} = 61{,}445.67$ (at 10%) and $t_1 = (Ia)_{\overline{10}|}/a_{\overline{10}|} = 4.725$, say 4.73.
   (2) Payments by investor: $V_2 = -30{,}000 \,(v^5 + v^{15}) = -25{,}809.40$ and $t_2 = (5v^5 + 15v^{15})/(v^5 + v^{15}) = 7.783$, say 7.78.

(ii) By above, $t = (V_1 t_1 + V_2 t_2)/(V_1 + V_2) = 2.511$, say 2.51. (Note that, since V_2 relates to payments *by* the investor, V_2 is negative.)

9.10 **(a)** Consider a nominal amount of £1 of stock. Income is 0.05 per annum, payable continuously until the stock is redeemed at time n. We calculate the volatility as

$$\frac{0.05 \,(\bar{I}\bar{a})_{\overline{n}|} + nv^n}{0.05 \bar{a}_{\overline{n}|} + v^n} \quad \text{at } force \text{ of interest } 0.07$$

Noting that, at force of interest $\delta, (\bar{I}\bar{a})_{\overline{n}|} = (\bar{a}_{\overline{n}|} - nv^n)/\delta$, we find that the volatility is 11.592 when $n = 20$ and 14.345 when $n = 60$.

(b) The maximum volatility (at force of interest 0.07) occurs when, at this force of interest,

$$0.05(n - \bar{a}_{\overline{n}}) + 0.07\bar{a}_{\overline{n}} = \frac{0.07}{0.07 - 0.05}$$

i.e., when

$$0.05n + 0.02\frac{1 - e^{-0.07n}}{0.07} = 3.5$$

By numerical methods, we obtain the solution as $n = 64.3486$. With this value of n, the volatility at force of interest 7% is 14.349.

9.11 (a) (i) Present value is

$$V = \sum_{t=1}^{4}(1{,}000 + 100t)v^{5t} = 2{,}751.54 \quad \text{(at 5%)}$$

(ii) Discounted mean term is

$$\frac{\sum_{t=1}^{4} 5t(1{,}000 + 100t)v^{5t}}{V} \quad \text{(at 5%)} \quad = \frac{31{,}609}{2{,}751.54} = 11.4877 \text{ years}$$

(b) Suppose that the amounts invested in the 10-year stock and in the 30-year stock are A and B, respectively. Then

$$A + B = 2751.54 \quad \text{(from (a))}$$

The cash flows arising from the 10-year stock have present value

$$V_1 = A(0.05a_{\overline{10}} + v^{10})$$

and discounted mean term

$$t_1 = \frac{0.05(Ia)_{\overline{10}} + 10v^{10}}{0.05a_{\overline{10}} + v^{10}}$$

Note that, at an interest rate of 5%, these values are simply

$$V_1 = A \quad \text{and} \quad t_1 = \ddot{a}_{\overline{10}0.05} = 8.10782$$

Similarly, at an interest rate of 5%, the cash flows arising from the 30-year stock have present value $V_2 = B$ and discounted mean term $t_2 = \ddot{a}_{\overline{30}0.05} = 16.14107$.
Using Exercise 9.9, we calculate the discounted mean term of the total asset-proceeds as

$$t = \frac{(A \times 8.10782) + (B \times 16.14107)}{A + B}$$

$$= 16.14107 - 0.002\ 919\ 55A \quad \text{(since } A + B = 2{,}751.54)$$

Equating this expression to 11.4877, we obtain $A = £1,593.87$ and hence $B = £1,157.67$.

9.12 Suppose that the purchaser buys nominal amounts $£A$ and $£B$ of the 5-year and 15-year stock, respectively. For "full" immunization, we require

$$Ae^{5\delta_0} + Be^{-5\delta_0} = 1,000,000$$

and

$$5Ae^{5\delta_0} = 5Be^{-5\delta_0}$$

These equations imply that $A = 500,000e^{-5\delta_0}$ and $B = 500,000e^{5\delta_0}$. Assume now that $\delta_0 = 0.05$. Then $A = 389,400$ and $B = 642,013$. Therefore, the investor should buy a nominal amount £389,400 of the 5-year stock and a nominal amount £642.013 of the 15-year stock (in each case at a cost of £303,265). The total amount invested, £606,530, is the present value (at a force of interest of 5% per annum) of the liability. The difference between the value, at force of interest δ, of the investor's assets and liabilities is

$$V_A - V_L = 389,400e^{-5\delta} + 642,013e^{-15\delta} - 1,000,000e^{-10\delta}$$

(a) When $\delta = 0.07$, this profit is £2,485.
(b) When $\delta = 0.03$, this profit is £3,707.

9.13 **(a)** The cash available for investment now is $100,000v^8$ at a force of interest of 5%, i.e., £67,032. Note that the price per unit nominal of the 20-year stock is v^{20} (at $\delta = 0.05$), i.e., 0.367 879.
Suppose that the company buys a nominal amount $£X$ of the 20-year stock (at a cost of 0.367879X). This means that an amount $(67,032 - 0.367\ 879X)$ is held in cash, which has a discounted mean term of zero. Using Exercise 9.9, we calculate the discounted mean term of the company's assets as

$$\frac{(0.367897X \times 20) + (67,032 - 0.367879X) \times 0}{0.367897X + (67,032 - 0.367897X)} = \frac{7.357580X}{67,032}$$

For this to equal eight (the term of the liability), we need $X = 72,885$. The company should buy £72,885 nominal of 20-year zero coupon bonds at a cost of £26,813 and hold £40,219 in cash.

(b) At force of interest δ per annum, the difference between the value of the company's assets and liabilities is

$$V_A - V_L = 72,885e^{-20\delta} + 40,219 - 100,000e^{-8\delta}$$

(i) When $\delta = 0.03$, this is £1,556.
(ii) When $\delta = 0.07$, this is £1,071.

9.14 **(a)** The single premium is $10{,}000v^{15}$ at 8%, i.e., £3,152.42. Let £X be the nominal amount of 20-year stock purchased. The purchase price is Xv^{20} at 8%, i.e., $0.214\,548X$. For the discounted mean term of the assets (at 8%) to equal that of the liability, we need (see Exercise 9.9)

$$\frac{(0.214548X)20 + (3{,}152.42 - 0.214548X)0}{3{,}152.42} = 15$$

from which we obtain $X = 11{,}019.98$, say 11,020. The company buys a nominal amount £11,020 of 20-year stock at a cost of £2,364.32 and retains cash of £788.10.

(b) **(i)** $V_A - V_L = 788.10 + 11{,}020v^{20} - 10{,}000v^{15}$ at 5%
$$= £131.25$$

say £131 (profit).

(ii) As in (i), but at 10%, i.e., £32.24, say £32 (profit).

(iii) We assume that, for $0 \le t \le 20$,

$$\delta(t) = \ln(1.05) + \frac{t}{20}\left[\ln(1.1) - \ln(1.05)\right]$$

This implies that, if

$$v(t) = \exp\left(-\int_0^t \delta(s)\,ds\right)$$

then $v(15) = 0.370\,268$ and $v(20) = 0.236\,690$. Hence,

$$V_A - V_L = 788.10 + 11{,}020v(20) - 10{,}000v(15)$$
$$= 306.26$$

that is, a *loss* of £306.

9.15 **(a)** For brevity, we write δ rather than δ_0.

(i) $Aae^{\delta a} = bBe^{-\delta b} = b(S - Ae^{\delta a}) \;\Rightarrow\; A = \dfrac{bS}{(a+b)e^{\delta a}}$

$Bbe^{-\delta b} = aAe^{\delta a} = a(S - Be^{-\delta b}) \;\Rightarrow\; \dfrac{aS}{(a+b)e^{-\delta b}}$

(ii) $Bbe^{-\delta b} = aAe^{\delta a} = a(S - Be^{-\delta b}) \;\Rightarrow\; a = \dfrac{Bbe^{-\delta b}}{S - Be^{-\delta b}}$

Then, using this value of a, calculate A as

$$A = \left[S - Be^{-\delta b}\right]e^{-\delta a}$$

(iii) $Aae^{\delta a} = bBe^{-\delta b} = b(S - Ae^{\delta a}) \;\Rightarrow\; b = \dfrac{Aae^{\delta a}}{S - Ae^{\delta a}}$

Then, using this value of b, calculate B as

$$B = \left[S - Ae^{\delta a}\right]e^{\delta b}$$

(iv) $S - Ae^{\delta a} = Be^{-\delta b} = \dfrac{Aae^{\delta a}}{b} \;\Rightarrow\; (a+b)e^{\delta a} = \dfrac{bS}{A}$
i.e.,

$$f(a) = \dfrac{bS}{A}$$

where

$$f(x) = (x+b)e^{\delta x}$$

Note that, for $x \geq 0$, $f(x)$ is an *increasing* function which tends to infinity as x tends to infinity. Also, $f(0) = b < bS/A$ (since, by hypothesis, $A < S$). Hence, there is a unique positive value of x such that $f(x) = bS/A$.

Having found a, calculate B as

$$B = \dfrac{Aae^{\delta a}}{be^{-\delta b}}$$

(b)

$$S - Be^{-\delta b} = Ae^{\delta a} = \dfrac{Bbe^{-\delta b}}{a} \;\Rightarrow\; (a+b)e^{-\delta b} = \dfrac{aS}{B}$$

i.e., $g(b) = aS/B$, where

$$g(x) = (a+x)e^{-\delta x}$$

If $\delta a < 1$, the graph of $g(x)$ is as in Figure S.9.1. In this case, the maximum value of $g(x)$ occurs when $x = (1/\delta) - a$ and is $[\delta e^{1-\delta a}]^{-1}$. If $\delta a < 1$ and

$$a < \dfrac{aS}{B} < \dfrac{1}{\delta e^{1-\delta a}}$$

i.e., if $\delta a < 1$ and

$$a\delta e^{1-\delta a} < \dfrac{B}{S} < 1$$

there are then *two* positive values of x for which $g(x) = aS/B$. When $\delta = 0.05$, $S = 1$, $a = 15$, and $B = 0.98$, the preceding conditions are satisfied. The solutions are

(i) $b = 1.489\ 885$ and $A = 0.042\ 679$;
(ii) $b = 8.976\ 051$ and $A = 0.176\ 843$.

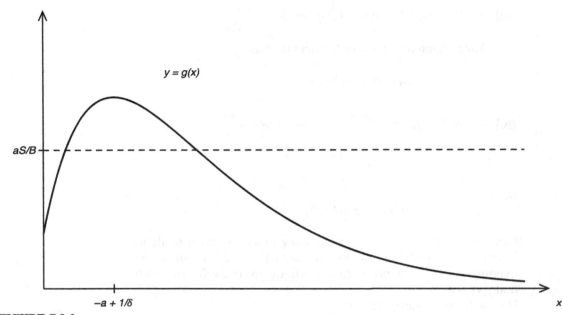

FIGURE S.9.1

$y = g(x)$ for Solution 9.15

CHAPTER 10 EXERCISES

10.1 The contract is settled in cash, so the settlement would be €20,000,000 × (0.875 − 0.90) = −$500,000. This amount would be paid by Sun Microsystems to the dealer. Sun would convert euros to dollars at the spot rate of $0.90, receiving €20,000,000× 0.90 = $18,000,000. The net cash receipt is $17,500,000, which results in an effective rate of $0.875.

10.2 (a) Assuming no arbitrage, the forward price should be

$$K = 200 \times (1.05)^{3/12} = \$202.45.$$

Because the forward contract offered by the dealer is overpriced, take the short position in the forward contract and buy the security now. Doing so will yield an arbitrage profit of $2.55, i.e. borrow $200 and buy the security. At the end of 3 months, repay $202.45. At the end of 3 months, deliver the security for $205.00. Arbitrage profit is then $2.55.

(b) At a price of $198, the contract offered is underpriced relative to the no-arbitrage forward price of $202.45. Enter the long position of the forward contract to buy in 3 months at £198. Short the stock now (i.e. borrow and sell it) and invest the proceeds. Doing so will yield an arbitrage profit of $4.45: i.e., short the security for $200 and invest the proceeds for 3 months ($202.45). At the end of 3 months, buy the security for $198 under the contract and return it. Arbitrage profit is then $4.45.

10.3 **(a)** We have $S_0 = \$150$, $T = 250/365$, and $i = 0.0525$ per annum (risk free).

The present value of proceeds is

$$I = \$1.25 \times \left(v^{30/365} + v^{120/365} + v^{210/365} \right) = \$3.69$$

and so the forward price is

$$K = (150 - 3.69) \times (1.0525)^{250/365} = \$151.53$$

(b) We now have $S_t = \$115$, $K = \$151.53$, $t = 100/365$, $T - t = 150/365$, and $i = 0.0525$ per annum (risk-free). After 100 days, two dividends remain: the first one in 20 days and second one in 110 days. The PV of these are

$$1.25 \times \left(v^{20/365} + v^{110/365} \right) = \$2.48$$

and so the value of the long position is

$$V_l(t) = \$115 - \$2.48 - \$151.53 v^{150/365} = -\$35.86$$

A negative value is a gain to the short position.

(c) We have $S_T = \pounds130$ and $K = \$151.53$. The value of the long contract at expiry is then $\$130.00 - \$151.53 = -\$21.53$. This represents a gain to the short position.

10.4 We have that $T = \dfrac{100}{365} = 0.274$.

(a) Under discrete compounding, we denote the US interest rate as $i = 6.3\%$ per annum and the British interest rate as $i^f = 5.8\%$. The futures price is

$$f_0(T) = \left(\frac{S_0}{(1 + i^f)^T} \right) (1 + i)^T$$

$$f_0(0.274) = \left(\frac{1.4390}{1.058^{0.274}} \right) 1.063^{0.274} = \$1.4409$$

(b) The actual futures price of $1.4650 is higher than the price computed above; the futures contract is therefore overpriced. To take advantage, a US arbitrageur needs to buy the foreign currency (£) and sell the futures contract. First, however, we must determine how many units of the currency to buy. Because we need to have 1 unit of currency, including interest, the number of units to buy is

$$\frac{1}{1.058^{0.274}} = 0.9847$$

So we buy 0.9847 units, which costs $0.9847 \times \$1.4390 = \$1.417$. We sell the futures at $1.4650 and hold until expiration. During that time, the accumulation of interest will cause the 0.9847 units of currency grow to 1 unit. Using the futures contract, at expiration, we convert this unit at the futures rate of $1.4660. The return per dollar invested is $1.4650/1.417 = 1.0339$, i.e., 3.39% over 100 days. The US annual risk-free rate is 6.3%, which is equivalent to a return per dollar invested of $(1.063)^{0.274} = 1.0169$ (i.e., 1.69%) over the 100-day period. Therefore, the return to the arbitrageur from the transactions described previously exceeds the risk-free return.

10.5 $f = S - I - Ke^{-r(T-t)}$
where
t is the present time;
T is the time of maturity of the forward contract;
r is the continuously compounded risk-free rate of interest for the interval from t to T;
S is the spot price of the security at time t;
I is the present value, at the risk-free interest rate, of the income generated by the security during the interval from t to T;
K is the delivery price of the forward contract;
f is the value of a long position in the forward contract.
Here, working with £100 nominal, $S = 95$, $K = 98$, $T - t = 1$, and $r = 0.052$:

$$I = 2.5\left(e^{-0.046 \times 0.5} + e^{-0.052 \times 1}\right) = 4.81648$$

$$\Rightarrow f = 95 - 4.81648 - 98e^{-0.052} = -2.85071$$

The value of the investor's short position in a forward contract on £1 million is therefore

$$\left(\frac{1,000,000}{100}\right) \times -f = 10,000 \times 2.85071$$

$$= £28,507$$

10.6 (a) The "no arbitrage" assumption means that neither of the following applies:

(i) an investor can make a deal that would give him an immediate profit, with no risk of future loss;

(ii) an investor can make a deal that has zero initial cost, no risk of future loss, and a non-zero probability of a future profit.

(b) In all states of the world, security B pays 80% of A. Therefore, its price must be 80% of A's price, or the investor could obtain a better payoff by purchasing only one security and make risk-free profits by selling one security short and buying the other. The price of B must therefore be 16p.

10.7 (a) The current value of the forward price of the old contract is

$$95 \times (1.03)^5 - 5(1.03)^{-2} - 6(1.03)^{-4}$$

whereas the current value of the forward price of a new contract is

$$145 - 5(1.03)^{-2} - 6(1.03)^{-4}$$

Hence, the current value of the old forward contract is

$$145 - 95(1.03)^5 = £34.87$$

(b) The current value of the forward price of the old contract is

$$95(1.02)^{-12}(1.03)^5 = 86.8376$$

whereas the current value of the forward price of a new contract is

$$145(1.02)^{-7} = 126.2312$$

therefore, the current value of the old forward contract is

$$126.23 - 86.84 = £39.39$$

10.8 Assuming no arbitrage:
Present value of dividends is (in £):

$$0.5v^{1/2}(\text{at } 5\%) + 0.5v(\text{at } 6\%) = 0.5(0.97590 + 0.94340) = 0.95965$$

Hence, the forward price is $F = (9 - 0.95965) \times 1.06 = £8.5228$.

10.9 The present value of the dividends, I, is

$$I = 0.5v^{1/12} + 0.5v^{7/12} = 0.5(0.99594 + 0.97194)$$
$$= 0.98394 \text{ calculated} \quad \text{at } i = 5\%$$

Hence, the forward price is (again calculated at $i = 5\%$)

$$F = (10 - 0.98394)(1 + i)^{11/12}$$
$$= 9.42845 = £9.43$$

10.10 The forward price at the outset of the contract was

$$\left(94.5 - 9v_{5\%}^{10} - 10v_{5\%}^{11}\right) \times (1.05)^{12} = 149.29$$

The forward price that should be offered now is

$$\left(143 - 9v_{5\%}^2 - 10v_{5\%}^3\right) \times (1.05)^4 = 153.39$$

Hence, the value of the contract now is

$$(153.39 - 149.29)v_{5\%}^4 = 3.37$$

Note that this result can also be obtained directly from

$$143 - 94.5 \times (1.05)^8 = 3.38$$

since the coupons are irrelevant in this calculation.

10.11 Assume no arbitrage.

(a) Buying the forward is exactly the same as buying the bond except that the forward will not pay coupons and the forward does not require immediate settlement.

Let the forward price $= F$. The equation of value is

$$F = 97(1.06) - 3.5 \times \frac{1.06}{(1.05)^{\frac{1}{2}}} - 3.5$$

$$= 102.82 - 3.62059 - 3.5 = 95.6994$$

(b) Let the 6-month forward interest rate $= f_{0.5,0.5} = \dfrac{1.06}{(1.05)^{\frac{1}{2}}} - 1 = 3.4454\%$

(c) $P = 2(1.05)^{-0.5} + 102(1.06)^{-1} = 1.9518 + 96.2264 = 98.1782$

(d) Gross redemption yield is i such that

$$98.1782 = 2(1+i)^{-0.5} + 102(1+i)^{-1}$$

Using the formula for solving a quadratic (interpolation will do): $(1+i)^{-0.5} = 0.97133$. Therefore, $i \approx 6\%$ (in fact, 5.99%).

(e) The answer is very close to 6% (the 1-year spot rate) because the payments from the bond are so heavily weighted towards the redemption time in 1 year.

10.12 $K = 60e^{0.06 \times 3/12} - 2.80e^{0.06 \times 1/12} = 60.90678 - 2.81404 = £58.09$

10.13 (a) The forward price is the accumulated value of the share less the accumulated value of the expected dividends:

$$F = 9.56 \, (1.03)^{9/12} - 0.2(1.03)^{8/12} - 0.2 \, (1.03)^{2/12}$$

$$= 9.7743 - 0.20398 - 0.20099$$

$$= £9.3693$$

(b) Although the share will be bought in 9 months, it is not necessary to take into account the expected share price. The current share price already makes an allowance for expected movements in the price, and the investor is simply buying an instrument that is (more or less) identical to the underlying share but with deferred payment. As such, under given assumptions, the forward can be priced from the underlying share.

CHAPTER 11 EXERCISES

11.1 We have $S_0 = \$1.05$ and $T = \dfrac{60}{365} = 0.1644$.

(a) $K = \$0.95$.

Call: The maximum value for the call is $S_0 = \$1.05$. The lower bound for the call is $S_0 - \dfrac{K}{(1+i)^T} = 1.05 - \dfrac{0.95}{1.055^{0.1644}} = \$0.11$.

We therefore have the range $\$0.11 \le C_0 \le \$1.05$.

Put: The maximum value for the put is

$\dfrac{K}{(1+i)^T} = \dfrac{0.95}{1.055^{0.1644}} = \$0.94$. The lower bound for the put is

$\dfrac{0.95}{1.055^{0.1644}} - 1.05 = -\$0.11$.

We therefore have the range $0 \le p_0 \le \$0.94$.

(b) $K = \$1.10$.

Call: The maximum value for the call is $S_0 = \$1.05$. The lower bound for the call is $1.05 - \dfrac{1.10}{1.055^{0.1644}} = -\$0.04$.

We therefore have the range $\$0 \le c_0 \le \$1.05$.

Put: The maximum value for the put is $\dfrac{1.10}{1.055^{0.1644}} = \$1.09$. The

lower bound for the put is $\dfrac{1.10}{1.055^{0.1644}} - 1.05 = \$0.04$.

We therefore have the range $\$0.04 \le p_0 \le \$1.09$.

11.2 We can illustrate put-call parity by showing that for the fiduciary call and protective put the current values and values at expiration are the same:

Call price, $c_0 = \$6.64$

Put price, $p_0 = \$2.75$

Exercise price, $K = \$30.00$

Risk free rate (discretely compounding), $i = 4\%$ per annum

Time to expiration, $T = 219/365 = 0.6$

Current stock price, $S_0 = \$33.19$

Current bond price, $K/(1+i)^T = 30/1.04^{0.6} = 29.30$

| | | Value at Expiration | |
Transaction	Current Value	$S_T = 20$	$S_T = 40$
Fiduciary call:			
Buy call	6.64	0	$40 - 30 = 10$
Buy bond	29.30	30	30
Total	35.94	30	40
Protective put:			
Buy put	2.75	$30 - 20 = 10$	0
Buy stock	33.19	20	40
Total	35.94	30	40

The values in the table show that the current values and values at expiration for the fiduciary call and the protective put are the same. That is, $c_0 + \dfrac{K}{(1+i)^T} = p_0 + S_0$ and put-call parity is satisfied.

11.3 (a) The payments at the beginning of the swap are as follows:
The British company (domestic party) pays the counterparty £75 million.
The counterparty pays the British company $105 million.

(b) The semi-annual payments are as follows.
The British company (domestic party) pays the counterparty $105,000,000 \times 0.06 \times 180/360 = \$3,150,000$.
The counterparty pays the British company £75,000,000 $\times 0.05 \times 180/360 = £1,875,000$.

(c) The payments at the end of the swap are as follows:
The British company (domestic party) pays the counterparty $105 million + $3,150,000.
The counterparty pays the British company £75 million + £1,875,000.

11.4 (a) We have $S_0 = \$99$, $T = 0.5$, $K = 95$, and $p_0 = 5$.

   (i) $p_T = \max(0, K - S_T) = \max(0, 95 - 100) = 0$
     Profit $= p_T - p_0 = -5$

   (ii) $p_T = \max(0, K - S_T) = \max(0, 95 - 95) = 0$
     Profit $= p_T - p_0 = -5$

   (iii) $p_T = \max(0, K - S_T) = \max(0, 95 - 93) = 2$
     Profit $= p_T - p_0 = 2 - 5 = -3$

   (iv) $p_T = \max(0, K - S_T) = \max(0, 95 - 90) = 5$
     Profit $= p_T - p_0 = 5 - 5 = 0$

   (v) $p_T = \max(0, K - S_T) = \max(0, 95 - 85) = 10$
     Profit $= p_T - p_0 = 10 - 5 = 5$

(b) Breakeven: $S_T^* = K - p_0 = 95 - 5 = 90$. Clearly, this result is consistent with our solution above where the profit is exactly zero in (a)(iv), in which the price at expiration is 90.

(c) Maximum profit (to put buyer) $= K - p_0 = 95 - 5 = 90$. This profit would be realized in the unlikely scenario of the price of the underlying falling to zero.

11.5 (a) We have $S_0 = \$77$, $K = 80$, and $c_0 = 5$.

  (i) $V_T = S_T - \max(0, S_T - K) = 70 - \max(0, 70 - 80) = 70$
  Profit $= V_T - V_0 = 70 - (S_0 - c_0) = 70 - (77 - 6) = -1$

  (ii) $V_T = S_T - \max(0, S_T - K) = 75 - \max(0, 75 - 80) = 70$
  Profit $= V_T - V_0 = 75 - (S_0 - c_0) = 75 - (77 - 6) = 4$

  (iii) $V_T = S_T - \max(0, S_T - K) = 80 - \max(0, 80 - 80) = 80$
  Profit $= V_T - V_0 = 80 - (S_0 - c_0) = 80 - (77 - 6) = 9$

  (iv) $V_T = S_T - \max(0, S_T - K) = 85 - \max(0, 85 - 80) = 80$
  Profit $= V_T - V_0 = 80 - (S_0 - c_0) = 80 - (77 - 6) = 9$

(b)
  (i) Maximum profit $= K - S_0 + c_0 = 80 - 77 + 6 = 9$.
  (ii) Maximum loss $= S_0 - c_0 = 77 - 6 = 71$.
  (iii) The maximum profit would be realized if the expiration price of the underlying is at or above the exercise price of $80.
  (iv) The maximum loss would be incurred if the underlying price drops to zero.

(c) Breakeven: $S_T^* = S_0 - c_0 = 77 - 6 = 71$

11.6 (a) The position is commonly called a bull spread.

(b) Let K_1 be the lower of the two strike prices and K_2 be the higher of the two strike prices.

  (i) $V_T = \max(0, S_T - K_1) - \max(0, S_T - K_2)$
  $= \max(0, 89 - 75) - \max(0, 89 - 85) = 14 - 4 = 10$
  Profit $= V_T - V_0 = V_T - (c_1 - c_2) = 10 - (10 - 2) = 2$

  (ii) $V_T = \max(0, S_T - K_1) - \max(0, S_T - K_2)$
  $= \max(0, 78 - 75) - \max(0, 78 - 85) = 3 - 0 = 3$
  Profit $= V_T - V_0 = V_T - (c_1 - c_2) = 3 - (10 - 2) = -5$

  (iii) $V_T = \max(0, S_T - K_1) - \max(0, S_T - K_2)$
  $= \max(0, 70 - 75) - \max(0, 70 - 85) = 0 - 0 = 0$
  Profit $= V_T - V_0 = V_T - (c_1 - c_2) = 0 - (10 - 2) = -8$

(c)
  (i) Maximum
    profit $= K_2 - K_1 - (c_1 - c_2) = 85 - 75 - (10 - 2) = 2$
  (ii) Maximum loss $= c_1 - c_2 = 10 - 2 = 8$

(d) Breakeven: $S_T^* = K_1 + (c_1 - c_2) = 75 + (10 - 2) = 83$

11.7 Let $K_1 = 110$, $K_2 = 115$, and $K_3 = 120$. Then
$V_0 = p_1 - 2p_2 + p_3 = 3.50 - 2 \times 6 + 9 = 0.50$.

(a)   **(i)** $V_T = \max(0, K_1 - S_T) - 2\max(0, K_2 - S_T) + \max(0, K_3 - S_T) = \max(0, 110 - 106) - 2\max(0, 115 - 106) + \max(0, 120 - 106) = 4 - 2 \times 9 + 14 = 0$

Profit $= V_T - V_0 = 0 - 0.50 = -0.50$

(ii) $V_T = \max(0, K_1 - S_T) - 2\max(0, K_2 - S_T) + \max(0, K_3 - S_T) = \max(0, 110 - 110) - 2\max(0, 115 - 110) + \max(0, 120 - 110) = 0 - 2 \times 5 + 10 = 0$

Profit $= V_T - V_0 = 0 - 0.50 = -0.50$

(iii) $V_T = \max(0, K_1 - S_T) - 2\max(0, K_2 - S_T) + \max(0, K_3 - S_T) = \max(0, 110 - 115) - 2\max(0, 115 - 115) + \max(0, 120 - 115) = 0 - 2 \times 0 + 5 = 5$

Profit $= V_T - V_0 = 5 - 0.50 = 4.50$

(iv) $V_T = \max(0, K_1 - S_T) - 2\max(0, K_2 - S_T) + \max(0, K_3 - S_T) = \max(0, 110 - 120) - 2\max(0, 115 - 120) + \max(0, 120 - 120) = 0 - 2 \times 0 + 0 = 0$

Profit $= V_T - V_0 = 0 - 0.50 = -0.50$

(v) $V_T = \max(0, K_1 - S_T) - 2\max(0, K_2 - S_T) + \max(0, K_3 - S_T) = \max(0, 110 - 123) - 2\max(0, 115 - 123) + \max(0, 120 - 123) = 4 - 2 \times 9 + 14 = 0$

Profit $= V_T - V_0 = 0 - 0.50 = -0.50$

(b)   **(i)** Maximum profit $= K_2 - K_1 - (p_1 - 2p_2 + p_3) = 115 - 110 - 0.50 = 4.50$. The maximum profit would be realized if the expiration price of the stock is at the exercise price of $115.

(ii) Maximum loss $= p_1 - 2p_2 + p_3 = 0.50$. The maximum loss would be incurred if the expiration price of the stock is at or below the exercise price of $110, or if the expiration price of the stock is at or above the exercise price of $120.

(c) Breakeven at $S_T^* = K_T + (p_1 - 2p_2 + p_3)$ and also $S_T^* = 2K_2 - K_1 - (p_1 - 2p_2 + p_3)$. So, $S_T^* = 110 + 0.50 = 110.50$ and also $S_T^* = 2 \times 115 - 110 - 0.50 = 119.50$.

11.8 (a) **(i)** $V_T = S_T + \max(0, K_1 - S_T) - \max(0, S_T - K_2)$

$= 92 + \max(0, 75 - 92) - \max(0, 92 - 90) = 92 + 0 - 2$

$= 90$

Profit $= V_T - V_0 = 90 - 80 = 10$

(ii) $V_T = S_T + \max(0, K_1 - S_T) - \max(0, S_T - K_2)$

$= 90 + \max(0, 75 - 90) - \max(0, 90 - 90) = 90 + 0 - 2$

$= 90$

Profit $= V_T - V_0 = 90 - 80 = 10$

(iii) $V_T = S_T + \max(0, K_1 - S_T) - \max(0, S_T - K_2)$

$= 82 + \max(0, 75 - 82) - \max(0, 82 - 90) = 82 + 0 - 0$

$= 82$

Profit $= V_T - V_0 = 82 - 80 = 2$

(iv) $V_T = S_T + \max(0, K_1 - S_T) - \max(0, S_T - K_2)$

$= 75 + \max(0, 75 - 75) - \max(0, 75 - 90) = 75 + 0 - 0$

$= 75$

Profit $= V_T - V_0 = 75 - 80 = -5$

(v) $V_T = S_T + \max(0, K_1 - S_T) - \max(0, S_T - K_2)$

$= 70 + \max(0, 75 - 70) - \max(0, 70 - 90) = 70 + 5 - 0$

$= 75$

Profit $= V_T - V_0 = 75 - 80 = -5$

(b) **(i)** Maximum profit $= K_2 - S_0 = 90 - 80 = 10$;

(ii) Maximum loss $= -(K_1 - S_0) = -(75 - 80) = 5$;

(iii) The maximum profit would be realized if the expiration price of the stock is at or above the exercise price of \$90;

(iv) The maximum loss would be incurred if the expiration price of the stock is at or below the exercise price of \$75.

(c) Breakeven: $S_T^* = 80$

11.9 (a) The position is commonly called a (long) straddle.

(b) **(i)** $V_T = \max(0, S_T - K) + \max(0, K - S_T)$

$= \max(0, 35 - 25) + \max(0, 25 - 35) = 10 + 0$

$= 10$

Profit $= V_T - (c_0 + p_0) = 10 - (4 + 1) = 5$

(ii) $V_T = \max(0, S_T - K) + \max(0, K - S_T)$

$= \max(0, 29 - 25) + \max(0, 25 - 29) = 4 + 0$

$= 4$

Profit $= V_T - (c_0 + p_0) = 4 - (4 + 1) = -1$

(iii) $V_T = \max(0, S_T - K) + \max(0, K - S_T)$

$= \max(0, 25 - 25) + \max(0, 25 - 25) = 0 + 0$

$= 0$

Profit $= V_T - (c_0 + p_0) = 0 - (4 + 1) = -5$

(iv) $V_T = \max(0, S_T - K) + \max(0, K - S_T)$

$= \max(0, 20 - 25) + \max(0, 25 - 20) = 0 + 5$

$= 5$

Profit $= V_T - (c_0 + p_0) = 5 - (4 + 1) = 0$

(v) $V_T = \max(0, S_T - K) + \max(0, K - S_T)$

$= \max(0, 15 - 25) + \max(0, 25 - 15) = 0 + 10$

$= 10$

Profit $= V_T - (c_0 + p_0) = 10 - (4 + 1) = 5$

    (c) **(i)** Maximum profit is unlimited.

          (ii) Maximum loss $= c_0 + p_0 = 4 + 1 = 5$

    (d) Breakeven: $S_T^* = X \pm (c_0 + p_0) = 25 \pm (4 + 1) = 30, 20$

11.10 **(a)** S&M (quoted $+0.75\%$ over sterling government bonds) has a poorer credit rating than BIM (quoted $+0.50\%$ over US\$ government bonds), as evidenced by the spread over corresponding 5-year government rates.

          To avoid any exchange rate risk on the exchange of interest rate payments, S&M will need to borrow at the 5-year fixed rate of 6% per annum in sterling and receive payments at a rate of 6% per annum fixed for 5 years from the global investment bank as part of the swap design.

          To avoid any exchange rate risk on the exchange of interest rate payments, BIM will need to borrow at the 5-year fixed rate of 5.25% per annum in US\$ and receive payments at a rate of 5.25% per annum fixed for 5 years from the global investment bank as part of the swap design.

          The difference between the US\$ payments by S&M to the global investment bank and the sterling payments by BIM to the global investment bank provides the margin for the global investment bank. However, the global investment bank will probably want to charge a higher rate of interest to S&M than to BIM to reflect the poorer credit rating of the former.

    (b) In arriving at its fee, the global investment bank would probably wish to tilt the charges to BIM and S&M to reflect their relative credit ratings. Thus, the global investment bank may wish to charge S&M a somewhat wider margin than BIM.

          One possibility would be to charge S&M a US\$ 5-year fixed rate of 5.80% (1.05% over 5-year US\$ government bonds) and BIM a sterling 5-year fixed rate of 5.90% (0.95% over 5-year sterling government bonds).

    (c) The global investment bank is left with a residual foreign exchange risk on each exchange of interest payments between the two parties.

          This risk could be hedged by forward foreign exchange contracts. The global investment bank is also left with credit risk.

          Credit risk could be hedged using credit derivatives.

11.11 **(a)** Marks should be awarded for the following features:

          Same axes and same scales—one rather than two diagrams;

          Common strike price;

          2:1 ratio of premiums (suggested on inspection by eye);

          General shape of each option's "hockey stick" diagram. See

Figures 11.3.1 and 11.3.2 (with modification to include the option premiums).

(b) Marks should be awarded for the following features:
- Same axes and same scales—one rather than two diagrams;
- Common strike price of the combined option portfolio;
- Maximum loss of 3 units [implied by 2:1 ratio of premiums for basic options in (a)] (suggested on inspection by eye);
- "V" shape of diagram. See Figure 11.6.4 (long straddle).

(c) The portfolio of options pays off at maturity when there is a large move either way in the price of the underlying stock over the life of the option. Conversely, investors in the portfolio of options will lose out if there is little or no movement in the price of the underlying stock by the maturity date of the option.

11.12 (a) The main uses of derivatives are as follows.
- *Speculation:* Exchange-traded derivatives could be used for speculation, effectively betting on a strong view of a particular market movement. The difference between speculation using options and speculation using the underlying asset is that buying the underlying asset requires an initial cash outlay equal to the total value of what is bought, whereas entering into a future contract or an option contract requires only a fraction of the initial cash outlay. Thus, a much higher level of leverage (gearing) can be achieved.
- *Arbitrage:* Arbitrage involves locking-in a riskless profit by simultaneously entering into two transactions in two or more markets. Using various combinations of options and the underlying instruments, portfolios with the same return but with different constituent parts can be created. Arbitrage opportunities can arise when the prices of these different portfolios get out of alignment and a riskless profit can be made. In practice, only very small arbitrage opportunities are observed in prices that are quoted in most financial markets. Also, transaction costs would probably eliminate the profit for all but the very large investment houses that face very low transaction costs.
- *Hedging:* Hedging allows a fund manager to reduce a risk that the fund already faces. Hedging using options, for example, involves taking a long or short position in a number of options contracts which is the opposite to the position held in the underlying asset. Conceptually, a loss made in the underlying asset will be offset by an approximately equal gain on the options position. This technique is very useful where, say,

a fund is going to sell its holding in 2 or 3 months and it wishes to avoid a fall in market values. However, if the market rises, there will be a loss on the futures position approximately equal in value to gain on the underlying equities, so the strategy does close off the opportunity for the fund to participate in any upward movements in the underlying assets while the hedge is in place.

- *Portfolio management:* Options can be used to manage the reallocation of assets from one market to another. For example, call options on equity indices can be used to gain exposure to upside movements in the markets; put options can be used to remove exposure to downside movements in markets. Calls and puts can be used to change a fund's exposure to an asset category or to change a fund's exposure within an asset category.

(b) **(i)** *Margin:* The amount of money to be maintained depending on how the market value of the contracts change.

(ii) *Initial Margin:* The initial payment put down to cover the risk of the contract.
Variation Margin: If there is no "margin account", the margin which is payable or received on a daily basis to mark to market. Otherwise, the margin required following a margin call.

(iii) The clearinghouse is removing the credit risk, and it needs some form of compensation to cover itself for this risk.

(c) **(i)** *Put:* The right to sell an underlying asset for a certain price by a certain date.
Call: The right to buy an underlying asset for a certain price by a certain date.

(ii) *American:* Exercised at any date to expiry. European can be exercised only on expiry date.

(d) *Call:* The loss on the 3 month is 5p to 80p when breakeven, then "in the money"; 6 month is 10p loss until 85p.
Put: The 3 month is in profit to 75p and then loss of option; the 6 month is in loss at 80p.

11.13 (a) The value of the promise can be thought as part of a call option contract. A call option consists of a contract to deliver a share in return for the payment of an exercise, where the share price exceeds the exercise price.
The promise made to the employees is the first part of the call option—the promise to deliver a share provided the price exceeds a certain level. Therefore, the first component of the Black—Scholes formula gives the value

$$\text{Value} = S.N.(d_1) \text{ where } d_1 = \left(\frac{\ln s/k + r + \tfrac{1}{2}\sigma^2}{\sigma\sqrt{T}} \right) T$$

$$= 1{,}000 \times N \left(\frac{\ln\left(\frac{1}{1.5}\right) + 0.04 + \tfrac{1}{2}0.3^2}{0.3} \right)$$

$$= N(-1.0682) \times 1{,}000 = \text{£}142.70$$

(b) Limiting the gain under the contract can be represented by a portfolio of the preceding promise less a call option with exercise price of £2.

The value of the call option is therefore

$$1{,}000(N(d_1) - 2e^{-0.04} N(d_2)) \quad d_1 \text{ as above } d_2 = d_1 - \sigma\sqrt{T}$$

where
$$d_1 = -2.027$$
$$d_2 = -2.327$$

The value of the call option is therefore

$$1{,}000 \times 0.00215 = \text{£}2.15$$

The value of the revised promise is therefore

$$\text{£}142.70 - \text{£}2.15 = \text{£}140.53$$

(c) **(i)** The employee has taken a view about the expected growth in share price. The result is a value that is not consistent with risk neutral pricing. This means that, if there were a market in these contracts, the price the employee has derived is not equal to the price the market would place on it.

    (ii) If the employer was willing to buy such promises at their suggested price, an arbitrageur would sell off £300 and hedge the position at £142.70, resulting in substantial risk-free profits.

11.14 **(a)** The bank has become a riskier investment; there may be an additional sector risk if this bank's performance impacts or infers a wider contagion. As the bank increases its risk, the expected return investors seek from the investment should also increase. However, the investment makes only 1% of portfolio, so although the bank has increased in risk, the impact at the total portfolio level should be minimal.

(b)
- Expect the share price to increase.
- Regret aversion—by maintaining existing arrangements, people minimise the pain associated with feeling of loss.
- Overconfidence in their ability.
- Status Quo bias—like to keep things the same.
- Good diversifier in overall portfolio.
- Could be part of an index-tracking portfolio.
- Tax considerations.
- Income may be appreciated.

(c) The individual is close to retirement and would expect to be in less risky investments where capital is more guaranteed. However, it this depends on exactly when the investor expects to retire.

The holding is a small part of overall portfolio, so it might be suitable if most other holdings are relatively safe.

If majority of investments are equity, then probably not suitable.

(d) Write options, for example put options for a lower price than the current share price for which the investor will receive a premium.

The closer to the current price, the higher the premium they will receive.

(e) Using example in (d) to answer (e) and (f):

Will keep premium, and current holdings will increase in value by 30%.

(f) Will make loss on put options and would have to buy additional shares at agreed price, leading to a higher loss and increasing exposure to bank. Existing holdings will decrease in value.

11.15 **(a)** A swap agreement in which the fixed rate receiver has the right to terminate the swap on one or more dates prior to its scheduled maturity. This early termination provision is designed to protect a party from adverse effects of large changes in fixed rates.

(b) The pension fund wants to enter into swaps to reduce risk, but the actual liabilities are subject to refinement, which might mean swap adjustment.

There is a yield pickup on the swap and, therefore, it is being held for tactical reasons and not as a long-term investment.

(c) Interest rate swap.

(d)

Period	Number of Days in Period	Annual Forward Interest	1/2 Year Interest Rate		
1	183	4.00%	2.03%	Value:	50,000,000
2	181	4.25%	2.14%	Term:	3
3	182	4.50%	2.28%	Payment:	semi-annual arrears
4	182	4.75%	2.40%	Days:	360
5	181	5.00%	2.51%		
6	183	5.25%	2.67%		

(i)

	PV	Discount	PV of Payments	Notional
1	1/[1+(days/360 × interest)]	0.9801	996,406	24,910,160
2	1/[1+(days/360 × interest) 2 periods]	0.9596	1,025,205	24,122,468
3	etc.	0.9382	1,067,229	23,716,197
4	etc.	0.9162	1,100,102	23,160,035
5	etc.	0.8938	1,123,398	22,467,962
6	etc.	0.8705	1,161,602	22,125,746
Total		**5.5584**	**6,473,942**	**140,502,569**

(ii)

PV of notional	140,502,569
PV of floating rate	6,473,942
Theoretical swap rate	4.61%

(e) PV of fixed rate = 4.75/4.61 × 6473942 = 6,670,520
Profit = 6,670,520 − 6,473,942 = £196,578 (or the "rounded" equivalent).
The fixed rate is higher than the theoretical swap rate, so assuming the payments reflect the assumed, then the pension fund will be in the money, i.e., calculation should show a profit, as they have been paid a higher amount from the bank (via fixed) than paid out.

(f) The higher interest rates would mean the pension fund would be paying out more than assumed and, therefore, the profit assumed would be reduced or turned into a loss (out of the money).

11.16 (a) (i) European: Right to purchase at set price at set date in future. American: Right to purchase at set price at any point before expiry date.

(ii) American because it has the added flexibility.

(b) The figures in Chapter 11 illustrate the basic shape of the payoff in each case.

(i) *Butterfly spread:* The investor does not believe a stock will rise or fall much before expiry; he thinks volatility will be low. He wants limited risk strategy but also limits profit.

(ii) *Straddle:* The investor believes the underlying price will change significantly but does not know which way it will go. Profit if volatility is high.

(iii) *Bear spreads:* A bear call spread is a limited profit, limited risk options trading strategy that can be used when the options trader is moderately bearish on the underlying security. He thinks the share price will fall.

(c) $O - S_t + K$, where S_t is greater than K; otherwise, zero. S_t = price of stock, K = exercise price, O is price of option.

(d) The chart would show £50 profit when share price starts at 0 until the exercise price is £1.50. The investor would then start to decrease the profit. At £2.00, the investor profit would be £0. At £2.50, the loss would be £50.

(e) (i) £0.75 = £50 profit.

(ii) £1.50 = £50 profit assuming the investor can buy stocks in market at zero cost.

(iii) £2.15 = £15 loss assuming the investor can buy stocks at zero cost. The loss on the purchase of shares is £65, and profit from the premium is £50.

(f) The initial margin is $0.2 \times 50\text{p} \times 100$ shares = £10. Then the investor has to post 100% of the movement, which is 5p. The additional margin is then £5 for the 100 shares, so the total margin = £15.

11.17 The proof of this result is an adaptation of that of the standard call-put parity. Two (self-financing) portfolios are considered.

- *Portfolio A:* Buying the call and selling the put at time t. Its value at time t is $P_t - C_t$ and at time T, it is $S_T - K$ in all states of the universe.

- *Portfolio B:* Buying a fraction $\exp(-\delta(T-t))$ of the underlying asset for $S_t\exp(-\delta(T-t))$ and borrowing $K\exp(-r(T-t))$ at time t. Its value at time t is then $K\exp(-r(T-t)) - S_t\exp(-\delta(T-t))$. Its value at maturity is then $S_T - K$ by taking into account the dividends which are paid continuously at rate δ.

  If we use the absence of arbitrage opportunity, both portfolios should have the same value at any intermediate time, in particular at time t. Hence,

  $$C_t - P_t = S_t\exp(-\delta(T-t)) - K\exp(-r(T-t)).$$

  Another proof can include the following portfolios:

- *Portfolio A:* At time t, buying a call option and lending $K\exp(-r(T-t))$;
- *Portfolio B:* At time t, buying the put option and buying one share.

11.18 **(a)** Try $\sigma=10\%$. The Black–Sholes formula gives a price of $p_{10}=44.05$.

Try $\sigma=40\%$. The Black–Sholes formula gives a price of $p_{40}=76.05$.

Interpolating gives a trial value of $(76.05-52.73)/(76.05-44.05)\times10 + (52.73-44.05)/(76.05-44.05)\times40 = 20.2\%$.

Evaluating gives $p_{20.2}=52.96$.

Interpolation with p_{40} give $\sigma = ((76.05-52.73)\times20.2 + (52.73-52.96)\times40)/(76.05-52.96) = 21.9\%$ (to the nearest 0.5%)

$p_{21.9}=54.75$.

The actual answer is 20%.

(b) The payoff is

$100\times\min(1,\max(S_T-320,0)) = 100\times(\max(S_T-320,0)-\max(S_T-321,0))$

so is 100 times the difference between two call options with the corresponding strikes. If we use the Black–Scholes formula, the price of the second call option is 52.06p, and, hence, the value of the derivative is $p=100\times(52.73-52.06)=67$p.

11.19 **(a)** Using put-call parity, $0=S-Ke^{-rT}$, so $K=Se^{rT}=306.06$p.

(b) $d_1=(\ln(S/K)+r+\frac12\sigma^2 T)/\sigma\sqrt{T}=\frac12\sigma$, while $d_2=(\ln(S/K)+r-\frac12\sigma^2 T)/\sigma\sqrt{T}=-\frac12\sigma$.

Therefore, $C=S\Phi(d_1)-Ke^{-rT}\Phi(d_2)=S\times(\Phi(\frac12\sigma)-\Phi(-\frac12\sigma))=300(2\Phi(\frac12\sigma)-1)$

so $\Phi(\frac12\sigma)=0.52$ so $\sigma=0.1003=10.0\%$.

(c) $\Phi(d_1)=0.52$, so the hedge is $5{,}000\times0.52=2{,}600$ shares and $600-2{,}600\times3=£7{,}200$ short in cash.

Appendix 3: Solutions to Exercises

CHAPTER 12 EXERCISES

12.1 **(a)** Let i_t be the (random) rate of interest in year t. Let S_{10} be the accumulation of the unit investment after 10 years:

$$E(S_{10}) = E[(1 + i_1)(1 + i_2)...(1 + i_{10})]$$

$$E(S_{10}) = E[1 + i_1]E[1 + i_2]...E[1 + i_{10}] \text{ as } \{i_t\} \text{ are independent}$$

$$E[i_t] = j$$

Therefore, $E(S_{10}) = (1 + j)^{10} = 1.07^{10} = 1.96715$

(b) $E(S_{10}^2) = E\big[[(1 + i_1)(1 + i_2)...(1 + i_{10})]^2\big]$

$$= E(1 + i_1)^2 E(1 + i_2)^2...E(1 + i_{10})^2 \text{ (using independence)}$$

$$= E(1 + 2i_1 + i_1^2)E(1 + 2i_2 + i_2^2)...E(1 + 2i_{10} + i_{10}^2)$$

$$= \big[E(1 + 2i_t + i_t^2)\big]^{10} = (1 + 2j + s^2 + j^2)^{10}$$

as $E[i_i^2] = V[i_t] + E[i_t]^2 = S^2 + j^2$

Therefore $\text{var}[S_n] = (1 + 2j + S^2 + j^2)^{10} - (1 + j)^{20}$

$$= (1 + 2 \times 0.07 + 0.016 + 0.07^2)^{10} - (1.07)^{20} = 0.5761$$

(c) If 1,000 units had been invested, the expected accumulation would have been 1,000 times bigger. The variance would have been 1,000,000 times bigger.

12.2 **(a)** Let S_3 = Accumulated fund after 3 years of investment of 1 at time 0

i_t = Interest rate for year t

Then, the fund after 3 years

$$= 80,000 \times S_3 = 80,000(1 + i_1)(1 + i_2)(1 + i_3)$$

$$E(i_1) = \frac{1}{3}(0.04 + 0.06 + 0.08) = 0.06$$

$$E(i_2) = 0.75 \times 0.07 + 0.25 \times 0.05 = 0.065$$

$$E(i_3) = 0.7 \times 0.06 + 0.3 \times 0.04 = 0.054$$

Then

$$E[80,000S_3] = 80,000\, E[S_3]$$

$$= E[80,000(1 + i_1)(1 + i_2)(1 + i_3)]$$

$$= 80,000\, E(1 + i_1)\, E(1 + i_2)\, E(1 + i_3)$$

since the i_t's are independent

$$= 80,000 \times 1.06 \times 1.065 \times 1.054 = \£95,188.85$$

(b)

$$\text{var}[80{,}000S_3] = 80{,}000^2 \times \text{var}[S_3]$$

$$\text{where var}[S_3] = E[S_3^2] - (E[S_3])^2$$

$$E[S_3^2] = E[(1+i_1)^2(1+i_2)^2(1+i_3)^2]$$

$$= E[(1+i_1)^2].E[(1+i_2)^2].E[(1+i_3)^2]$$

using independence

$$(1 + 2E[i_1] + E[i_1^2]).(1 + 2E[i_2] + E[i_2^2]).(1 + 2E[i_3] + E[i_3^2])$$

Now

$$E(i_1^2) = \frac{1}{3}(0.04^2 + 0.06^2 + 0.08^2) = 0.0038667$$

$$E(i_2^2) = 0.75 \times 0.07^2 + 0.25 \times 0.05^2 = 0.0043$$

$$E(i_3^2) = 0.7 \times 0.06^2 + 0.3 \times 0.04^2 = 0.0030$$

Hence, $E[S_3^2] = (1 + 2 \times 0.06 + 0.0038667)$

$$\times(1 + 2 \times 0.065 + 0.0043)$$

$$\times(1 + 2 \times 0.054 + 0.003)$$

$$= 1.41631$$

Hence var$[80{,}000S_3] = 80{,}000^2$ var$[S_3]80{,}000^2(1.41631$

$$- (1.18986)^2)$$

$$= 3{,}476{,}355$$

(c) Note that $80{,}000 \times 1.08 \times 1.07 \times 1.06 = 97{,}995 > 97{,}000$. But if, in any year, the highest interest rate for the year is not achieved, then the fund after 3 years falls below £97,000. Hence, the answer is the probability that the highest interest rate is achieved in each year

$$\frac{1}{3} \times 0.75 \times 0.7 = 0.175$$

12.3 (a)

$$E(1+i) = e^{\mu + \frac{1}{2}\sigma^2}$$

$$= e^{0.05 + \frac{1}{2} \times 0.004}$$

$$= 1.0533757$$

Therefore, $E[i] = 0.0533757$ since $E(1+i) = 1 + E(i)$

Let A be the accumulation at the end of 25 years of £3,000 paid annually in advance for 25 years.

Then

$$E[A] = 3,000\ddot{s}_{\overline{20}|} \quad \text{at rate } j = 0.0533757$$

$$= 3,000\frac{\left((1+j)^{25} - 1\right)}{j} \times (1+j)$$

$$= 3,000\frac{(1.0533757^{25} - 1)}{0.0533757} \times 1.0533757$$

$$= £158,036.43$$

(b) Let the accumulation be S_{20}; then S_{20} has a log-normal distribution with parameters 20μ and $20\sigma^2$:

$$\therefore E[S_{20}] = e^{20\mu + 1/2 \times 20\sigma^2}\left\{\text{or}(1+j)^{20}\right\}$$

$$= \exp(20 \times 0.05 + 10 \times 0.004) = e^{1.04} = 2.829217$$

$$\ln S_{20} \sim N(20\mu, 20\sigma^2) \text{ and so } \ln S_{20} \sim N(1, 0.08)$$

$$\Pr(S_{20} > 2.829217) = \Pr(\ln S_{20} > \ln 2.829217)$$

$$= \Pr\left(Z > \frac{\ln 2.829217 - 1}{\sqrt{0.08}}\right) \text{ where } Z \sim N(0, 1)$$

$$= \Pr(Z > 0.14) = 1 - \Phi(0.14)$$

$$= 1 - 0.55567$$

$$= 0.44433, \text{ i.e., } 44.4\%$$

12.4 (a) $E[i] = \frac{1}{3}(0.03) + \frac{1}{3}(0.06) + \frac{1}{3}(0.09) = 0.06 \text{ or } 6\%$

If the annual premium was calculated at this mean rate of interest, it would be $10,000/\ddot{s}_{\overline{20}|0.06} = £256.46$.

(b) **(i)** The expected value of the accumulated profit is

$$\frac{1}{3}(P\ddot{s}_{\overline{20}|0.03} - 10,000) + \frac{1}{3}(P\ddot{s}_{\overline{20}|0.06} - 10,000)$$

$$+ \frac{1}{3}(P\ddot{s}_{\overline{20}|0.09} - 10,000)$$

which equals $40.811 \ 25P - 10,000$. This will be zero if $P = 245.03$.

If $P = 256.46$ [as in (a)], the expected value of the accumulated profit is £466.45.

(ii) The expected value of the net present value of the policy immediately before the first premium is paid is

$$\frac{1}{3}\left(P\ddot{a}_{\overline{20}|0.03} - 10{,}000v_{0.03}^{20}\right) + \frac{1}{3}\left(P\ddot{a}_{\overline{20}|0.06} - 10{,}000v_{0.06}^{20}\right)$$
$$+ \frac{1}{3}\left(P\ddot{a}_{\overline{20}|0.09} - 10{,}000v_{0.09}^{20}\right)$$

which equals $12.477\ 34P - 3{,}479.70$. This will be zero if $P = £278.88$.

If $P = 256.46$ [as in (a)], the expected value of the net present value of the policy is $-£279.76$.

12.5 (a) (i) We note that, if θ denotes the accumulated profit at the maturity date, then

$$\theta = PS_{10} - 1{,}000$$

Hence,

$$E[\theta] = PE[S_{10}] - 1{,}000$$
$$= P(1 + E[i])^{10} - 1{,}000$$

Let $E[i] = j$ and $\text{var}[i] = s^2$. Then

standard deviation of $\theta = P(\text{standard deviation of } S_{10})$

$$= P\left\{\left[(1+j)^2 + s^2\right]^{10} - (1+j)^{20}\right\}^{\frac{1}{2}}$$

(ii) Let ϕ be the net present value of the policy immediately after it is effected. Then

$$\phi = P - 1{,}000(1 + i_1)^{-1}(1 + i_2)^{-1}...(1 + i_{10})^{-1}$$

where i_t denotes the yield in the tth year. Therefore,

$$E[\phi] = P - 1{,}000E\left[(1 + i_1)...(1 + i_{10})^{-1}\right]$$

Since (by assumption) $i_1, i_2,..., i_{10}$ are independently and identically distributed, the last equation implies that

$$E[\phi] = P - 1{,}000\left(E\left[\frac{1}{1+i}\right]\right)^{10}$$

Also, since P is a constant, the standard deviation of ϕ
$= 1{,}000 \times$ standard deviation of $[(P - \phi)/1{,}000]$
$= 1{,}000 \times$ standard deviation of $[(1 + i_i)^{-1}...(1 + i_{10})^{-1}]$
Using the fact that, for any random variable X, $\text{var }[X], = E[X^2]$

$- (E[X])^2$, we immediately obtain from the last equation: the standard deviation of ϕ

$$= 1{,}000\left\{ \left(E\left[\frac{1}{(1+i)^2}\right]\right)^{10} - \left(E\left[\frac{1}{1+i}\right]\right)^{20}\right\}^{\frac{1}{2}}$$

(b) The reader should verify the following results for each of the given models:

Model I $j = E[i] = 0.04$

$$s^2 = \text{var}[i] = \frac{1}{3}(-0.02)^2 + \frac{1}{3}(0)^2 + \frac{1}{3}(0.02)^2 = \frac{8}{3}10^{-4}$$

$$E\left[\frac{1}{1+i}\right] = \frac{1}{3}\left(\frac{1}{1.02}\right) + \frac{1}{3}\left(\frac{1}{1.04}\right) + \frac{1}{3}\left(\frac{1}{1.06}\right) = 0.961\,657\,01$$

Therefore,

$$\left(E\left[\frac{1}{1+i}\right]\right)^{10} = 0.677\,232\,23$$

$$E\left[\frac{1}{(1+i)^2}\right] = \frac{1}{3}\left(\frac{1}{1.02}\right)^2 + \frac{1}{3}\left(\frac{1}{1.04}\right)^2 + \frac{1}{3}\left(\frac{1}{1.06}\right)^2$$

$$= 0.92\,524\,048$$

and so

$$\left(E\left[\frac{1}{(1+i)^2}\right]\right)^{10} = 0.459\,775\,94$$

Model II $j = E[i] = 0.04$

$$s^2 = \text{var}[i] = \frac{4}{3}10^{-4}$$

$$E\left[\frac{1}{1+i}\right] = \int_{0.02}^{0.06} \frac{1}{1+x}25dx = 0.961\,657\,01$$

and therefore

$$\left(E\left[\frac{1}{1+i}\right]\right)^{10} = 0.676\,397\,61$$

$$E\left[\frac{1}{(1+i)^2}\right] = \int_{0.02}^{0.06} \frac{1}{(1+x)^2}25dx = 0.924\,898\,26$$

and so

$$\left(E\left[\frac{1}{(1+i)^2}\right]\right)^{10} = 0.458\,078\,21$$

Model III $j = E[i] = 0.04$

$$s^2 = \text{var}[i] = \frac{2}{3}10^{-4}$$

Note that the probability density function of i is

$$f(x) = \begin{cases} 2{,}500(x - 0.02) & \text{if } 0.02 \leq x < 0.04 \\ 2{,}500(0.06 - x) & \text{if } 0.04 \leq x \leq 0.06 \end{cases}$$

$$E\left[\frac{1}{1+i}\right] = \int_{0.02}^{0.06} \frac{1}{1+x} f(x)dx = 0.961\,597\,48$$

and therefore

$$\left(E\left[\frac{1}{1+i}\right]\right)^{10} = 0.67597890$$

$$E\left[\frac{1}{(1+i)^2}\right] = \int_{0.02}^{0.06} \frac{1}{(1+x)^2} f(x)dx = 0.924\,728\,15$$

and therefore

$$\left(E\left[\frac{1}{(1+i)^2}\right]\right)^{10} = 0.457\,236\,39$$

Combining these results with those of (a), we obtain the following values:

	Value of P for Which the Expected Value of the Accumulated Profit is Zero (£)	Standard Deviation of Accumulated Profit (P as in Previous Column) (£)	Value of P for Which the Expected Value of the Net Present Value at Outset is Zero (£)	Standard Deviation of Net Present Value (£)
Model I	675.56	49.68	677.23	33.65
Model II	675.56	35.12	676.40	23.76
Model III	675.56	24.83	675.98	17.00

12.6 (a)

$$j = E[i] = \frac{1}{3}(0.03) + \frac{1}{3}(0.06) + \frac{1}{3}(0.09) = 0.06$$

$$s^2 = \text{var}[i] = \frac{1}{3}(-0.03)^2 + \frac{1}{3}(0)^2 + \frac{1}{3}(0.03)^2 = 0.0006$$

(b) Hence, we obtain

$$E[S_n] = 1.06^n$$
$$\text{var}[S_n] = 1.1242^n - 1.06^{2n}$$

This gives the following values:

n	5	10	15	20
$E[S_n]$	1.338 23	1.790 85	2.396 56	3.207 14
$\sqrt{(\text{var}[S_n])}$	0.069 19	0.131 02	0.214 89	0.332 28

(c)

$$E[A_n] = \ddot{S}_{\overline{n}|0.06}$$

and we calculate the following values:

n	$\sqrt{(\text{var}[A_n])}$	n	$\sqrt{(\text{var}[A_n])}$	n	$\sqrt{(\text{var}[A_n])}$
1	0.024 49	6	0.280 93	11	0.802 84
2	0.056 75	7	0.361 94	12	0.946 24
3	0.098 50	8	0.453 92	13	1.104 76
4	0.149 63	9	0.557 55	14	1.279 48
5	0.210 32	10	0.673 58	15	1.471 55

(d)

n	5	10	15
$\sqrt{(\text{var}[A_n])}/E[A_n]$	0.0352	0.0482	0.0596

Note that, as n increases, the value of the standard deviation increases relative to the value of the mean.

12.7 (a) $j = E[i] = 0.3(0.02) + 0.5(0.04) + 0.2(0.07) = 0.04$
$s^2 = \text{var}\,[i] = 0.3(-0.02)^2 + 0.5(0)^2 + 0.2(0.03)^2 = 0.0003$
We have
$1{,}000\, E[S_{15}] = 1{,}000(1.04)^{15} = 1{,}800.94$
$1{,}000\sqrt{(\text{var}[S_{15}])} = 1{,}000(1.08\,19^{\,15} - 1.04^{30})^{1/2} = 116.28$
We carried out 10,000 simulations. Of these, 254 gave an

accumulation of less than £1,600; 468 gave an accumulation greater than £2,000. Our estimated probabilities and corresponding 95% confidence intervals are

(i) 0.0254; (0.0223, 0.0285)
(ii) 0.0468; (0.0427, 0.0509)

(b)

$$100E[A_{15}] = 100\ddot{s}_{\overline{15}|0.04} = 2{,}082.45$$

$$100\sqrt{(\text{var}[A_{15}])} = 86.87$$

Note that of the 10,000 simulations, 60 gave an accumulation of less than £1,900, and 90 gave an accumulation greater than £2,300. The estimated probabilities and 95% confidence intervals are therefore

(i) 0.006 (0.0045, 0.0075);
(ii) 0.009 (0.0071, 0.0109).

12.8 (a) $V_n = (1+i_1)^{-1}(1+i_2)^{-1}\dots(1+i_n)^{-1}$

Hence,

$$\ln V_n = -\ln(1+i_1) - \ln(1+i_2) - \dots - \ln(1+i_n).$$

Since, for each value of t, $\ln(1+i_t)$ is normally distributed with mean μ and variance σ^2, each term of the right side of the last equation is normally distributed with mean $-\mu$ and variance σ^2. Also, the terms are independently distributed. This means that $\ln V_n$ is normally distributed with mean $-n\mu$ and variance $n\sigma^2$, i.e., V_n has log-normal distribution with parameters $-n\mu$ and $n\sigma^2$.

(b) We are given that $E[i] = 0.08$ and $\text{var}[i] = (0\cdot05)^2 = 0.0025$. This implies that $\mu = 0.075\,890\,52$ and $\sigma^2 = 0.002\,141\,05$.
Let $m' = -n\mu$ and $\sigma'^2 = n\sigma^2$. Then

$$E[V_n] = \exp\left(\mu' + \frac{1}{2}\sigma'^2\right) = \exp\left(-n\mu + \frac{1}{2}n\sigma^2\right)$$

$$= \left[\exp\left(-\mu + \frac{1}{2}\sigma^2\right)\right]^n$$

and

$$E\big[(V_n)^2\big] = \exp(2\mu' + 2\sigma'^2) = \exp(-2n\mu + 2n\sigma^2)$$

$$= \left[\exp(-\mu + \sigma^2)\right]^{2n}$$

The standard deviation of V_n is $\{E[(V_n)^2] - (E[V_n])^2\}^{\frac{1}{2}}$.

Using the last two equations (with μ and σ^2 as calculated previously), we may readily calculate the following values:

n	$1,000E[V_n]$	$1,000\sqrt{(\text{var}[V_n])}$
5	687.91	71.37
10	473.22	69.62
15	325.53	58.81
20	223.93	46.84

12.9 **(a)** For $i = 1, 2, 3$, we have

$$E[X_i] = \frac{1}{2} \quad \text{and} \quad \text{var}[X_i] = \frac{1}{12}$$

Therefore,

$$E[X] = \frac{1}{3}\sum_{i=1}^{3} E[X_i] = \frac{1}{2} \quad \text{and} \quad \text{var}[X] = \left(\frac{1}{3}\right)^2 \sum_{i=1}^{3} \text{var}[X_i] = \frac{1}{36}$$

Hence, the standard deviation of X is $1/6$.

(b) *Exact* values for the probabilities are

    (i) $9/128 \simeq 0.070\ 313$;
    (ii) 0.036;
    (iii) $\frac{2}{3} \simeq 0.666\ 667$;
    (iv) 0.1625.

Readers should compare their estimated values with the preceding true values.

(c) *Exact* values of t are

    (i) $[\sqrt[3]{90}]^{-1} \simeq 0.223\ 144$;

    (ii) $1 - \sqrt[3]{(2/900)} \simeq 0.869\ 504$;

    (iii) The unique real root of the equation
$$36t^3 - 54t^2 + 18t - 1 = 0.$$

This root is approximately $0.382\ 380$.

Again, readers should compare these values with their estimates.

Appendix 4: Compound Interest Tables

Compound Interest Tables								
Constants					**1 Per Cent**			
Function	**Value**	**n**	**$(1 + i)^n$**	**v^n**	**$s_{\overline{n}\|}$**	**$a_{\overline{n}\|}$**	**$(Ia)_{\overline{n}\|}$**	**n**
i	0.010 000	1	1.010 00	0.990 10	1.000 0	0.990 1	0.990 1	1
$i^{(2)}$	0.009 975	2	1.020 10	0.980 30	2.010 0	1.970 4	2.950 7	2
$i^{(4)}$	0.009 963	3	1.030 30	0.970 59	3.030 1	2.941 0	5.862 5	3
$i^{(12)}$	0.009 954	4	1.040 60	0.960 98	4.060 4	3.902 0	9.706 4	4
δ	0.009 950	5	1.051 01	0.951 47	5.101 0	4.853 4	14.463 7	5
$(1 + i)$	1.010 000	6	1.061 52	0.942 05	6.152 0	5.795 5	20.116 0	6
$(1 + i)^{1/2}$	1.004 988	7	1.072 14	0.932 72	7.213 5	6.728 2	26.645 0	7
$(1 + i)^{1/4}$	1.002 491	8	1.082 86	0.923 48	8.285 7	7.651 7	34.032 9	8
$(1 + i)^{1/12}$	1.000 830	9	1.093 69	0.914 34	9.368 5	8.566 0	42.261 9	9
		10	1.104 62	0.905 29	10.462 2	9.471 3	51.314 8	10
v	0.990 099	11	1.115 67	0.896 32	11.566 8	10.367 6	61.174 4	11
$v^{1/2}$	0.995 037	12	1.126 83	0.887 45	12.682 5	11.255 1	71.823 8	12
$v^{1/4}$	0.997 516	13	1.138 09	0.878 66	13.809 3	12.133 7	83.246 4	13
$v^{1/12}$	0.999 171	14	1.149 47	0.869 96	14.947 4	13.003 7	95.425 8	14
		15	1.160 97	0.861 35	16.096 9	13.865 1	108.346 1	15
d	0.009 901	16	1.172 58	0.852 82	17.257 9	14.717 9	121.991 2	16
$d^{(2)}$	0.009 926	17	1.184 30	0.844 38	18.430 4	15.562 3	136.345 6	17
$d^{(4)}$	0.009 938	18	1.196 15	0.836 02	19.614 7	16.398 3	151.394 0	18
$d^{(12)}$	0.009 946	19	1.208 11	0.827 74	20.810 9	17.226 0	167.121 0	19
		20	1.220 19	0.819 54	22.019 0	18.045 6	183.511 9	20
$i/i^{(2)}$	1.002 494	21	1.232 39	0.811 43	23.239 2	18.857 0	200.551 9	21
$i/i^{(4)}$	1.003 742	22	1.244 72	0.803 40	24.471 6	19.660 4	218.226 7	22
$i/i^{(12)}$	1.004 575	23	1.257 16	0.795 44	25.716 3	20.455 8	236.521 8	23

(Continued)

(Continued)

Compound Interest Tables								
Constants					**1 Per Cent**			
Function	Value	n	$(1 + i)^n$	v^n	$s_{\overline{n}}$	$a_{\overline{n}}$	$(Ia)_{\overline{n}}$	n
		24	1.269 73	0.787 57	26.973 5	21.243 4	255.423 4	24
		25	1.282 43	0.779 77	28.243 2	22.023 2	274.917 6	25
i/δ	1.004 992	26	1.295 26	0.772 05	29.525 6	22.795 2	294.990 9	26
		27	1.308 21	0.764 40	30.820 9	23.559 6	315.629 8	27
		28	1.321 29	0.756 84	32.129 1	24.316 4	336.821 2	28
		29	1.334 50	0.749 34	33.450 4	25.065 8	358.552 1	29
		30	1.347 85	0.741 92	34.784 9	25.807 7	380.809 8	30
$i/d^{(2)}$	1.007 494	31	1.361 33	0.734 58	36.132 7	26.542 3	403.581 7	31
$i/d^{(4)}$	1.006 242	32	1.374 94	0.727 30	37.494 1	27.269 6	426.855 4	32
$i/d^{(12)}$	1.005 408	33	1.388 69	0.720 10	38.869 0	27.989 7	450.618 8	33
		34	1.402 58	0.712 97	40.257 7	28.702 7	474.859 9	34
		35	1.416 60	0.705 91	41.660 3	29.408 6	499.566 9	35
		36	1.430 77	0.698 92	43.076 9	30.107 5	524.728 2	36
		37	1.445 08	0.692 00	44.507 6	30.799 5	550.332 4	37
		38	1.459 53	0.685 15	45.952 7	31.484 7	576.368 2	38
		39	1.474 12	0.678 37	47.412 3	32.163 0	602.824 6	39
		40	1.488 86	0.671 65	48.886 4	32.834 7	629.690 7	40
		41	1.503 75	0.665 00	50.375 2	33.499 7	656.955 9	41
		42	1.518 79	0.658 42	51.879 0	34.158 1	684.609 5	42
		43	1.533 98	0.651 90	53.397 8	34.810 0	712.641 2	43
		44	1.549 32	0.645 45	54.931 8	35.455 5	741.040 8	44
		45	1.564 81	0.639 05	56.481 1	36.094 5	769.798 2	45
		46	1.580 46	0.632 73	58.045 9	36.727 2	798.903 7	46
		47	1.596 26	0.626 46	59.626 3	37.353 7	828.347 5	47
		48	1.612 23	0.620 26	61.222 6	37.974 0	858.120 0	48
		49	1.628 35	0.614 12	62.834 8	38.588 1	888.211 8	49
		50	1.644 63	0.608 04	64.463 2	39.196 1	918.613 7	50

Compound Interest Tables								
Constants					**2 Per Cent**			
Function	**Value**	**n**	**$(1 + i)^n$**	**v^n**	**$s_{\overline{n}\rvert}$**	**$a_{\overline{n}\rvert}$**	**$(Ia)_{\overline{n}\rvert}$**	**n**
i	0.020 000	1	1.020 00	0.980 39	1.000 0	0.980 4	0.980 4	1
$i^{(2)}$	0.019 901	2	1.040 40	0.961 17	2.020 0	1.941 6	2.902 7	2
$i^{(4)}$	0.019 852	3	1.061 21	0.942 32	3.060 4	2.883 9	5.729 7	3
$i^{(12)}$	0.019 819	4	1.082 43	0.923 85	4.121 6	3.807 7	9.425 1	4
δ	0.019 803	5	1.104 08	0.905 73	5.204 0	4.713 5	13.953 7	5
$(1 + i)$	1.020 000	6	1.126 16	0.887 97	6.308 1	5.601 4	19.281 6	6
$(1 + i)^{1/2}$	1.009 950	7	1.148 69	0.870 56	7.434 3	6.472 0	25.375 5	7
$(1 + i)^{1/4}$	1.004 963	8	1.171 66	0.853 49	8.583 0	7.325 5	32.203 4	8
$(1 + i)^{1/12}$	1.001 652	9	1.195 09	0.836 76	9.754 6	8.162 2	39.734 2	9
		10	1.218 99	0.820 35	10.949 7	8.982 6	47.937 7	10
v	0.980 392	11	1.243 37	0.804 26	12.168 7	9.786 8	56.784 6	11
$v^{1/2}$	0.990 148	12	1.268 24	0.788 49	13.412 1	10.575 3	66.246 5	12
$v^{1/4}$	0.995 062	13	1.293 61	0.773 03	14.680 3	11.348 4	76.295 9	13
$v^{1/12}$	0.998 351	14	1.319 48	0.757 88	15.973 9	12.106 2	86.906 2	14
		15	1.345 87	0.743 01	17.293 4	12.849 3	98.051 4	15
d	0.019 608	16	1.372 79	0.728 45	18.639 3	13.577 7	109.706 5	16
$d^{(2)}$	0.019 705	17	1.400 24	0.714 16	20.012 1	14.291 9	121.847 3	17
$d^{(4)}$	0.019 754	18	1.428 25	0.700 16	21.412 3	14.992 0	134.450 2	18
$d^{(12)}$	0.019 786	19	1.456 81	0.686 43	22.840 6	15.678 5	147.492 3	19
		20	1.485 95	0.672 97	24.297 4	16.351 4	160.951 8	20
$i/i^{(2)}$	1.004 975	21	1.515 67	0.659 78	25.783 3	17.011 2	174.807 1	21
$i/i^{(4)}$	1.007 469	22	1.545 98	0.646 84	27.299 0	17.658 0	189.037 5	22
$i/i^{(12)}$	1.009 134	23	1.576 90	0.634 16	28.845 0	18.292 2	203.623 1	23
		24	1.608 44	0.621 72	30.421 9	18.913 9	218.544 4	24
		25	1.640 61	0.609 53	32.030 3	19.523 5	233.782 7	25
i/δ	1.009 967	26	1.673 42	0.597 58	33.670 9	20.121 0	249.319 8	26
		27	1.706 89	0.585 86	35.344 3	20.706 9	265.138 0	27
		28	1.741 02	0.574 37	37.051 2	21.281 3	281.220 5	28
		29	1.775 84	0.563 11	38.792 2	21.844 4	297.550 8	29
		30	1.811 36	0.552 07	40.568 1	22.396 5	314.112 9	30
$i/d^{(2)}$	1.014 975	31	1.847 59	0.541 25	42.379 4	22.937 7	330.891 5	31
$i/d^{(4)}$	1.012 469	32	1.884 54	0.530 63	44.227 0	23.468 3	347.871 8	32
$i/d^{(12)}$	1.010 801	33	1.922 23	0.520 23	46.111 6	23.988 6	365.039 3	33

(Continued)

(Continued)

Compound Interest Tables

Constants							2 Per Cent		
Function	Value	n	$(1+i)^n$	v^n	$s_{\overline{n}\rvert}$	$a_{\overline{n}\rvert}$	$(Ia)_{\overline{n}\rvert}$	n
		34	1.960 68	0.510 03	48.033 8	24.498 6	382.380 3	34
		35	1.999 89	0.500 03	49.994 5	24.998 6	399.881 3	35
		36	2.039 89	0.490 22	51.994 4	25.488 8	417.529 3	36
		37	2.080 69	0.480 61	54.034 3	25.969 5	435.311 9	37
		38	2.122 30	0.471 19	56.114 9	26.440 6	453.217 0	38
		39	2.164 74	0.461 95	58.237 2	26.902 6	471.233 0	39
		40	2.208 04	0.452 89	60.402 0	27.355 5	489.348 6	40
		41	2.252 20	0.444 01	62.610 0	27.799 5	507.553 0	41
		42	2.297 24	0.435 30	64.862 2	28.234 8	525.835 8	42
		43	2.343 19	0.426 77	67.159 5	28.661 6	544.186 9	43
		44	2.390 05	0.418 40	69.502 7	29.080 0	562.596 5	44
		45	2.437 85	0.410 20	71.892 7	29.490 2	581.055 3	45
		46	2.486 61	0.402 15	74.330 6	29.892 3	599.554 4	46
		47	2.536 34	0.394 27	76.817 2	30.286 6	618.085 0	47
		48	2.587 07	0.386 54	79.353 5	30.673 1	636.638 8	48
		49	2.638 81	0.378 96	81.940 6	31.052 1	655.207 8	49
		50	2.691 59	0.371 53	84.579 4	31.423 6	673.784 2	50

Compound Interest Tables

Constants							3 Per Cent		
Function	Value	n	$(1+i)^n$	v^n	$s_{\overline{n}\rvert}$	$a_{\overline{n}\rvert}$	$(Ia)_{\overline{n}\rvert}$	n
i	0.030 000	1	1.030 00	0.970 87	1.000 0	0.970 9	0.970 9	1
$i^{(2)}$	0.029 778	2	1.060 90	0.942 60	2.030 0	1.913 5	2.856 1	2
$i^{(4)}$	0.029 668	3	1.092 73	0.915 14	3.090 9	2.828 6	5.601 5	3
$i^{(12)}$	0.029 595	4	1.125 51	0.888 49	4.183 6	3.717 1	9.155 4	4
δ	0.029 559	5	1.159 27	0.862 61	5.309 1	4.579 7	13.468 5	5
$(1+i)$	1.030 000	6	1.194 05	0.837 48	6.468 4	5.417 2	18.493 4	6
$(1+i)^{1/2}$	1.014 889	7	1.229 87	0.813 09	7.662 5	6.230 3	24.185 0	7
$(1+i)^{1/4}$	1.007 417	8	1.266 77	0.789 41	8.892 3	7.019 7	30.500 3	8

(Continued)

Compound Interest Tables							
Constants					**3 Per Cent**		
Function	**Value**	**n**	$(1 + i)^n$	v^n	$s_{\overline{n}}$	$a_{\overline{n}}$	$(Ia)_{\overline{n}}$	**n**
$(1 + i)^{1/12}$	1.002 466	9	1.304 77	0.766 42	10.159 1	7.786 1	37.398 1	9
		10	1.343 92	0.744 09	11.463 9	8.530 2	44.839 0	10
v	0.970 874	11	1.384 23	0.722 42	12.807 8	9.252 6	52.785 6	11
$v^{1/2}$	0.985 329	12	1.425 76	0.701 38	14.192 0	9.954 0	61.202 2	12
$v^{1/4}$	0.992 638	13	1.468 53	0.680 95	15.617 8	10.635 0	70.054 6	13
$v^{1/12}$	0.997 540	14	1.512 59	0.661 12	17.086 3	11.296 1	79.310 2	14
		15	1.557 97	0.64186	18.598 9	11.937 9	88.938 1	15
d	0.029 126	16	1.604 71	0.623 17	20.156 9	12.561 1	98.908 8	16
$d^{(2)}$	0.029 341	17	1.652 85	0.605 02	21.761 6	13.166 1	109.194 1	17
$d^{(4)}$	0.029 450	18	1.702 43	0.587 39	23.414 4	13.753 5	119.767 2	18
$d^{(12)}$	0.029 522	19	1.753 51	0.570 29	25.116 9	14.323 8	130.602 6	19
		20	1.806 11	0.553 68	26.870 4	14.877 5	141.676 1	20
$i/i^{(2)}$	1.007 445	21	1.860 29	0.537 55	28.676 5	15.415 0	152.964 7	21
$i/i^{(4)}$	1.011 181	22	1.916 10	0.521 89	30.536 8	15.936 9	164.446 3	22
$i/i^{(12)}$	1.013 677	23	1.973 59	0.506 69	32.452 9	16.443 6	176.100 2	23
		24	2.032 79	0.491 93	34.426 5	16.935 5	187.906 6	24
		25	2.093 78	0.477 61	36.459 3	17.413 1	199.846 8	25
i/δ	1.014 926	26	2.156 59	0.463 69	38.553 0	17.876 8	211.902 8	26
		27	2.221 29	0.450 19	40.709 6	18.327 0	224.057 9	27
		28	2.287 93	0.437 08	42.930 9	18.764 1	236.296 1	28
		29	2.356 57	0.424 35	45.218 9	19.188 5	248.602 1	29
		30	2.427 26	0.411 99	47.575 4	19.600 4	260.961 7	30
$i/d^{(2)}$	1.022 445	31	2.500 08	0.399 99	50.002 7	20.000 4	273.361 3	31
$i/d^{(4)}$	1.018 681	32	2.575 08	0.388 34	52.502 8	20.388 8	285.788 1	32
$i/d^{(12)}$	1.016 177	33	2.652 34	0.377 03	55.077 8	20.765 8	298.230 0	33
		34	2.731 91	0.366 04	57.730 2	21.131 8	310.675 5	34
		35	2.813 86	0.355 38	60.462 1	21.487 2	323.113 9	35
		36	2.898 28	0.345 03	63.275 9	21.832 3	335.535 1	36
		37	2.985 23	0.334 98	66.174 2	22.167 2	347.929 5	37
		38	3.074 78	0.325 23	69.159 4	22.492 5	360.288 1	38
		39	3.167 03	0.315 75	72.234 2	22.808 2	372.602 4	39
		40	3.262 04	0.306 56	75.401 3	23.114 8	384.864 7	40

(Continued)

(Continued)

Compound Interest Tables								
Constants				**3 Per Cent**				
Function	Value	n	$(1 + i)^n$	v^n	$s_{\overline{n}}$	$a_{\overline{n}}$	$(Ia)_{\overline{n}}$	n
		41	3.359 90	0.297 63	78.663 3	23.412 4	397.067 5	41
		42	3.460 70	0.288 96	82.023 2	23.701 4	409.203 8	42
		43	3.564 52	0.280 54	85.483 9	23.981 9	421.267 1	43
		44	3.671 45	0.272 37	89.048 4	24.254 3	433.251 5	44
		45	3.781 60	0.264 44	92.719 9	24.518 7	445.151 2	45
		46	3.895 04	0.256 74	96.501 5	24.775 4	456.961 1	46
		47	4.011 90	0.249 26	100.396 5	25.024 7	468.676 2	47
		48	4.132 25	0.242 00	104.408 4	25.266 7	480.292 2	48
		49	4.256 22	0.234 95	108.540 6	25.501 7	491.804 7	49
		50	4.383 91	0.228 11	112.796 9	25.729 8	503.210 1	50

Compound Interest Tables								
Constants				**4 Per Cent**				
Function	Value	n	$(1 + i)^n$	v^n	$s_{\overline{n}}$	$a_{\overline{n}}$	$(Ia)_{\overline{n}}$	n
i	0.040 000	1	1.040 00	0.961 54	1.000 0	0.961 5	0.961 5	1
$i^{(2)}$	0.039 608	2	1.081 60	0.924 56	2.040 0	1.886 1	2.810 7	2
$i^{(4)}$	0.039 414	3	1.124 86	0.889 00	3.121 6	2.775 1	5.477 6	3
$i^{(12)}$	0.039 285	4	1.169 86	0.854 80	4.246 5	3.629 9	8.896 9	4
δ	0.039 221	5	1.216 65	0.821 93	5.416 3	4.451 8	13.006 5	5
$(1 + i)$	1.040 000	6	1.265 32	0.790 31	6.633 0	5.242 1	17.748 4	6
$(1 + i)^{1/2}$	1.019 804	7	1.315 93	0.759 92	7.898 3	6.002 1	23.067 8	7
$(1 + i)^{1/4}$	1.009 853	8	1.368 57	0.730 69	9.214 2	6.732 7	28.913 3	8
$(1 + i)^{1/12}$	1.003 274	9	1.423 31	0.702 59	10.582 8	7.435 3	35.236 6	9
		10	1.480 24	0.675 56	12.006 1	8.110 9	41.992 2	10
v	0.961 538	11	1.539 45	0.649 58	13.486 4	8.760 5	49.137 6	11
$v^{1/2}$	0.980 581	12	1.601 03	0.624 60	15.025 8	9.385 1	56.632 8	12
$v^{1/4}$	0.990 243	13	1.665 07	0.600 57	16.626 8	9.985 6	64.440 3	13
$v^{1/12}$	0.996 737	14	1.731 68	0.577 48	18.291 9	10.563 1	72.524 9	14
		15	1.800 94	0.555 26	20.023 6	11.118 4	80.853 9	15
d	0.038 462	16	1.872 98	0.533 91	21.824 5	11.652 3	89.396 4	16
$d^{(2)}$	0.038 839	17	1.947 90	0.513 37	23.697 5	12.165 7	98.123 8	17
$d^{(4)}$	0.039 029	18	2.025 82	0.493 63	25.645 4	12.659 3	107.009 1	18

(Continued)

Compound Interest Tables								
Constants					**4 Per Cent**			
Function	**Value**	**n**	**$(1 + i)^n$**	**v^n**	**$s_{\overline{n}|}$**	**$a_{\overline{n}|}$**	**$(Ia)_{\overline{n}|}$**	**n**
$d^{(12)}$	0.039 157	19	2.106 85	0.474 64	27.671 2	13.133 9	116.027 3	19
		20	2.191 12	0.456 39	29.778 1	13.590 3	125.155 0	20
$i/i^{(2)}$	1.009 902	21	2.278 77	0.438 83	31.969 2	14.029 2	134.370 5	21
$i/i^{(4)}$	1.014 877	22	2.369 92	0.421 96	34.248 0	14.451 1	143.653 5	22
$i/i^{(12)}$	1.018 204	23	2.464 72	0.405 73	36.617 9	14.856 8	152.985 2	23
		24	2.563 30	0.390 12	39.082 6	15.247 0	162.348 2	24
		25	2.665 84	0.375 12	41.645 9	15.622 1	171.726 1	25
i/δ	1.019 869	26	2.772 47	0.360 69	44.311 7	15.982 8	181.104 0	26
		27	2.883 37	0.346 82	47.084 2	16.329 6	190.468 0	27
		28	2.998 70	0.333 48	49.967 6	16.663 1	199.805 4	28
		29	3.118 65	0.320 65	52.966 3	16.983 7	209.104 3	29
		30	3.243 40	0.308 32	56.084 9	17.292 0	218.353 9	30
$i/d^{(2)}$	1.029 902	31	3.373 13	0.296 46	59.328 3	17.588 5	227.544 1	31
$i/d^{(4)}$	1.024 877	32	3.508 06	0.285 06	62.701 5	17.873 6	236.666 0	32
$i/d^{(12)}$	1.021 537	33	3.648 38	0.274 09	66.209 5	18.147 6	245.711 1	33
		34	3.794 32	0.263 55	69.857 9	18.411 2	254.671 9	34
		35	3.946 09	0.253 42	73.652 2	18.664 6	263.541 4	35
		36	4.103 93	0.243 67	77.598 3	18.908 3	272.313 5	36
		37	4.268 09	0.234 30	81.702 2	19.142 6	280.982 5	37
		38	4.438 81	0.225 29	85.970 3	19.367 9	289.543 3	38
		39	4.616 37	0.216 62	90.409 1	19.584 5	297.991 5	39
		40	4.801 02	0.208 29	95.025 5	19.792 8	306.323 1	40
		41	4.993 06	0.200 28	99.826 5	19.993 1	314.534 5	41
		42	5.192 78	0.192 57	104.819 6	20.185 6	322.622 6	42
		43	5.400 50	0.185 17	110.012 4	20.370 8	330.584 9	43
		44	5.616 52	0.178 05	115.412 9	20.548 8	338.418 9	44
		45	5.841 18	0.171 20	121.029 4	20.720 0	346.122 8	45
		46	6.074 82	0.164 61	126.870 6	20.884 7	353.695 1	46
		47	6.317 82	0.158 28	132.945 4	21.042 9	361.134 3	47
		48	6.570 53	0.152 19	139.263 2	21.195 1	368.439 7	48
		49	6.833 35	0.146 34	145.833 7	21.341 5	375.610 4	49
		50	7.106 68	0.140 71	152.667 1	21.482 2	382.646 0	50

Compound Interest Tables								
Constants			**5 Per Cent**					
Function	Value	n	$(1+i)^n$	v^n	$s_{\overline{n}}$	$a_{\overline{n}}$	$(Ia)_{\overline{n}}$	n
i	0.050 000	1	1.050 00	0.952 38	1.000 0	0.952 4	0.952 4	1
$i^{(2)}$	0.049 390	2	1.102 50	0.907 03	2.050 0	1.859 4	2.766 4	2
$i^{(4)}$	0.049 089	3	1.157 62	0.863 84	3.152 5	2.723 2	5.358 0	3
$i^{(12)}$	0.048 889	4	1.215 51	0.822 70	4.310 1	3.546 0	8.648 8	4
δ	0.048 790	5	1.276 28	0.783 53	5.525 6	4.329 5	12.566 4	5
$(1+i)$	1.050 000	6	1.340 10	0.746 22	6.801 9	5.075 7	17.043 7	6
$(1+i)^{1/2}$	1.024 695	7	1.407 10	0.710 68	8.142 0	5.786 4	22.018 5	7
$(1+i)^{1/4}$	1.012 272	8	1.477 46	0.676 84	9.549 1	6.463 2	27.433 2	8
$(1+i)^{1/12}$	1.004 074	9	1.551 33	0.644 61	11.026 6	7.107 8	33.234 7	9
		10	1.628 89	0.613 91	12.577 9	7.721 7	39.373 8	10
v	0.952 381	11	1.710 34	0.584 68	14.206 8	8.306 4	45.805 3	11
$v^{1/2}$	0.975 900	12	1.795 86	0.556 84	15.917 1	8.863 3	52.487 3	12
$v^{1/4}$	0.987 877	13	1.885 65	0.530 32	17.713 0	9.393 6	59.381 5	13
$v^{1/12}$	0.995 942	14	1.979 93	0.505 07	19.598 6	9.898 6	66.452 4	14
		15	2.078 93	0.481 02	21.578 6	10.379 7	73.667 7	15
d	0.047 619	16	2.182 87	0.458 11	23.657 5	10.837 8	80.997 5	16
$d^{(2)}$	0.048 200	17	2.292 02	0.436 30	25.840 4	11.274 1	88.414 5	17
$d^{(4)}$	0.048 494	18	2.406 62	0.415 52	28.132 4	11.689 6	95.893 9	18
$d^{(12)}$	0.048 691	19	2.526 95	0.395 73	30.539 0	12.085 3	103.412 8	19
		20	2.653 30	0.376 89	33.066 0	12.462 2	110.950 6	20
$i/i^{(2)}$	1.012 348	21	2.785 96	0.358 94	35.719 3	12.821 2	118.488 4	21
$i/i^{(4)}$	1.018 559	22	2.925 26	0.341 85	38.505 2	13.163 0	126.009 1	22
$i/i^{(12)}$	1.022 715	23	3.071 52	0.325 57	41.430 5	13.488 6	133.497 3	23
		24	3.225 10	0.310 07	44.502 0	13.798 6	140.938 9	24
		25	3.386 35	0.295 30	47.727 1	14.093 9	148.321 5	25
i/δ	1.024 797	26	3.555 67	0.281 24	51.113 5	14.375 2	155.633 7	26
		27	3.733 46	0.267 85	54.669 1	14.643 0	162.865 6	27
		28	3.920 13	0.255 09	58.402 6	14.898 1	170.008 2	28
		29	4.116 14	0.242 95	62.322 7	15.141 1	177.053 7	29
		30	4.321 94	0.231 38	66.438 8	15.372 5	183.995 0	30
$i/d^{(2)}$	1.037 348	31	4.538 04	0.220 36	70.760 8	15.592 8	190.826 1	31
$i/d^{(4)}$	1.031 059	32	4.764 94	0.209 87	75.298 8	15.802 7	197.541 9	32
$i/d^{(12)}$	1.026 881	33	5.003 19	0.199 87	80.063 8	16.002 5	204.137 7	33
		34	5.253 35	0.190 35	85.067 0	16.192 9	210.609 7	34
		35	5.516 02	0.181 29	90.320 3	16.374 2	216.954 9	35

(Continued)

Compound Interest Tables

Constants							5 Per Cent			
Function	Value	n	$(1 + i)^n$	v^n	$s_{\overline{n}	}$	$a_{\overline{n}	}$	$(Ia)_{\overline{n}	}$	n
		36	5.791 82	0.172 66	95.836 3	16.546 9	223.170 5	36
		37	6.081 41	0.164 44	101.628 1	16.711 3	229.254 7	37
		38	6.385 48	0.156 61	107.709 5	16.867 9	235.205 7	38
		39	6.704 75	0.149 15	114.095 0	17.017 0	241.022 4	39
		40	7.039 99	0.142 05	120.799 8	17.159 1	246.704 3	40
		41	7.391 99	0.135 28	127.839 8	17.294 4	252.250 8	41
		42	7.761 59	0.128 84	135.231 8	17.423 2	257.662 1	42
		43	8.149 67	0.122 70	142.993 3	17.545 9	262.938 4	43
		44	8.557 15	0.116 86	151.143 0	17.662 8	268.080 3	44
		45	8.985 01	0.111 30	159.700 2	17.774 1	273.088 6	45
		46	9.434 26	0.106 00	168.685 2	17.880 1	277.964 5	46
		47	9.905 97	0.100 95	178.119 4	17.981 0	282.709 1	47
		48	10.401 27	0.096 14	188.025 4	18.077 2	287.323 9	48
		49	10.921 33	0.091 56	198.426 7	18.168 7	291.810 5	49
		50	11.467 40	0.087 20	209.348 0	18.255 9	296.170 7	50

Compound Interest Tables

Constants							6 Per Cent			
Function	Value	n	$(1 + i)^n$	v^n	$s_{\overline{n}	}$	$a_{\overline{n}	}$	$(Ia)_{\overline{n}	}$	n
i	0.060 000	1	1.060 00	0.943 40	1.000 0	0.943 4	0.943 4	1
$i^{(2)}$	0.059 126	2	1.123 60	0.890 00	2.060 0	1.833 4	2.723 4	2
$i^{(4)}$	0.058 695	3	1.191 02	0.839 62	3.183 6	2.673 0	5.242 2	3
$i^{(12)}$	0.058 411	4	1.262 40	0.792 09	4.374 6	3.465 1	8.410 6	4
δ	0.058 269	5	1.338 23	0.747 26	5.637 1	4.212 4	12.146 9	5
$(1 + i)$	1.060 000	6	1.418 52	0.704 96	6.975 3	4.917 3	16.376 7	6
$(1 + i)^{1/2}$	1.029 563	7	1.503 63	0.665 06	8.393 8	5.582 4	21.032 1	7
$(1 + i)^{1/4}$	1.014 674	8	1.593 85	0.627 41	9.897 5	6.209 8	26.051 4	8
$(1 + i)^{1/12}$	1.004 868	9	1.689 48	0.591 90	11.491 3	6.801 7	31.378 5	9
		10	1.790 85	0.558 39	13.180 8	7.360 1	36.962 4	10

(Continued)

(Continued)

Compound Interest Tables

Constants							6 Per Cent		
Function	Value	n	$(1+i)^n$	v^n	$s_{\overline{n}\rceil}$	$a_{\overline{n}\rceil}$	$(Ia)_{\overline{n}\rceil}$	n
v	0.943 396	11	1.898 30	0.526 79	14.971 6	7.886 9	42.757 1	11
$v^{1/2}$	0.971 286	12	2.012 20	0.496 97	16.869 9	8.383 8	48.720 7	12
$v^{1/4}$	0.985 538	13	2.132 93	0.468 84	18.882 1	8.852 7	54.815 6	13
$v^{1/12}$	0.995 156	14	2.260 90	0.442 30	21.015 1	9.295 0	61.007 8	14
		15	2.396 56	0.417 27	23.276 0	9.712 2	67.266 8	15
d	0.056 604	16	2.540 35	0.393 65	25.672 5	10.105 9	73.565 1	16
$d^{(2)}$	0.057 428	17	2.692 77	0.371 36	28.212 9	10.477 3	79.878 3	17
$d^{(4)}$	0.057 847	18	2.854 34	0.350 34	30.905 7	10.827 6	86.184 5	18
$d^{(12)}$	0.058 128	19	3.025 60	0.330 51	33.760 0	11.158 1	92.464 3	19
		20	3.207 14	0.311 80	36.785 6	11.469 9	98.700 4	20
$i/i^{(2)}$	1.014 782	21	3.399 56	0.294 16	39.992 7	11.764 1	104.877 6	21
$i/i^{(4)}$	1.022 227	22	3.603 54	0.277 51	43.392 3	12.041 6	110.982 7	22
$i/i^{(12)}$	1.027 211	23	3.819 75	0.261 80	46.995 8	12.303 4	117.004 1	23
		24	4.048 93	0.246 98	50.815 6	12.550 4	122.931 6	24
		25	4.291 87	0.233 00	54.864 5	12.783 4	128.756 5	25
i/δ	1.029 709	26	4.549 38	0.219 81	59.156 4	13.003 2	134.471 6	26
		27	4.822 35	0.207 37	63.705 8	13.210 5	140.070 5	27
		28	5.111 69	0.195 63	68.528 1	13.406 2	145.548 2	28
		29	5.418 39	0.184 56	73.639 8	13.590 7	150.900 3	29
		30	5.743 49	0.174 11	79.058 2	13.764 8	156.123 6	30
$i/d^{(2)}$	1.044 782	31	6.088 10	0.164 25	84.801 7	13.929 1	161.215 5	31
$i/d^{(4)}$	1.037 227	32	6.453 39	0.154 96	90.889 8	14.084 0	166.174 2	32
$i/d^{(12)}$	1.032 211	33	6.840 59	0.146 19	97.343 2	14.230 2	170.998 3	33
		34	7.251 03	0.137 91	104.183 8	14.368 1	175.687 3	34
		35	7.686 09	0.130 11	111.434 8	14.498 2	180.241 0	35
		36	8.147 25	0.122 74	119.120 9	14.621 0	184.659 6	36
		37	8.636 09	0.115 79	127.268 1	14.736 8	188.944 0	37
		38	9.154 25	0.109 24	135.904 2	14.846 0	193.095 1	38
		39	9.703 51	0.103 06	145.058 5	14.949 1	197.114 2	39
		40	10.285 72	0.097 22	154.762 0	15.046 3	201.003 1	40
		41	10.902 86	0.091 72	165.047 7	15.138 0	204.763 6	41
		42	11.557 03	0.086 53	175.950 5	15.224 5	208.397 8	42

(Continued)

Compound Interest Tables								
Constants		\multicolumn{7}{c}{**6 Per Cent**}							
Function	**Value**	n	$(1+i)^n$	v^n	$s_{\overline{n}|}$	$a_{\overline{n}|}$	$(Ia)_{\overline{n}|}$	n
		43	12.250 45	0.081 63	187.507 6	15.306 2	211.907 8	43
		44	12.985 48	0.077 01	199.758 0	15.383 2	215.296 2	44
		45	13.764 61	0.072 65	212.743 5	15.455 8	218.565 5	45
		46	14.590 49	0.068 54	226.508 1	15.524 4	221.718 2	46
		47	15.465 92	0.064 66	241.098 6	15.589 0	224.757 2	47
		48	16.393 87	0.061 00	256.564 5	15.650 0	227.685 1	48
		49	17.377 50	0.057 55	272.958 4	15.707 6	230.504 8	49
		50	18.420 15	0.054 29	290.335 9	15.761 9	233.219 2	50

Compound Interest Tables								
Constants		\multicolumn{7}{c}{**7 Per Cent**}							
Function	**Value**	n	$(1+i)^n$	v^n	$s_{\overline{n}|}$	$a_{\overline{n}|}$	$(Ia)_{\overline{n}|}$	n
i	0.070 000	1	1.070 00	0.934 58	1.000 0	0.934 6	0.934 6	1
$i^{(2)}$	0.068 816	2	1.144 90	0.873 44	2.070 0	1.808 0	2.681 5	2
$i^{(4)}$	0.068 234	3	1.225 04	0.816 30	3.214 9	2.624 3	5.130 4	3
$i^{(12)}$	0.067 850	4	1.310 80	0.762 90	4.439 9	3.387 2	8.181 9	4
δ	0.067 659	5	1.402 55	0.712 99	5.750 7	4.100 2	11.746 9	5
$(1+i)$	1.070 000	6	1.500 73	0.666 34	7.153 3	4.766 5	15.744 9	6
$(1+i)^{1/2}$	1.034 408	7	1.605 78	0.622 75	8.654 0	5.389 3	20.104 2	7
$(1+i)^{1/4}$	1.017 059	8	1.718 19	0.582 01	10.259 8	5.971 3	24.760 2	8
$(1+i)^{1/12}$	1.005 654	9	1.838 46	0.543 93	11.978 0	6.515 2	29.655 6	9
		10	1.967 15	0.508 35	13.816 4	7.023 6	34.739 1	10
v	0.934 579	11	2.104 85	0.475 09	15.783 6	7.498 7	39.965 2	11
$v^{1/2}$	0.966 736	12	2.252 19	0.444 01	17.888 5	7.942 7	45.293 3	12
$v^{1/4}$	0.983 228	13	2.409 85	0.414 96	20.140 6	8.357 7	50.687 8	13
$v^{1/12}$	0.994 378	14	2.578 53	0.387 82	22.550 5	8.745 5	56.117 3	14
		15	2.759 03	0.362 45	25.129 0	9.107 9	61.554 0	15
d	0.065 421	16	2.952 16	0.338 73	27.888 1	9.446 6	66.973 7	16
$d^{(2)}$	0.066 527	17	3.158 82	0.316 57	30.840 2	9.763 2	72.355 5	17
$d^{(4)}$	0.067 090	18	3.379 93	0.295 86	33.999 0	10.059 1	77.681 0	18
$d^{(12)}$	0.067 468	19	3.616 53	0.276 51	37.379 0	10.335 6	82.934 7	19

(Continued)

(Continued)

Compound Interest Tables								
Constants					7 Per Cent			
Function	**Value**	n	$(1+i)^n$	v^n	$s_{\overline{n}\|}$	$a_{\overline{n}\|}$	$(Ia)_{\overline{n}\|}$	n
		20	3.869 68	0.258 42	40.995 5	10.594 0	88.103 1	20
$i/i^{(2)}$	1.017 204	21	4.140 56	0.241 51	44.865 2	10.835 5	93.174 8	21
$i/i^{(4)}$	1.025 880	22	4.430 40	0.225 71	49.005 7	11.061 2	98.140 5	22
$i/i^{(12)}$	1.031 691	23	4.740 53	0.210 95	53.436 1	11.272 2	102.992 3	23
		24	5.072 37	0.197 15	58.176 7	11.469 3	107.723 8	24
		25	5.427 43	0.184 25	63.249 0	11.653 6	112.330 1	25
i/δ	1.034 605	26	5.807 35	0.172 20	68.676 5	11.825 8	116.807 1	26
		27	6.213 87	0.160 93	74.483 8	11.986 7	121.152 3	27
		28	6.648 84	0.150 40	80.697 7	12.137 1	125.363 5	28
		29	7.114 26	0.140 56	87.346 5	12.277 7	129.439 9	29
		30	7.612 26	0.131 37	94.460 8	12.409 0	133.380 9	30
$i/d^{(2)}$	1.052 204	31	8.145 11	0.122 77	102.073 0	12.531 8	137.186 8	31
$i/d^{(4)}$	1.043 380	32	8.715 27	0.114 74	110.218 2	12.646 6	140.858 5	32
$i/d^{(12)}$	1.037 525	33	9.325 34	0.107 23	118.933 4	12.753 8	144.397 3	33
		34	9.978 11	0.100 22	128.258 8	12.854 0	147.804 7	34
		35	10.676 58	0.093 66	138.236 9	12.947 7	151.082 9	35
		36	11.423 94	0.087 54	148.913 5	13.035 2	154.234 2	36
		37	12.223 62	0.081 81	160.337 4	13.117 0	157.261 2	37
		38	13.079 27	0.076 46	172.561 0	13.193 5	160.166 5	38
		39	13.994 82	0.071 46	185.640 3	13.264 9	162.953 3	39
		40	14.974 46	0.066 78	199.635 1	13.331 7	165.624 5	40
		41	16.022 67	0.062 41	214.609 6	13.394 1	168.183 3	41
		42	17.144 26	0.058 33	230.632 2	13.452 4	170.633 1	42
		43	18.344 35	0.054 51	247.776 5	13.507 0	172.977 2	43
		44	19.628 46	0.050 95	266.120 9	13.557 9	175.218 8	44
		45	21.002 45	0.047 61	285.749 3	13.605 5	177.361 4	45
		46	22.472 62	0.044 50	306.751 8	13.650 0	179.408 4	46
		47	24.045 71	0.041 59	329.224 4	13.691 6	181.363 0	47
		48	25.728 91	0.038 87	353.270 1	13.730 5	183.228 6	48
		49	27.529 93	0.036 32	378.999 0	13.766 8	185.008 5	49
		50	29.457 03	0.033 95	406.528 9	13.800 7	186.705 9	50

Compound Interest Tables								
Constants				8 Per Cent				
Function	Value	n	$(1+i)^n$	v^n	$s_{\overline{n}\|}$	$a_{\overline{n}\|}$	$(Ia)_{\overline{n}\|}$	n
i	0.080 000	1	1.080 00	0.925 93	1.000 0	0.925 9	0.925 9	1
$i^{(2)}$	0.078 461	2	1.166 40	0.857 34	2.080 0	1.783 3	2.640 6	2
$i^{(4)}$	0.077 706	3	1.259 71	0.793 83	3.246 4	2.577 1	5.022 1	3
$i^{(12)}$	0.077 208	4	1.360 49	0.735 03	4.506 1	3.312 1	7.962 2	4
δ	0.076 961	5	1.469 33	0.680 58	5.866 6	3.992 7	11.365 1	5
$(1+i)$	1.080 000	6	1.586 87	0.630 17	7.335 9	4.622 9	15.146 2	6
$(1+i)^{1/2}$	1.039 230	7	1.713 82	0.583 49	8.922 8	5.206 4	19.230 6	7
$(1+i)^{1/4}$	1.019 427	8	1.850 93	0.540 27	10.636 6	5.746 6	23.552 7	8
$(1+i)^{1/12}$	1.006 434	9	1.999 00	0.500 25	12.487 6	6.246 9	28.055 0	9
		10	2.158 92	0.463 19	14.486 6	6.710 1	32.686 9	10
v	0.925 926	11	2.331 64	0.428 88	16.645 5	7.139 0	37.404 6	11
$v^{1/2}$	0.962 250	12	2.518 17	0.397 11	18.977 1	7.536 1	42.170 0	12
$v^{1/4}$	0.980 944	13	2.719 62	0.367 70	21.495 3	7.903 8	46.950 1	13
$v^{1/12}$	0.993 607	14	2.937 19	0.340 46	24.214 9	8.244 2	51.716 5	14
		15	3.172 17	0.315 24	27.152 1	8.559 5	56.445 1	15
d	0.074 074	16	3.425 94	0.291 89	30.324 3	8.851 4	61.115 4	16
$d^{(2)}$	0.075 499	17	3.700 02	0.270 27	33.750 2	9.121 6	65.710 0	17
$d^{(4)}$	0.076 225	18	3.996 02	0.250 25	37.450 2	9.371 9	70.214 4	18
$d^{(12)}$	0.076 715	19	4.315 70	0.231 71	41.446 3	9.603 6	74.617 0	19
		20	4.660 96	0.214 55	45.762 0	9.818 1	78.907 9	20
$i/i^{(2)}$	1.019 615	21	5.033 83	0.198 66	50.422 9	10.016 8	83.079 7	21
$i/i^{(4)}$	1.029 519	22	5.436 54	0.183 94	55.456 8	10.200 7	87.126 4	22
$i/i^{(12)}$	1.036 157	23	5.871 46	0.170 32	60.893 3	10.371 1	91.043 7	23
		24	6.341 18	0.157 70	66.764 8	10.528 8	94.828 4	24
		25	6.848 48	0.146 02	73.105 9	10.674 8	98.478 9	25
i/δ	1.039 487	26	7.396 35	0.135 20	79.954 4	10.810 0	101.994 1	26
		27	7.988 06	0.125 19	87.350 8	10.935 2	105.374 2	27
		28	8.627 11	0.115 91	95.338 8	11.051 1	108.619 8	28
		29	9.317 27	0.107 33	103.965 9	11.158 4	111.732 3	29
		30	10.062 66	0.099 38	113.283 2	11.257 8	114.713 6	30
$i/d^{(2)}$	1.059 615	31	10.867 67	0.092 02	123.345 9	11.349 8	117.566 1	31
$i/d^{(4)}$	1.049 519	32	11.737 08	0.085 20	134.213 5	11.435 0	120.292 5	32
$i/d^{(12)}$	1.042 824	33	12.676 05	0.078 89	145.950 6	11.513 9	122.895 8	33

(Continued)

(Continued)

Compound Interest Tables

Constants							8 Per Cent			
Function	Value	n	$(1 + i)^n$	v^n	$s_{\overline{n}}$	$a_{\overline{n}}$	$(Ia)_{\overline{n}}$	n
		34	13.690 13	0.073 05	158.626 7	11.586 9	125.379 3	34
		35	14.785 34	0.067 63	172.316 8	11.654 6	127.746 6	35
		36	15.968 17	0.062 62	187.102 1	11.717 2	130.001 0	36
		37	17.245 63	0.057 99	203.070 3	11.775 2	132.146 5	37
		38	18.625 28	0.053 69	220.315 9	11.828 9	134.186 8	38
		39	20.115 30	0.049 71	238.941 2	11.878 6	136.125 6	39
		40	21.724 52	0.046 03	259.056 5	11.924 6	137.966 8	40
		41	23.462 48	0.042 62	280.781 0	11.967 2	139.714 3	41
		42	25.339 48	0.039 46	304.243 5	12.006 7	141.371 8	42
		43	27.366 64	0.036 54	329.583 0	12.043 2	142.943 0	43
		44	29.555 97	0.033 83	356.949 6	12.077 1	144.431 7	44
		45	31.920 45	0.031 33	386.505 6	12.108 4	145.841 5	45
		46	34.474 09	0.029 01	418.426 1	12.137 4	147.175 8	46
		47	37.232 01	0.026 86	452.900 2	12.164 3	148.438 2	47
		48	40.210 57	0.024 87	490.132 2	12.189 1	149.631 9	48
		49	43.427 42	0.023 03	530.342 7	12.212 2	150.760 2	49
		50	46.901 61	0.021 32	573.770 2	12.233 5	151.826 3	50

Compound Interest Tables

Constants							9 Per Cent			
Function	Value	n	$(1 + i)^n$	v^n	$s_{\overline{n}}$	$a_{\overline{n}}$	$(Ia)_{\overline{n}}$	n
i	0.090 000	1	1.090 00	0.917 43	1.000 0	0.917 4	0.917 4	1
$i^{(2)}$	0.088 061	2	1.188 10	0.841 68	2.090 0	1.759 1	2.600 8	2
$i^{(4)}$	0.087 113	3	1.295 03	0.772 18	3.278 1	2.531 3	4.917 3	3
$i^{(12)}$	0.086 488	4	1.411 58	0.708 43	4.573 1	3.239 7	7.751 0	4
δ	0.086 178	5	1.538 62	0.649 93	5.984 7	3.889 7	11.000 7	5
$(1 + i)$	1.090 000	6	1.677 10	0.596 27	7.523 3	4.485 9	14.578 3	6
$(1 + i)^{1/2}$	1.044 031	7	1.828 04	0.547 03	9.200 4	5.033 0	18.407 5	7
$(1 + i)^{1/4}$	1.021 778	8	1.992 56	0.501 87	11.028 5	5.534 8	22.422 5	8
$(1 + i)^{1/12}$	1.007 207	9	2.171 89	0.460 43	13.021 0	5.995 2	26.566 3	9
		10	2.367 36	0.422 41	15.192 9	6.417 7	30.790 4	10

(Continued)

Compound Interest Tables								
Constants					9 Per Cent			
Function	**Value**	**n**	**$(1 + i)^n$**	**v^n**	**$s_{\overline{n}|}$**	**$a_{\overline{n}|}$**	**$(Ia)_{\overline{n}|}$**	**n**
v	0.917 431	11	2.580 43	0.387 53	17.560 3	6.805 2	35.053 3	11
$v^{1/2}$	0.957 826	12	2.812 66	0.355 53	20.140 7	7.160 7	39.319 7	12
$v^{1/4}$	0.978 686	13	3.065 80	0.326 18	22.953 4	7.486 9	43.560 0	13
$v^{1/12}$	0.992 844	14	3.341 73	0.299 25	26.019 2	7.786 2	47.749 5	14
		15	3.642 48	0.274 54	29.360 9	8.060 7	51.867 6	15
d	0.082 569	16	3.970 31	0.251 87	33.003 4	8.312 6	55.897 5	16
$d^{(2)}$	0.084 347	17	4.327 63	0.231 07	36.973 7	8.543 6	59.825 7	17
$d^{(4)}$	0.085 256	18	4.717 12	0.211 99	41.301 3	8.755 6	63.641 6	18
$d^{(12)}$	0.085 869	19	5.141 66	0.194 49	46.018 5	8.950 1	67.336 9	19
		20	5.604 41	0.178 43	51.160 1	9.128 5	70.905 5	20
$i/i^{(2)}$	1.022 015	21	6.108 81	0.163 70	56.764 5	9.292 2	74.343 2	21
$i/i^{(4)}$	1.033 144	22	6.658 60	0.150 18	62.873 3	9.442 4	77.647 2	22
$i/i^{(12)}$	1.040 608	23	7.257 87	0.137 78	69.531 9	9.580 2	80.816 2	23
		24	7.911 08	0.126 40	76.789 8	9.706 6	83.849 9	24
		25	8.623 08	0.115 97	84.700 9	9.822 6	86.749 1	25
i/δ	1.044 354	26	9.399 16	0.106 39	93.324 0	9.929 0	89.515 3	26
		27	10.245 08	0.097 61	102.723 1	10.026 6	92.150 7	27
		28	11.167 14	0.089 55	112.968 2	10.116 1	94.658 0	28
		29	12.172 18	0.082 15	124.135 4	10.198 3	97.040 5	29
		30	13.267 68	0.075 37	136.307 5	10.273 7	99.301 7	30
$i/d^{(2)}$	1.067 015	31	14.461 77	0.069 15	149.575 2	10.342 8	101.445 2	31
$i/d^{(4)}$	1.055 644	32	15.763 33	0.063 44	164.037 0	10.406 2	103.475 3	32
$i/d^{(12)}$	1.048 108	33	17.182 03	0.058 20	179.800 3	10.464 4	105.395 9	33
		34	18.728 41	0.053 39	196.982 3	10.517 8	107.211 3	34
		35	20.413 97	0.048 99	215.710 8	10.566 8	108.925 8	35
		36	22.251 23	0.044 94	236.124 7	10.611 8	110.543 7	36
		37	24.253 84	0.041 23	258.375 9	10.653 0	112.069 2	37
		38	26.436 68	0.037 83	282.629 8	10.690 8	113.506 6	38
		39	28.815 98	0.034 70	309.066 5	10.725 5	114.860 0	39
		40	31.409 42	0.031 84	337.882 4	10.757 4	116.133 5	40
		41	34.236 27	0.029 21	369.291 9	10.786 6	117.331 1	41
		42	37.317 53	0.026 80	403.528 1	10.813 4	118.456 6	42

(Continued)

(Continued)

Compound Interest Tables								
Constants						**9 Per Cent**		
Function	Value	n	$(1 + i)^n$	v^n	$s_{\overline{n}|}$	$a_{\overline{n}|}$	$(Ia)_{\overline{n}|}$	n
		43	40.676 11	0.024 58	440.845 7	10.838 0	119.513 7	43
		44	44.336 96	0.022 55	481.521 8	10.860 5	120.506 1	44
		45	48.327 29	0.020 69	525.858 7	10.881 2	121.437 3	45
		46	52.676 74	0.018 98	574.186 0	10.900 2	122.310 5	46
		47	57.417 65	0.017 42	626.862 8	10.917 6	123.129 1	47
		48	62.585 24	0.015 98	684.280 4	10.933 6	123.896 0	48
		49	68.217 91	0.014 66	746.865 6	10.948 2	124.614 3	49
		50	74.357 52	0.013 45	815.083 6	10.961 7	125.286 7	50

Compound Interest Tables								
Constants						**10 Per Cent**		
Function	Value	n	$(1 + i)^n$	v^n	$s_{\overline{n}|}$	$a_{\overline{n}|}$	$(Ia)_{\overline{n}|}$	n
i	0.100 000	1	1.100 00	0.909 09	1.000 0	0.909 1	0.909 1	1
$i^{(2)}$	0.097 618	2	1.210 00	0.826 45	2.100 0	1.735 5	2.562 0	2
$i^{(4)}$	0.096 455	3	1.331 00	0.751 31	3.310 0	2.486 9	4.815 9	3
$i^{(12)}$	0.095 690	4	1.464 10	0.683 01	4.641 0	3.169 9	7.548 0	4
δ	0.095 310	5	1.610 51	0.620 92	6.105 1	3.790 8	10.652 6	5
$(1 + i)$	1.100 000	6	1.771 56	0.564 47	7.715 6	4.355 3	14.039 4	6
$(1 + i)^{1/2}$	1.048 809	7	1.948 72	0.513 16	9.487 2	4.868 4	17.631 5	7
$(1 + i)^{1/4}$	1.024 114	8	2.143 59	0.466 51	11.435 9	5.334 9	21.363 6	8
$(1 + i)^{1/12}$	1.007 974	9	2.357 95	0.424 10	13.579 5	5.759 0	25.180 5	9
		10	2.593 74	0.385 54	15.937 4	6.144 6	29.035 9	10
v	0.909 091	11	2.853 12	0.350 49	18.531 2	6.495 1	32.891 3	11
$v^{1/2}$	0.953 463	12	3.138 43	0.318 63	21.384 3	6.813 7	36.714 9	12
$v^{1/4}$	0.976 454	13	3.452 27	0.289 66	24.522 7	7.103 4	40.480 5	13
$v^{1/12}$	0.992 089	14	3.797 50	0.263 33	27.975 0	7.366 7	44.167 2	14
		15	4.177 25	0.239 39	31.772 5	7.606 1	47.758 1	15
d	0.090 909	16	4.594 97	0.217 63	35.949 7	7.823 7	51.240 1	16
$d^{(2)}$	0.093 075	17	5.054 47	0.197 84	40.544 7	8.021 6	54.603 5	17
$d^{(4)}$	0.094 184	18	5.559 92	0.179 86	45.599 2	8.201 4	57.841 0	18
$d^{(12)}$	0.094 933	19	6.115 91	0.163 51	51.159 1	8.364 9	60.947 6	19
		20	6.727 50	0.148 64	57.275 0	8.513 6	63.920 5	20

(Continued)

| Compound Interest Tables | | | | | | | | |
| Constants | | | 10 Per Cent | | | | | |
Function	Value	n	$(1 + i)^n$	v^n	$s_{\overline{n}}$	$a_{\overline{n}}$	$(Ia)_{\overline{n}}$	n
$i/i^{(2)}$	1.024 404	21	7.400 25	0.135 13	64.002 5	8.648 7	66.758 2	21
$i/i^{(4)}$	1.036 756	22	8.140 27	0.122 85	71.402 7	8.771 5	69.460 8	22
$i/i^{(12)}$	1.045 045	23	8.954 30	0.111 68	79.543 0	8.883 2	72.029 4	23
		24	9.849 73	0.101 53	88.497 3	8.984 7	74.466 0	24
		25	10.834 71	0.092 30	98.347 1	9.077 0	76.773 4	25
i/δ	1.049 206	26	11.918 18	0.083 91	109.181 8	9.160 9	78.955 0	26
		27	13.109 99	0.076 28	121.099 9	9.237 2	81.014 5	27
		28	14.420 99	0.069 34	134.209 9	9.306 6	82.956 1	28
		29	15.863 09	0.063 04	148.630 9	9.369 6	84.784 2	29
		30	17.449 40	0.057 31	164.494 0	9.426 9	86.503 5	30
$i/d^{(2)}$	1.074 404	31	19.194 34	0.052 10	181.943 4	9.479 0	88.118 6	31
$i/d^{(4)}$	1.061 756	32	21.113 78	0.047 36	201.137 8	9.526 4	89.634 2	32
$i/d^{(12)}$	1.053 378	33	23.225 15	0.043 06	222.251 5	9.569 4	91.055 0	33
		34	25.547 67	0.039 14	245.476 7	9.608 6	92.385 9	34
		35	28.102 44	0.035 58	271.024 4	9.644 2	93.631 3	35
		36	30.912 68	0.032 35	299.126 8	9.676 5	94.795 9	36
		37	34.003 95	0.029 41	330.039 5	9.705 9	95.884 0	37
		38	37.404 34	0.026 73	364.043 4	9.732 7	96.899 9	38
		39	41.144 78	0.024 30	401.447 8	9.757 0	97.847 8	39
		40	45.259 26	0.022 09	442.592 6	9.779 1	98.731 6	40
		41	49.785 18	0.020 09	487.851 8	9.799 1	99.555 1	41
		42	54.763 70	0.018 26	537.637 0	9.817 4	100.322 1	42
		43	60.240 07	0.016 60	592.400 7	9.834 0	101.035 9	43
		44	66.264 08	0.015 09	652.640 8	9.849 1	101.699 9	44
		45	72.890 48	0.013 72	718.904 8	9.862 8	102.317 2	45
		46	80.179 53	0.012 47	791.795 3	9.875 3	102.891 0	46
		47	88.197 49	0.011 34	871.974 9	9.886 6	103.423 8	47
		48	97.017 23	0.010 31	960.172 3	9.896 9	103.918 6	48
		49	106.718 96	0.009 37	1057.189 6	9.906 3	104.377 8	49
		50	117.390 85	0.008 52	1163.908 5	9.914 8	104.803 7	50

Compound Interest Tables								
Constants			11 Per Cent					
Function	Value	n	$(1+i)^n$	v^n	$s_{\overline{n}}$	$a_{\overline{n}}$	$(Ia)_{\overline{n}}$	n
i	0.110 000	1	1.110 00	0.900 90	1.000 0	0.900 9	0.900 9	1
$i^{(2)}$	0.107 131	2	1.232 10	0.811 62	2.110 0	1.712 5	2.524 1	2
$i^{(4)}$	0.105 733	3	1.367 63	0.731 19	3.342 1	2.443 7	4.717 7	3
$i^{(12)}$	0.104 815	4	1.518 07	0.658 73	4.709 7	3.102 4	7.352 6	4
δ	0.104 360	5	1.685 06	0.593 45	6.227 8	3.695 9	10.319 9	5
$(1+i)$	1.110 000	6	1.870 41	0.534 64	7.912 9	4.230 5	13.527 7	6
$(1+i)^{1/2}$	1.053 565	7	2.076 16	0.481 66	9.783 3	4.712 2	16.899 4	7
$(1+i)^{1/4}$	1.026 433	8	2.304 54	0.433 93	11.859 4	5.146 1	20.370 8	8
$(1+i)^{1/12}$	1.008 735	9	2.558 04	0.390 92	14.164 0	5.537 0	23.889 1	9
		10	2.839 42	0.352 18	16.722 0	5.889 2	27.410 9	10
v	0.900 901	11	3.151 76	0.317 28	19.561 4	6.206 5	30.901 1	11
$v^{1/2}$	0.949 158	12	3.498 45	0.285 84	22.713 2	6.492 4	34.331 1	12
$v^{1/4}$	0.974 247	13	3.883 28	0.257 51	26.211 6	6.749 9	37.678 8	13
$v^{1/12}$	0.991 341	14	4.310 44	0.231 99	30.094 9	6.981 9	40.926 8	14
		15	4.784 59	0.209 00	34.405 4	7.190 9	44.061 8	15
d	0.099 099	16	5.310 89	0.188 29	39.189 9	7.379 2	47.074 5	16
$d^{(2)}$	0.101 684	17	5.895 09	0.169 63	44.500 8	7.548 8	49.958 2	17
$d^{(4)}$	0.103 010	18	6.543 55	0.152 82	50.395 9	7.701 6	52.709 0	18
$d^{(12)}$	0.103 908	19	7.263 34	0.137 68	56.939 5	7.839 3	55.324 9	19
		20	8.062 31	0.124 03	64.202 8	7.963 3	57.805 6	20
$i/i^{(2)}$	1.026 783	21	8.949 17	0.111 74	72.265 1	8.075 1	60.152 2	21
$i/i^{(4)}$	1.040 353	22	9.933 57	0.100 67	81.214 3	8.175 7	62.366 9	22
$i/i^{(12)}$	1.049 467	23	11.026 27	0.090 69	91.147 9	8.266 4	64.452 8	23
		24	12.239 16	0.081 70	102.174 2	8.348 1	66.413 7	24
		25	13.585 46	0.073 61	114.413 3	8.421 7	68.253 9	25
i/δ	1.054 044	26	15.079 86	0.066 31	127.998 8	8.488 1	69.978 1	26
		27	16.738 65	0.059 74	143.078 6	8.547 8	71.591 1	27
		28	18.579 90	0.053 82	159.817 3	8.601 6	73.098 1	28
		29	20.623 69	0.048 49	178.397 2	8.650 1	74.504 3	29
		30	22.892 30	0.043 68	199.020 9	8.693 8	75.814 8	30
$i/d^{(2)}$	1.081 783	31	25.410 45	0.039 35	221.913 2	8.733 1	77.034 7	31
$i/d^{(4)}$	1.067 853	32	28.205 60	0.035 45	247.323 6	8.768 6	78.169 3	32
$i/d^{(12)}$	1.058 634	33	31.308 21	0.031 94	275.529 2	8.800 5	79.223 3	33
		34	34.752 12	0.028 78	306.837 4	8.829 3	80.201 7	34
		35	38.574 85	0.025 92	341.589 6	8.855 2	81.109 0	35

(Continued)

Compound Interest Tables

Constants								
Function	**Value**				**11 Per Cent**			
		n	$(1 + i)^n$	v^n	$s_{\overline{n}}$	$a_{\overline{n}}$	$(Ia)_{\overline{n}}$	n
		36	42.818 08	0.023 35	380.164 4	8.878 6	81.949 8	36
		37	47.528 07	0.021 04	422.982 5	8.899 6	82.728 2	37
		38	52.756 16	0.018 96	470.510 6	8.918 6	83.448 5	38
		39	58.559 34	0.017 08	523.266 7	8.935 7	84.114 5	39
		40	65.000 87	0.015 38	581.826 1	8.951 1	84.729 9	40
		41	72.150 96	0.013 86	646.826 9	8.964 9	85.298 2	41
		42	80.087 57	0.012 49	718.977 9	8.977 4	85.822 6	42
		43	88.897 20	0.011 25	799.065 5	8.988 6	86.306 3	43
		44	98.675 89	0.010 13	887.962 7	8.998 8	86.752 2	44
		45	109.530 24	0.009 13	986.638 6	9.007 9	87.163 0	45
		46	121.578 57	0.008 23	1096.168 8	9.016 1	87.541 4	46
		47	134.952 21	0.007 41	1217.747 4	9.023 5	87.889 7	47
		48	149.796 95	0.006 68	1352.699 6	9.030 2	88.210 1	48
		49	166.274 62	0.006 01	1502.496 5	9.036 2	88.504 8	49
		50	184.564 83	0.005 42	1668.771 2	9.041 7	88.775 7	50

Compound Interest Tables

Constants								
Function	**Value**				**12 Per Cent**			
		n	$(1 + i)^n$	v^n	$s_{\overline{n}}$	$a_{\overline{n}}$	$(Ia)_{\overline{n}}$	n
i	0.120 000	1	1.120 00	0.892 86	1.000 0	0.892 9	0.892 9	1
$i^{(2)}$	0.116 601	2	1.254 40	0.797 19	2.120 0	1.690 1	2.487 2	2
$i^{(4)}$	0.114 949	3	1.404 93	0.711 78	3.374 4	2.401 8	4.622 6	3
$i^{(12)}$	0.113 866	4	1.573 52	0.635 52	4.779 3	3.037 3	7.164 7	4
δ	0.113 329	5	1.762 34	0.567 43	6.352 8	3.604 8	10.001 8	5
$(1 + i)$	1.120 000	6	1.973 82	0.506 63	8.115 2	4.111 4	13.041 6	6
$(1 + i)^{1/2}$	1.058 301	7	2.210 68	0.452 35	10.089 0	4.563 8	16.208 0	7
$(1 + i)^{1/4}$	1.028 737	8	2.475 96	0.403 88	12.299 7	4.967 6	19.439 1	8
$(1 + i)^{1/12}$	1.009 489	9	2.773 08	0.360 61	14.775 7	5.328 2	22.684 6	9
		10	3.105 85	0.321 97	17.548 7	5.650 2	25.904 3	10
v	0.892 857	11	3.478 55	0.287 48	20.654 6	5.937 7	29.066 5	11
$v^{1/2}$	0.944 911	12	3.895 98	0.256 68	24.133 1	6.194 4	32.146 7	12
$v^{1/4}$	0.972 065	13	4.363 49	0.229 17	28.029 1	6.423 5	35.125 9	13
$v^{1/12}$	0.990 600	14	4.887 11	0.204 62	32.392 6	6.628 2	37.990 6	14
		15	5.473 57	0.182 70	37.279 7	6.810 9	40.731 0	15

(Continued)

(Continued)

Compound Interest Tables

Constants									
Function	**Value**	**n**	**$(1+i)^n$**	**v^n**	**$s_{\overline{n}}$**	**$a_{\overline{n}}$**	**$(Ia)_{\overline{n}}$**	**n**	

12 Per Cent

Function	Value	n	$(1+i)^n$	v^n	$s_{\overline{n}}$	$a_{\overline{n}}$	$(Ia)_{\overline{n}}$	n
d	0.107 143	16	6.130 39	0.163 12	42.753 3	6.974 0	43.341 0	16
$d^{(2)}$	0.110 178	17	6.866 04	0.145 64	48.883 7	7.119 6	45.816 9	17
$d^{(4)}$	0.111 738	18	7.689 97	0.130 04	55.749 7	7.249 7	48.157 6	18
$d^{(12)}$	0.112 795	19	8.612 76	0.116 11	63.439 7	7.365 8	50.363 7	19
		20	9.646 29	0.103 67	72.052 4	7.469 4	52.437 0	20
$i/i^{(2)}$	1.029 150	21	10.803 85	0.092 56	81.698 7	7.562 0	54.380 8	21
$i/i^{(4)}$	1.043 938	22	12.100 31	0.082 64	92.502 6	7.644 6	56.198 9	22
$i/i^{(12)}$	1.053 875	23	13.552 35	0.073 79	104.602 9	7.718 4	57.896 0	23
		24	15.178 63	0.065 88	118.155 2	7.784 3	59.477 2	24
		25	17.000 06	0.058 82	133.333 9	7.843 1	60.947 8	25
i/δ	1.058 867	26	19.040 07	0.052 52	150.333 9	7.895 7	62.313 3	26
		27	21.324 88	0.046 89	169.374 0	7.942 6	63.579 4	27
		28	23.883 87	0.041 87	190.698 9	7.984 4	64.751 8	28
		29	26.749 93	0.037 38	214.582 8	8.021 8	65.835 9	29
		30	29.959 92	0.033 38	241.332 7	8.055 2	66.837 2	30
$i/d^{(2)}$	1.089 150	31	33.555 11	0.029 80	271.292 6	8.085 0	67.761 1	31
$i/d^{(4)}$	1.073 938	32	37.581 73	0.026 61	304.847 7	8.111 6	68.612 6	32
$i/d^{(12)}$	1.063 875	33	42.091 53	0.023 76	342.429 4	8.135 4	69.396 6	33
		34	47.142 52	0.021 21	384.521 0	8.156 6	70.117 8	34
		35	52.799 62	0.018 94	431.663 5	8.175 5	70.780 7	35
		36	59.135 57	0.016 91	484.463 1	8.192 4	71.389 4	36
		37	66.231 84	0.015 10	543.598 7	8.207 5	71.948 1	37
		38	74.179 66	0.013 48	609.830 5	8.221 0	72.460 4	38
		39	83.081 22	0.012 04	684.010 2	8.233 0	72.929 8	39
		40	93.050 97	0.010 75	767.091 4	8.243 8	73.359 6	40
		41	104.217 09	0.009 60	860.142 4	8.253 4	73.753 1	41
		42	116.723 14	0.008 57	964.359 5	8.261 9	74.112 9	42
		43	130.729 91	0.007 65	1081.082 6	8.269 6	74.441 8	43
		44	146.417 50	0.006 83	1211.812 5	8.276 4	74.742 3	44
		45	163.987 60	0.006 10	1358.230 0	8.282 5	75.016 7	45
		46	183.666 12	0.005 44	1522.217 6	8.288 0	75.267 2	46
		47	205.706 05	0.004 86	1705.883 8	8.292 8	75.495 7	47
		48	230.390 78	0.004 34	1911.589 8	8.297 2	75.704 0	48
		49	258.037 67	0.003 88	2141.980 6	8.301 0	75.893 9	49
		50	289.002 19	0.003 46	2400.018 2	8.304 5	76.066 9	50

Compound Interest Tables

Constants			13 Per Cent					
Function	Value	n	$(1+i)^n$	v^n	$s_{\overline{n}\rvert}$	$a_{\overline{n}\rvert}$	$(Ia)_{\overline{n}\rvert}$	n
i	0.130 000	1	1.130 00	0.884 96	1.000 0	0.885 0	0.885 0	1
$i^{(2)}$	0.126 029	2	1.276 90	0.783 15	2.130 0	1.668 1	2.451 2	2
$i^{(4)}$	0.124 104	3	1.442 90	0.693 05	3.406 9	2.361 2	4.530 4	3
$i^{(12)}$	0.122 842	4	1.630 47	0.613 32	4.849 8	2.974 5	6.983 7	4
δ	0.122 218	5	1.842 44	0.542 76	6.480 3	3.517 2	9.697 5	5
$(1+i)$	1.130 000	6	2.081 95	0.480 32	8.322 7	3.997 5	12.579 4	6
$(1+i)^{1/2}$	1.063 015	7	2.352 61	0.425 06	10.404 7	4.422 6	15.554 8	7
$(1+i)^{1/4}$	1.031 026	8	2.658 44	0.376 16	12.757 3	4.798 8	18.564 1	8
$(1+i)^{1/12}$	1.010 237	9	3.004 04	0.332 88	15.415 7	5.131 7	21.560 1	9
		10	3.394 57	0.294 59	18.419 7	5.426 2	24.505 9	10
v	0.884 956	11	3.835 86	0.260 70	21.814 3	5.686 9	27.373 6	11
$v^{1/2}$	0.940 721	12	4.334 52	0.230 71	25.650 2	5.917 6	30.142 1	12
$v^{1/4}$	0.969 908	13	4.898 01	0.204 16	29.984 7	6.121 8	32.796 2	13
$v^{1/12}$	0.989 867	14	5.534 75	0.180 68	34.882 7	6.302 5	35.325 7	14
		15	6.254 27	0.159 89	40.417 5	6.462 4	37.724 1	15
d	0.115 044	16	7.067 33	0.141 50	46.671 7	6.603 9	39.988 0	16
$d^{(2)}$	0.118 558	17	7.986 08	0.125 22	53.739 1	6.729 1	42.116 7	17
$d^{(4)}$	0.120 369	18	9.024 27	0.110 81	61.725 1	6.839 9	44.111 3	18
$d^{(12)}$	0.121 597	19	10.197 42	0.098 06	70.749 4	6.938 0	45.974 5	19
		20	11.523 09	0.086 78	80.946 8	7.024 8	47.710 2	20
$i/i^{(2)}$	1.031 507	21	13.021 09	0.076 80	92.469 9	7.101 6	49.322 9	21
$i/i^{(4)}$	1.047 509	22	14.713 83	0.067 96	105.491 0	7.169 5	50.818 1	22
$i/i^{(12)}$	1.058 269	23	16.626 63	0.060 14	120.204 8	7.229 7	52.201 5	23
		24	18.788 09	0.053 23	136.831 5	7.282 9	53.478 9	24
		25	21.230 54	0.047 10	155.619 6	7.330 0	54.656 4	25
i/δ	1.063 676	26	23.990 51	0.041 68	176.850 1	7.371 7	55.740 2	26
		27	27.109 28	0.036 89	200.840 6	7.408 6	56.736 1	27
		28	30.633 49	0.032 64	227.949 9	7.441 2	57.650 2	28
		29	34.615 84	0.028 89	258.583 4	7.470 1	58.487 9	29
		30	39.115 90	0.025 57	293.199 2	7.495 7	59.254 9	30
$i/d^{(2)}$	1.096 507	31	44.200 96	0.022 62	332.315 1	7.518 3	59.956 2	31
$i/d^{(4)}$	1.080 009	32	49.947 09	0.020 02	376.516 1	7.538 3	60.596 9	32
$i/d^{(12)}$	1.069 102	33	56.440 21	0.017 72	426.463 2	7.556 0	61.181 6	33

(Continued)

(Continued)

Compound Interest Tables

Constants — **13 Per Cent**

Function	Value	n	$(1+i)^n$	v^n	$s_{\overline{n}}$	$a_{\overline{n}}$	$(Ia)_{\overline{n}}$	n
		34	63.777 44	0.015 68	482.903 4	7.571 7	61.714 7	34
		35	72.068 51	0.013 88	546.680 8	7.585 6	62.200 4	35
		36	81.437 41	0.012 28	618.749 3	7.597 9	62.642 4	36
		37	92.024 28	0.010 87	700.186 7	7.608 7	63.044 5	37
		38	103.987 43	0.009 62	792.211 0	7.618 3	63.409 9	38
		39	117.505 80	0.008 51	896.198 4	7.626 8	63.741 8	39
		40	132.781 55	0.007 53	1013.704 2	7.634 4	64.043 1	40
		41	150.043 15	0.006 66	1146.485 8	7.641 0	64.316 3	41
		42	169.548 76	0.005 90	1296.528 9	7.646 9	64.564 0	42
		43	191.590 10	0.005 22	1466.077 7	7.652 2	64.788 5	43
		44	216.496 82	0.004 62	1657.667 8	7.656 8	64.991 7	44
		45	244.641 40	0.004 09	1874.164 6	7.660 9	65.175 6	45
		46	276.444 78	0.003 62	2118.806 0	7.664 5	65.342 0	46
		47	312.382 61	0.003 20	2395.250 8	7.667 7	65.492 5	47
		48	352.992 34	0.002 83	2707.633 4	7.670 5	65.628 5	48
		49	398.881 35	0.002 51	3060.625 8	7.673 0	65.751 3	49
		50	450.735 93	0.002 22	3459.507 1	7.675 2	65.862 3	50

Compound Interest Tables

Constants — **14 Per Cent**

Function	Value	n	$(1+i)^n$	v^n	$s_{\overline{n}}$	$a_{\overline{n}}$	$(Ia)_{\overline{n}}$	n
i	0.140 000	1	1.140 00	0.877 19	1.000 0	0.877 2	0.877 2	1
$i^{(2)}$	0.135 416	2	1.299 60	0.769 47	2.140 0	1.646 7	2.416 1	2
$i^{(4)}$	0.133 198	3	1.481 54	0.674 97	3.439 6	2.321 6	4.441 0	3
$i^{(12)}$	0.131 746	4	1.688 96	0.592 08	4.921 1	2.913 7	6.809 4	4
δ	0.131 028	5	1.925 41	0.519 37	6.610 1	3.433 1	9.406 2	5
$(1+i)$	1.140 000	6	2.194 97	0.455 59	8.535 5	3.888 7	12.139 7	6
$(1+i)^{1/2}$	1.067 708	7	2.502 27	0.399 64	10.730 5	4.288 3	14.937 2	7
$(1+i)^{1/4}$	1.033 299	8	2.852 59	0.350 56	13.232 8	4.638 9	17.741 7	8
$(1+i)^{1/12}$	1.010 979	9	3.251 95	0.307 51	16.085 3	4.946 4	20.509 2	9
		10	3.707 22	0.269 74	19.337 3	5.216 1	23.206 7	10

(Continued)

Compound Interest Tables									
Constants			**14 Per Cent**						
Function	Value	n	$(1 + i)^n$	v^n	$s_{\overline{n}}$	$a_{\overline{n}}$	$(Ia)_{\overline{n}}$	n
v	0.877 193	11	4.226 23	0.236 62	23.044 5	5.452 7	25.809 5	11
$v^{1/2}$	0.936 586	12	4.817 90	0.207 56	27.270 7	5.660 3	28.300 2	12
$v^{1/4}$	0.967 774	13	5.492 41	0.182 07	32.088 7	5.842 4	30.667 1	13
$v^{1/12}$	0.989 140	14	6.261 35	0.159 71	37.581 1	6.002 1	32.903 0	14
		15	7.137 94	0.140 10	43.842 4	6.142 2	35.004 5	15
d	0.122 807	16	8.137 25	0.122 89	50.980 4	6.265 1	36.970 7	16
$d^{(2)}$	0.126 828	17	9.276 46	0.107 80	59.117 6	6.372 9	38.803 3	17
$d^{(4)}$	0.128 905	18	10.575 17	0.094 56	68.394 1	6.467 4	40.505 4	18
$d^{(12)}$	0.130 316	19	12.055 69	0.082 95	78.969 2	6.550 4	42.081 4	19
		20	13.743 49	0.072 76	91.024 9	6.623 1	43.536 7	20
$i/i^{(2)}$	1.033 854	21	15.667 58	0.063 83	104.768 4	6.687 0	44.877 0	21
$i/i^{(4)}$	1.051 067	22	17.861 04	0.055 99	120.436 0	6.742 9	46.108 8	22
$i/i^{(12)}$	1.062 649	23	20.361 58	0.049 11	138.297 0	6.792 1	47.238 3	23
		24	23.212 21	0.043 08	158.658 6	6.835 1	48.272 3	24
		25	26.461 92	0.037 79	181.870 8	6.872 9	49.217 0	25
i/δ	1.068 472	26	30.166 58	0.033 15	208.332 7	6.906 1	50.078 9	26
		27	34.389 91	0.029 08	238.499 3	6.935 2	50.864 0	27
		28	39.204 49	0.025 51	272.889 2	6.960 7	51.578 2	28
		29	44.693 12	0.022 37	312.093 7	6.983 0	52.227 1	29
		30	50.950 16	0.019 63	356.786 8	7.002 7	52.815 9	30
$i/d^{(2)}$	1.103 854	31	58.083 18	0.017 22	407.737 0	7.019 9	53.349 6	31
$i/d^{(4)}$	1.086 067	32	66.214 83	0.015 10	465.820 2	7.035 0	53.832 9	32
$i/d^{(12)}$	1.074 316	33	75.484 90	0.013 25	532.035 0	7.048 2	54.270 1	33
		34	86.052 79	0.011 62	607.519 9	7.059 9	54.665 2	34
		35	98.100 18	0.010 19	693.572 7	7.070 0	55.022 0	35
		36	111.834 20	0.008 94	791.672 9	7.079 0	55.343 9	36
		37	127.490 99	0.007 84	903.507 1	7.086 8	55.634 1	37
		38	145.339 73	0.006 88	1030.998 1	7.093 7	55.895 5	38
		39	165.687 29	0.006 04	1176.337 8	7.099 7	56.130 9	39
		40	188.883 51	0.005 29	1342.025 1	7.105 0	56.342 7	40
		41	215.327 21	0.004 64	1530.908 6	7.109 7	56.533 1	41
		42	245.473 01	0.004 07	1746.235 8	7.113 8	56.704 2	42
		43	279.839 24	0.003 57	1991.708 8	7.117 3	56.857 9	43

(Continued)

(Continued)

Compound Interest Tables								
Constants		— **14 Per Cent**						
Function	Value	n	$(1 + i)^n$	v^n	$s_{\overline{n}\rceil}$	$a_{\overline{n}\rceil}$	$(Ia)_{\overline{n}\rceil}$	n
		44	319.016 73	0.003 13	2271.548 1	7.120 5	56.995 8	44
		45	363.679 07	0.002 75	2590.564 8	7.123 2	57.119 5	45
		46	414.594 14	0.002 41	2954.243 9	7.125 6	57.230 5	46
		47	472.637 32	0.002 12	3368.838 0	7.127 7	57.329 9	47
		48	538.806 55	0.001 86	3841.475 3	7.129 6	57.419 0	48
		49	614.239 46	0.001 63	4380.281 9	7.131 2	57.498 8	49
		50	700.232 99	0.001 43	4994.521 3	7.132 7	57.570 2	50

Compound Interest Tables								
Constants		**15 Per Cent**						
Function	Value	n	$(1 + i)^n$	v^n	$s_{\overline{n}\rceil}$	$a_{\overline{n}\rceil}$	$(Ia)_{\overline{n}\rceil}$	n
i	0.150 000	1	1.150 00	0.869 57	1.000 0	0.869 6	0.869 6	1
$i^{(2)}$	0.144 761	2	1.322 50	0.756 14	2.150 0	1.625 7	2.381 9	2
$i^{(4)}$	0.142 232	3	1.520 87	0.657 52	3.472 5	2.283 2	4.354 4	3
$i^{(12)}$	0.140 579	4	1.749 01	0.571 75	4.993 4	2.855 0	6.641 4	4
δ	0.139 762	5	2.011 36	0.497 18	6.742 4	3.352 2	9.127 3	5
$(1 + i)$	1.150 000	6	2.313 06	0.432 33	8.753 7	3.784 5	11.721 3	6
$(1 + i)^{1/2}$	1.072 381	7	2.660 02	0.375 94	11.066 8	4.160 4	14.352 8	7
$(1 + i)^{1/4}$	1.035 558	8	3.059 02	0.326 90	13.726 8	4.487 3	16.968 0	8
$(1 + i)^{1/12}$	1.011 715	9	3.517 88	0.284 26	16.785 8	4.771 6	19.526 4	9
		10	4.045 56	0.247 18	20.303 7	5.018 8	21.998 2	10
v	0.869 565	11	4.652 39	0.214 94	24.349 3	5.233 7	24.362 6	11
$v^{1/2}$	0.932 505	12	5.350 25	0.186 91	29.001 7	5.420 6	26.605 5	12
$v^{1/4}$	0.965 663	13	6.152 79	0.162 53	34.351 9	5.583 1	28.718 4	13
$v^{1/12}$	0.988 421	14	7.075 71	0.141 33	40.504 7	5.724 5	30.697 0	14
		15	8.137 06	0.122 89	47.580 4	5.847 4	32.540 4	15
d	0.130 435	16	9.357 62	0.106 86	55.717 5	5.954 2	34.250 2	16
$d^{(2)}$	0.134 990	17	10.761 26	0.092 93	65.075 1	6.047 2	35.830 0	17
$d^{(4)}$	0.137 348	18	12.375 45	0.080 81	75.836 4	6.128 0	37.284 5	18
$d^{(12)}$	0.138 951	19	14.231 77	0.070 27	88.211 8	6.198 2	38.619 5	19
		20	16.366 54	0.061 10	102.443 6	6.259 3	39.841 5	20

(Continued)

Compound Interest Tables								
Constants			**15 Per Cent**					
Function	Value	n	$(1 + i)^n$	v^n	$s_{\overline{n}\rvert}$	$a_{\overline{n}\rvert}$	$(Ia)_{\overline{n}\rvert}$	n
$i/i^{(2)}$	1.036 190	21	18.821 52	0.053 13	118.810 1	6.312 5	40.957 2	21
$i/i^{(4)}$	1.054 613	22	21.644 75	0.046 20	137.631 6	6.358 7	41.973 7	22
$i/i^{(12)}$	1.067 016	23	24.891 46	0.040 17	159.276 4	6.398 8	42.897 7	23
		24	28.625 18	0.034 93	184.167 8	6.433 8	43.736 1	24
		25	32.918 95	0.030 38	212.793 0	6.464 1	44.495 5	25
i/δ	1.073 254	26	37.856 80	0.026 42	245.712 0	6.490 6	45.182 3	26
		27	43.535 31	0.022 97	283.568 8	6.513 5	45.802 5	27
		28	50.065 61	0.019 97	327.104 1	6.533 5	46.361 8	28
		29	57.575 45	0.017 37	377.169 7	6.550 9	46.865 5	29
		30	66.211 77	0.015 10	434.745 1	6.566 0	47.318 6	30
$i/d^{(2)}$	1.111 190	31	76.143 54	0.013 13	500.956 9	6.579 1	47.725 7	31
$i/d^{(4)}$	1.092 113	32	87.565 07	0.011 42	577.100 5	6.590 5	48.091 1	32
$i/d^{(12)}$	1.079 516	33	100.699 83	0.009 93	664.665 5	6.600 5	48.418 8	33
		34	115.804 80	0.008 64	765.365 4	6.609 1	48.712 4	34
		35	133.175 52	0.007 51	881.170 2	6.616 6	48.975 2	35
		36	153.151 85	0.006 53	1014.345 7	6.623 1	49.210 3	36
		37	176.124 63	0.005 68	1167.497 5	6.628 8	49.420 4	37
		38	202.543 32	0.004 94	1343.622 2	6.633 8	49.608 0	38
		39	232.924 82	0.004 29	1546.165 5	6.638 0	49.775 4	39
		40	267.863 55	0.003 73	1779.090 3	6.641 8	49.924 8	40
		41	308.043 08	0.003 25	2046.953 9	6.645 0	50.057 9	41
		42	354.249 54	0.002 82	2354.996 9	6.647 8	50.176 4	42
		43	407.386 97	0.002 45	2709.246 5	6.650 3	50.282 0	43
		44	468.495 02	0.00213	3116.633 4	6.652 4	50.375 9	44
		45	538.769 27	0.001 86	3585.128 5	6.654 3	50.459 4	45
		46	619.584 66	0.001 61	4123.897 7	6.655 9	50.533 7	46
		47	712.522 36	0.001 40	4743.482 4	6.657 3	50.599 6	47
		48	819.400 71	0.001 22	5456.004 7	6.658 5	50.658 2	48
		49	942.310 82	0.001 06	6275.405 5	6.659 6	50.710 2	49
		50	1083.657 44	0.000 92	7217.716 3	6.660 5	50.756 3	50

Compound Interest Tables		20 Per Cent						
Constants								
Function	Value	n	$(1+i)^n$	v^n	$s_{\overline{n}}$	$a_{\overline{n}}$	$(Ia)_{\overline{n}}$	n
i	0.200 000	1	1.200 00	0.833 33	1.000 0	0.833 3	0.833 3	1
$i^{(2)}$	0.190 890	2	1.440 00	0.694 44	2.200 0	1.527 8	2.222 2	2
$i^{(4)}$	0.186 541	3	1.728 00	0.578 70	3.640 0	2.106 5	3.958 3	3
$i^{(12)}$	0.183 714	4	2.073 60	0.482 25	5.368 0	2.588 7	5.887 3	4
δ	0.182 322	5	2.488 32	0.401 88	7.441 6	2.990 6	7.896 7	5
$(1+i)$	1.200 000	6	2.985 98	0.334 90	9.929 9	3.325 5	9.906 1	6
$(1+i)^{1/2}$	1.095 445	7	3.583 18	0.279 08	12.915 9	3.604 6	11.859 7	7
$(1+i)^{1/4}$	1.046 635	8	4.299 82	0.232 57	16.499 1	3.837 2	13.720 2	8
$(1+i)^{1/12}$	1.015 309	9	5.159 78	0.193 81	20.798 9	4.031 0	15.464 5	9
		10	6.191 74	0.161 51	25.958 7	4.192 5	17.079 6	10
v	0.833 333	11	7.430 08	0.134 59	32.150 4	4.327 1	18.560 0	11
$v^{1/2}$	0.912 871	12	8.916 10	0.112 16	39.580 5	4.439 2	19.905 9	12
$v^{1/4}$	0.955 443	13	10.699 32	0.093 46	48.496 6	4.532 7	21.120 9	13
$v^{1/12}$	0.984 921	14	12.839 18	0.077 89	59.195 9	4.610 6	22.211 3	14
		15	15.407 02	0.064 91	72.035 1	4.675 5	23.184 9	15
d	0.166 667	16	18.488 43	0.054 09	87.442 1	4.729 6	24.050 3	16
$d^{(2)}$	0.174 258	17	22.186 11	0.045 07	105.930 6	4.774 6	24.816 6	17
$d^{(4)}$	0.178 229	18	26.623 33	0.037 56	128.116 7	4.812 2	25.492 7	18
$d^{(12)}$	0.180 943	19	31.948 00	0.031 30	154.740 0	4.843 5	26.087 4	19
		20	38.337 60	0.026 08	186.688 0	4.869 6	26.609 1	20
$i/i^{(2)}$	1.047 723	21	46.005 12	0.021 74	225.025 6	4.891 3	27.065 5	21
$i/i^{(4)}$	1.072 153	22	55.206 14	0.018 11	271.030 7	4.909 4	27.464 1	22
$i/i^{(12)}$	1.088 651	23	66.247 37	0.015 09	326.236 9	4.924 5	27.811 2	23
		24	79.496 85	0.012 58	392.484 2	4.937 1	28.113 1	24
		25	95.396 22	0.010 48	471.981 1	4.947 6	28.375 2	25
i/δ	1.096 963							
$i/d^{(2)}$	1.147 723							
$i/d^{(4)}$	1.122 153							
$i/d^{(12)}$	1.105 317							

| Compound Interest Tables | | | | | | | | |
| **Constants** | | 25 Per Cent | | | | | | |
Function	Value	n	$(1 + i)^n$	v^n	$s_{\overline{n}|}$	$a_{\overline{n}|}$	$(Ia)_{\overline{n}|}$	n
i	0.250 000	1	1.250 00	0.800 00	1.000 0	0.800 0	0.800 0	1
$i^{(2)}$	0.236 068	2	1.562 50	0.640 00	2.250 0	1.440 0	2.080 0	2
$i^{(4)}$	0.229 485	3	1.953 13	0.512 00	3.812 5	1.952 0	3.616 0	3
$i^{(12)}$	0.225 231	4	2.441 41	0.409 60	5.765 6	2.361 6	5.254 4	4
δ	0.223 144	5	3.051 76	0.327 68	8.207 0	2.689 3	6.892 8	5
$(1 + i)$	1.250 000	6	3.814 70	0.262 14	11.258 8	2.951 4	8.465 7	6
$(1 + i)^{1/2}$	1.118 034	7	4.768 37	0.209 72	15.073 5	3.161 1	9.933 7	7
$(1 + i)^{1/4}$	1.057 371	8	5.960 46	0.167 77	19.841 9	3.328 9	11.275 8	8
$(1 + i)^{1/12}$	1.018 769	9	7.450 58	0.134 22	25.802 3	3.463 1	12.483 8	9
		10	9.313 23	0.107 37	33.252 9	3.570 5	13.557 5	10
v	0.800 000	11	11.641 53	0.085 90	42.566 1	3.656 4	14.502 4	11
$v^{1/2}$	0.894 427	12	14.551 92	0.068 72	54.207 7	3.725 1	15.327 1	12
$v^{1/4}$	0.945 742	13	18.189 89	0.054 98	68.759 6	3.780 1	16.041 8	13
$v^{1/12}$	0.981 577	14	22.737 37	0.043 98	86.949 5	3.824 1	16.657 5	14
		15	28.421 71	0.035 18	109.686 8	3.859 3	17.185 3	15
d	0.200 000	16	35.527 14	0.028 15	138.108 5	3.887 4	17.635 6	16
$d^{(2)}$	0.211 146	17	44.408 92	0.022 52	173.635 7	3.909 9	18.018 4	17
$d^{(4)}$	0.217 034	18	55.511 15	0.018 01	218.044 6	3.927 9	18.342 7	18
$d^{(12)}$	0.221 082	19	69.388 94	0.014 41	273.555 8	3.942 4	18.616 5	19
		20	86.736 17	0.011 53	342.944 7	3.953 9	18.847 1	20
$i/i^{(2)}$	1.059 017	21	108.420 22	0.009 22	429.680 9	3.963 1	19.040 8	21
$i/i^{(4)}$	1.089 396	22	135.525 27	0.007 38	538.101 1	3.970 5	19.203 1	22
$i/i^{(12)}$	1.109 971	23	169.406 59	0.005 90	673.626 4	3.976 4	19.338 9	23
		24	211.758 24	0.004 72	843.032 9	3.981 1	19.452 2	24
		25	264.697 80	0.003 78	1054.791 2	3.984 9	19.546 7	25
i/δ	1.120 355							
$i/d^{(2)}$	1.184 017							
$i/d^{(4)}$	1.151 896							
$i/d^{(12)}$	1.130 804							

Additional Reading

Reading around and ahead of a subject is always recommended. Although this book can be considered as self-contained, we provide a list of potential additional reading here. Some texts are intended to give an alternative perspective on the material presented in this book, and some texts are intended as further reading.

The subject of *mathematical finance/financial mathematics* is ever growing, and this short list is certainty not intended to be exhaustive.

Baxter, M., Rennie, A., 1996. Financial Calculus: An Introduction to Derivative Pricing. Cambridge University Press, Cambridge.

Brigham, E.F., Houston, J.F., 2009. Fundamentals of Financial Management, 12th ed. South-Western College Pub, Mason.

Cairns, A.J.G., 2004. Interest Rate Models: An Introduction. Princeton University Press, Princeton.

Cochrane, J.H., 2005. Asset Pricing, Revised ed. Princeton University Press, Princeton.

Copeland, T.E., Weston, J.F., Shastri, K., 2003. Financial Theory and Corporate Policy, 4th ed. Pearson, London.

Cuthbertson, K., Nitzsche, D., 2001. Financial Engineering: Derivative & Risk Management. John Wiley & Sons, Chichester.

DeFusco, R.A., et al., 2010. Quantitative Investment Analysis, 2nd ed. John Wiley & Sons, Chichester.

Dodds, J.C., Ford, J.L., 1974. Expectations, Uncertainty and the Term Structure of Interest Rates. Martin Robertson & Co. Ltd., Oxford.

Glasserman, P., 2010. Monte Carlo Methods in Financial Engineering. Springer, New York.

Guerard Jr., J.B., Schwartz, E., 2007. Quantitative Corporate Finance. Springer, New York.

Homer, S., Sylla, R., 2005. A History of Interest Rates, 4th ed. Wiley, Hoboken.

Hull, J., 2011. Options, Futures and Other Derivatives, 11th ed. Pearson Education, Harlow.

Levy, H., Sarnat, M., 1994. Capital Investment and Financial Decisions, 5th ed. Prentice-Hall International, Edinburgh.

Mao, J.C.T., 1969. Quantitative Analysis of Financial Decisions. Macmillan, New York.

Merrett, A.J., Sykes, A., 1963. Finance and Analysis of Capital Projects. Longman, New York.

Michaelsen, J.B., 1973. The Term Structure of Interest Rates. Intext Educational Publishers, London.

Pinto, J.E., Henry, E., Robinson, T.R., Stowe, J.D., 2010. Equity Asset Valuation. John Wiley & Sons, Hoboken.

Redington, F.M., 1952. Review of the principles of life office valuations. Journal of the Institute of Actuaries 78 (3), 286–315.

Renwick, F.B., 1971. Introduction to Investments and Finance. Macmillan, New York.

Russell, A.M., Rickard, J.A., 1982. Uniqueness of non-negative internal rate of return. Journal of the Institute of Actuaries 109 (3), 435–445.

Taylor, G.C., 1980. Determination of internal rates of return in respect of an arbitrary cash flow. Journal of the Institute of Actuaries 107 (4), 487–497.

Van Horne, J.C., 1970. Function and Analysis of Capital Markets. Prentice-Hall, Englewood Cliffs, N J.

Wilmott, P., 2007. Paul Wilmott Introduces Quantitative Finance. John Wiley & Sons, Chichester.

Index

Printed in the United States
By Bookmasters